Enhancing Communication and Decision-Making With AI

Arul Kumar Natarajan
Samarkand International University of Technology, Uzbekistan

Mohammad Gouse Galety
Samarkand International University of Technology, Uzbekistan

Celestine Iwendi
University of Bolton, UK

Deepthi Das
Christ University, India

Achyut Shankar
University of Warwick, UK

IGI Global
Publishing Tomorrow's Research Today

Vice President of Editorial Melissa Wagner
Managing Editor of Acquisitions Mikaela Felty
Managing Editor of Book Development Jocelynn Hessler
Production Manager Mike Brehm
Cover Design Phillip Shickler

Published in the United States of America by
IGI Global Scientific Publishing
701 East Chocolate Avenue
Hershey, PA, 17033, USA
Tel: 717-533-8845
Fax: 717-533-8661
Website: https://www.igi-global.com E-mail: cust@igi-global.com

Library of Congress Cataloging-in-Publication Data

Names: Natarajan, Arulkumar, 1986- editor.
Title: Enhancing communication and decision-making with AI / edited by Arul Kumar Natarajan, Mohammad Gouse Galety, Celestine Iwendi, Deepthi Das, Achyut Shankar.
Description: Hershey, PA : IGI Global Scientific Publishing, [2025] | Includes bibliographical references and index. | Summary: "This edited book is structured to provide readers with a thorough understanding of how AI can be leveraged to improve communication strategies and decision-making processes"-- Provided by publisher.
Identifiers: LCCN 2024060865 (print) | LCCN 2024060866 (ebook) | ISBN 9798369392461 (hardcover) | ISBN 9798369392478 (paperback) | ISBN 9798369392485 (ebook)
Subjects: LCSH: Decision making--Social aspects. | Communication--Data processing.
Classification: LCC HD30.23 .E523 2025 (print) | LCC HD30.23 (ebook) | DDC 658.4/03--dc23/eng/20250307
LC record available at https://lccn.loc.gov/2024060865
LC ebook record available at https://lccn.loc.gov/2024060866

British Cataloguing in Publication Data
A Cataloguing in Publication record for this book is available from the British Library.

Table of Contents

Detailed Table of Contents

Generative artificial intelligence (GenAI) is one part of the diverse AI universe adapting to the growing needs in the business world. This adaptability covers other forms including images, writing codes, and further imitating emotional quotient as well as human-level creativity by implementing a toolkit called natural language processing (NLP), a monumental toolkit that equips the computers to understand, manipulate, and generate human-like language. This chapter explores how GenAI is impacting businesses and can use it to boost productivity to gain an advantage over their competition. To understand the opportunities and challenges posed by GenAI and to offer a new view, published literature as well as official reports by companies and government were studied. It is advised that the managers incorporate AI in their strategies not as mere IT but as a fundamental part of the company. Integrating GenAI will benefit companies by increasing their value proposition. Before the edge provided by GenAI becomes less distinct, businesses should adopt it to prevent themselves from lagging.

Home solutions that unite the internet of things (IoT) and artificial intelligence (AI) offer a ground-breaking opportunity. This chapter discloses how IoT devices and AI algorithms can change household inventory management and security. The main focus is on sensors and AI-based analytics for optimizing data collection, analysis, and decision-making processes. Predictive maintenance becomes possible when sensors allow real-time monitoring of stock levels. On the other hand, integrated artificial

the utilization of user-generated content on social media platforms like Facebook, Instagram, TikTok, and Twitter, businesses can obtain up-to-date insights into the assumptions and preferences of consumers regarding items. Advanced analytical techniques like machine learning, normal dialect preparation, and estimate research help firms sort through massive volumes of social media data to uncover patterns, identify designs, and predict future consumer behaviour. Additionally, these social media platforms provide a coordinated way for brands and businesses to engage with their audience, fostering meaningful relationships, soliciting feedback, and establishing brand credibility.

Previously, when looking towards the development of surveys to help us improve aftershave products, we relied on the traditional and often costly means of conducting comprehensive and expensive marketing surveys. Social media plays the role of a powerful tool for realizing consumer behavior and obtaining opinion of consumers of products that is referred to as consumer sentiment analysis. It observes and checks consumer preferences by investigating the content they are sharing, where they can be any images, videos, or simple text. Catchers get a new insight into target market needs and service adequacy. By tailoring to the consumers' needs as these demands keep changing, companies can make their operations better. Social media handling companies can gain rich insights from social media applications by using latest technologies like machine learning and natural language processing, especially through lexicon-based methods. Then, they can randomly select a few customers and conduct sentiment analysis on their feedback of the product which is known as opinion mining.

Combining NLP technology into communication platforms offers noteworthy opportunities for businesses to gain actionable insights into consumer behaviour and

The first objective examines AI's efficacy in detecting fake news, with Sandeep Amar discussing AI models. The second objective focuses on enhancing fact-checking through AI, with Pratyush Ranjan emphasising the importance of critical thinking alongside AI tools. The third objective explores government policy's role, with Nikhil Pahwa analysing Indian regulations and their impact on AI deployment for fact-checking. The study contributes to debates on addressing fake news, offering insights for policymakers, researchers, and practitioners. It highlights the need for a hybrid approach integrating AI's capabilities with human judgment to ensure information integrity.

Fuzzy soft sets find an extensive assortment of applications in many decision-making problems. Fuzzy sets are applied to overcome uncertainty, and soft sets came into existence as an extension. It is called soft because the boundary depends on the parameters. Soft sets are the generalization of fuzzy sets. Parameter reduction in soft sets shows an energetic role in reducing the complexity of any decision-making problems. Fuzzy soft sets are applied to identify a word. The set of alphabets is taken as the universal set, and the words formed using these alphabets give the parameter set. The length of the word is first calculated. The first three letters of the word to be identified are given as the input. The parameter set consists of words of the same length, and the first three letters are generated using NLG. The word is identified using machine learning of word classification and prediction. The results are compared, and the final accuracy is calculated.

The legal domain generates an immense volume of textual data daily, ranging from court opinions and statutes to contracts and legal briefs. As this volume grows, so does the need for effective methods to organize, categorize, and extract insights from these documents. Natural language processing (NLP) has emerged as a powerful tool to address these challenges by automating the analysis of legal texts. This chapter

Foreword

The rapid advancements in Artificial Intelligence (AI) have ushered in a new era of transformative possibilities for communication and decision-making across various industries. This edited volume, Enhancing Communication and Decision-Making With AI, brings together a collection of chapters exploring AI's multifaceted applications in enhancing business strategies, consumer insights, financial predictions, and public engagement.

Chapter 1 opens the discussion by examining how generative AI shapes the next frontier of competitive advantage, setting the stage for a deeper exploration of AI's potential in modern enterprises. The subsequent chapter, Chapter 2, bridges the Internet of Things (IoT) and AI, presenting a compelling case of how smart home solutions revolutionize business intelligence.

Social media, a critical platform for communication and influence, is rigorously analyzed in Chapters 3 through 6. Chapter 3 delves into the journey of social media influence, offering a comprehensive study of the techniques and strategies that drive effective diffusion. Building on this, Chapter 4 harnesses social media for trend analysis and behavioral understanding, unlocking valuable consumer insights. Chapter 5 continues this exploration by focusing on consumer sentiment analysis on social media, demonstrating its pivotal role in business intelligence. Chapter 6 introduces an innovative AI-NLP bot designed for predicting consumer vision across various communication platforms, showcasing the practical integration of AI in understanding consumer preferences.

Chapters 7 and 8 address the financial sector, meticulously reviewing AI's impact on predicting market movements and financial distress. Chapter 7 investigates the relationship between Twitter sentiment and Bitcoin price movements, providing insights into the intersection of social media and financial markets. Chapter 8 offers a scoping review of financial distress prediction in India using modern machine learning methods, highlighting the scope for future research in this critical area.

Chapter 9 focuses on public service delivery and citizen engagement and reviews the role of volunteered geographic information in enhancing public services. A qualitative investigation follows it in Chapter 10, which explores AI-driven solutions for combating fake news through industry expert interviews, shedding light on one of the most pressing challenges of the digital age.

The final chapters, Chapters 11 and 12, delve into the technical applications of AI in language processing. Chapter 11 presents word prediction using fuzzy sets and computational intelligence, illustrating the advancements in predictive algorithms. The book concludes with Chapter 12, which discusses natural language processing for legal document classification, underscoring the importance of AI in streamlining legal processes.

This volume provides a comprehensive overview of AI's diverse applications in enhancing communication and decision-making, offering valuable insights for academics, industry professionals, and policymakers. The chapters collectively emphasize AI's critical role in shaping the future of various sectors, making this book a timely and essential contribution to the field.

Subrata Chowdhury
Sreenivasa Institute of Technology and Management, India

Preface

Integrating Artificial Intelligence (AI) into communication and decision-making processes marks a significant shift in how organizations operate and engage with their environments. AI's ability to process vast amounts of data, generate actionable insights, and automate complex tasks has positioned it as a crucial tool in the modern world, driving innovation and efficiency across numerous sectors.

Enhancing Communication and Decision-Making with AI is a curated collection of research highlighting AI's transformative potential in these domains. The chapters within this volume explore the diverse applications of AI, from enhancing competitive advantage and business intelligence to understanding consumer behavior and predicting financial outcomes.

This edited book is structured to provide readers with a thorough understanding of how AI can be leveraged to improve communication strategies and decision-making processes. Each chapter offers a deep dive into specific use cases, methodologies, and emerging trends supported by rigorous analysis and real-world examples. The topics covered include generative AI, the intersection of IoT and AI, social media analytics, financial prediction models, citizen engagement in public services, and advancements in natural language processing.

This book aims to serve as a valuable resource for researchers, professionals, and students interested in the evolving role of AI in communication and decision-making. By presenting a broad spectrum of AI applications, the book demonstrates the current capabilities of AI technologies and provides insights into future possibilities and challenges.

In compiling this volume, contributions from leading experts and practitioners in the field have been gathered, each offering unique perspectives on the application of AI. This collaborative effort has resulted in a comprehensive guide that reflects the latest advancements and ongoing research in AI-driven communication and decision-making.

Hopefully, this book will inspire further exploration and innovation in AI, fostering a deeper understanding of its impact on our world and encouraging the development of new solutions that enhance how we communicate and make decisions in both personal and professional contexts.

Arul Kumar Natarajan
Samarkand International University of Technology, Uzbekistan

Mohammad Gouse Galety
Samarkand International University of Technology, Uzbekistan

Celestine Iwendi
University of Bolton, UK

Deepthi Das
Christ University, India

Achyut Shankar
University of Warwick, UK

Chapter 1
Generative AI:
Next Frontier for Competitive Advantage

Ipsa Bharti
https://orcid.org/0009-0000-4762-2390
Jamia Millia Islamia, India

Kavita Chauhan
Jamia Millia Islamia, India

Priyanka Aggarwal
https://orcid.org/0000-0002-7486-2012
Jamia Millia Islamia, India

ABSTRACT

Generative artificial intelligence (GenAI) is one part of the diverse AI universe adapting to the growing needs in the business world. This adaptability covers other forms including images, writing codes, and further imitating emotional quotient as well as human-level creativity by implementing a toolkit called natural language processing (NLP), a monumental toolkit that equips the computers to understand, manipulate, and generate human-like language. This chapter explores how GenAI is impacting businesses and can use it to boost productivity to gain an advantage over their competition. To understand the opportunities and challenges posed by GenAI and to offer a new view, published literature as well as official reports by companies and government were studied. It is advised that the managers incorporate AI in their strategies not as mere IT but as a fundamental part of the company. Integrating GenAI will benefit companies by increasing their value proposition. Before the edge provided by GenAI becomes less distinct, businesses should adopt it to prevent themselves from lagging.

DOI: 10.4018/979-8-3693-9246-1.ch001

INTRODUCTION

"AI is one of the most important things humanity is working on. It is more profound than, I dunno, electricity or fire." — Sundar Pichai, CEO of Google (Clifford, 2018).

In the ever-evolving field of technology, generative artificial intelligence, also known as GenAI, emerges as a notable symbol of both creativity and controversy. We are undergoing a rapid transformation in how we engage with technology. With the help of the training data, GenAI uses machine learning and neural networks to identify patterns and generate original content like texts and music. The inception of machines imitating the creative cognitive processes of humans has led to the synthesis of customized content, carrying substantial implications for both enterprises and individuals. As organizations and societies contend with the profound implications of GenAI, the conversation frequently alternates between idealistic visions and pessimistic apprehensions. GenAI has the potential to revolutionize various aspects of how a business operates. Imagine a world where machines not only assist in routine tasks but also generate creative new content, design products, and even strategize solutions to complex problems. This is not a distant future but a burgeoning reality.

GenAI has the potential to disrupt the various aspects of our professional accomplishments. It possesses the remarkable capability to generate content that greatly enhances and complements human expertise and skills. For instance, GenAI can be utilized for tasks such as crafting written memos and reports, designing visually appealing graphics for websites, devising personalized marketing strategies, and curating comprehensive employee learning programs ("Generative AI and the Future of Work", 2023). The applications of GenAI across different industries are virtually limitless, exemplifying the extensive range of work that can be enhanced through the utilization of this technology. Unlike traditional AI, which follows a set of predefined rules, GenAI learns the pattern from the training data and generates new outputs. This capability has opened the doors for new opportunities for businesses to innovate and stay ahead of the competition.

GenAI is booming as an addictive setup for individuals across the globe by expanding its boundaries of creativity, automation, and human-machine interaction. GenAI is one part of the diverse AI universe, which includes models like "Generative Adversarial Networks (GANs)," "Variational Autoencoders (VAEs)," and GPT-3, adapting to the growing needs of the users by offering its features beyond text. This adaptability covers other forms, including images, writing codes, and further imitating emotional quotients, as well as human-level creativity.

The continuing evolution of GenAI based on user feedback, market needs, and technological vision is successfully taking ownership of tasks across industries. Some leading examples of these tasks can be quoted as content creation personalized to the individual's expectations, software development automated to a certain extent to reduce laborious tasks, and innovative recommendations on design and similar functions. The outcome of these advancements has the potential to contribute to the global economy by driving economic growth, enhancing task-level efficiency independent of industry or domain, and fostering new stages in creative self-expression expression (Bond-Taylor et al., 2022; Floridi et al., 2018).

However, the continuous development and incorporation of genAI technologies by businesses also raise crucial ethical and societal concerns. Debates among scholars, policymakers, and the tech community revolve around issues of data privacy, copyright infringement, misinformation, and the perpetuation of societal biases within AI-generated content. These discussions emphasize the necessity for robust ethical frameworks and regulatory measures to ensure the responsible development and use of generative AI (Hao, 2022; Rahwan, 2018).

Furthermore, to minimize the risk of job displacement in the labor market, careful consideration and proactive strategies are required. GenAI must be incorporated into the workplace in a way that maximizes its potential to enhance human capabilities while also safeguarding against negative consequences for employment and social equality (Susskind, 2020; West et al., 2019). There is much discussion about the existential risks that GenAI poses, which range from ethical conundrums to employment displacement. However, amid these worries, there is a unique chance for companies to overcome these obstacles and turn them into success.

Ultimately, the goal is to study that GenAI is not just a technological novelty but a strategic asset. Businesses that recognize and utilize its potential can transform their operations and redefine industry standards. As we navigate through the various aspects of GenAI, it will become evident that the key to future competitiveness lies in the intelligent integration of this disruptive technology.

BACKGROUND OF THE STUDY

To understand the impact of GenAI and the transformations in industries, it is essential to consider its historical development and technological advancements. ELIZA, the first computer chatbot using decomposition and assembly rules, was developed in 1964-66 by Joseph Weizenbaum at MIT. The Chatbot mimicked a psychotherapist, creating an illusion of understanding (Weizenbaum, 1966). OpenAI, founded in 2015, advanced this field significantly and released different versions of GPTs: GPT-1, GPT-2, GPT 3.5 (ChatGPT), and GPT-4 in 2018, 2019, 2022, and 2023, respectively. OpenAI, after receiving investments from Microsoft, also integrated GPTs into products like Bing, Edge, and Windows, making it possible to showcase GPT's practical application (OpenAI & Pilipiszyn, 2021; Toews, 2024; Novet, 2023). Figure 1 shows the timeline and journey of GenAI.

Figure 1. Generative AI Journey

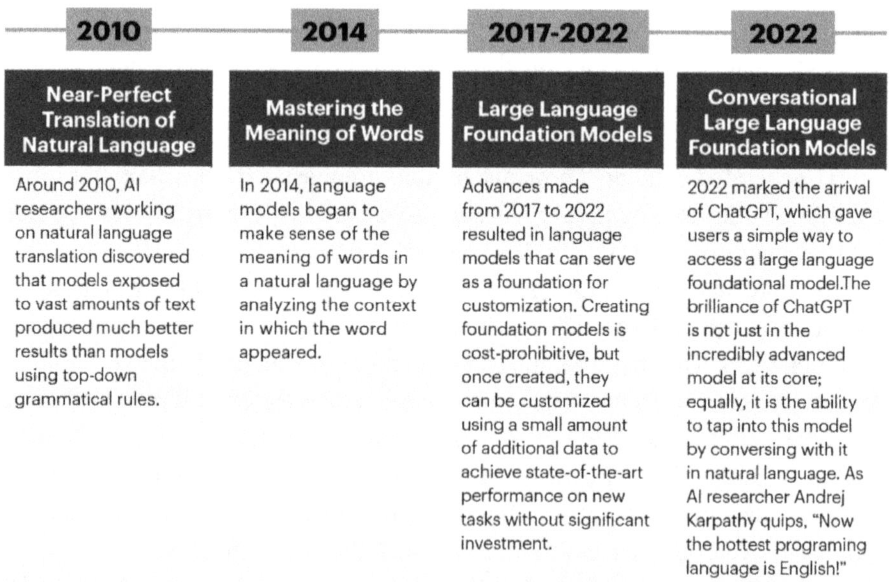

Source: (Generative AI: From Human to Machine, to Artificial Intelligence, n.d.)

Earlier, the language models aimed to replicate human language understanding. Due to Significant breakthroughs in the field, statistical approaches such as the Hidden Markov Model (HMMs) and n-grams focusing on character collocation rather than deep linguistic understanding came. These models enabled content-aware and flexible language processing, improved part-of-speech tagging, and also named entity recognition (Szudarski, 2023; Shukla, 2024). In the 1970s, IBM researchers Church and Mercer used statistical text analysis for machine learning, shifting the focus from traditional linguistic methods (Brown et al., 1990). Despite several debates regarding whether GenAI systems exhibit behavior that is beyond procedural collections, their practical applications, such as cognitive tools in various sectors, are evident. GenAI tools, like GPT, help in image creation, software code, sound, and content creation, thereby assisting with job application writing and personalized recommendations (Lo & Ross, 2024; Wittenberg et al., 2024). The capabilities demonstrated by these GenAI models are derived from the massive datasets they are trained on, enabling them to understand and generate human-like responses. However, the intelligence exhibited by these models varies significantly from that of human intelligence, highlighting their distinct abilities as well as inherent constraints.

MAIN FOCUS OF THE STUDY

The main focus of this chapter is to present how Artificial Intelligence (AI) has started to shape a new era of the business world, possibly replacing employees, as has been pointed out by AI experts due to its capabilities. Big corporations have trained AI for over a decade with massive amounts of data in diverse formats, leading to impressive features such as human-like interaction and logical reasoning capabilities to benefit diverse business functions. The trending AI product, GenAI, has been trained to converse like humans, along with significant features such as question-answering skills, analyzing sentiments in text, generating creative content, images, videos, and more. A major benefit of these features is that they can help reduce human effort by taking charge of repetitive tasks, eventually saving time and resources. These capabilities of GenAI to produce content, write codes, and generate images and videos based on instructions can help businesses and their leadership avoid cognitive biases, extract insights from large datasets efficiently, and quicken the decision-making process, to mention a few aspects.

WHAT IS GENERATIVE ARTIFICIAL INTELLIGENCE (GENAI)?

GenAI is an awe-inspiring final product emerging from extensive data fed to the computer over the last few decades, commonly known as Deep Learning. GenAI is a simple yet powerful text-based application capable of generating content in both written and visual forms. The content generated can be text, an image, a code, and so on. When trained, the models identify the patterns from the large dataset. Based on the patterns, it generates new and similar outputs. GenAI includes various applications, such as writing software code and creating data that mimics real-world data. It is also capable of developing conversational chatbots. These models include techniques like "variational autoencoders (VAEs)," "generative adversarial networks (GANs)," and "transformers," enabling the models to generate realistic images or translate languages (Martineau, 2024; *What Is Generative AI?*, 2024; *Explained: Generative AI*, 2023; Davenport, 2023).

There are different techniques through which GenAI can be trained: Supervised Learning (Labeled Data), Unsupervised Learning (Unlabeled Data), Reinforcement Learning (Labeled+Unlabeled Data), and Deep Learning (Extensive Information fed into the system) (Benedikt Frey & Osborne, 2023; Davenport, 2023).

Types of GenAI Models

GenAI has a variety of models, each with unique applications. These models can create new and engaging data.

Variational Autoencoders (VAE): VAEs are a type of neural network that uses encoder-decoder architecture to generate new data points by learning a latent representation of input data. VAE comprises an encoder that maps the data into a probability distribution within a latent space (lower-dimensional space) and a decoder that reconstructs the original input based on this distribution (Skalic et al., 2019).

Training VAEs centers around minimizing a loss function. A loss function is composed of a combination of reconstruction loss and a regularization term, specifically the Kullback-Leibler (KL) divergence. Even though VAEs exhibit generative capabilities, they sometimes generate blurry outputs. These issues are now being addressed by Vector Quantized Variational Autoencoder (VQ-VAE) and Recurrent Variational Autoencoder (RVAE) (Skalic et al., 2019).

Generative Adversarial Networks (GAN): GAN was first introduced by Ian Goodfellow (Goodfellow et al., 2020). This type of model consists of two neural networks, a generator and a discriminator. A generator and a discriminator are both trained simultaneously through adversarial learning.

A generator learns the patterns, identifies the statistical properties to understand the relationships within a dataset, and uses that knowledge to create synthetic data that resembles the training data. A discriminative AI model excels at sorting existing data into categories like spam vs. not spam emails and cat vs. dog images. The core function of a discriminator is to identify distinguishing features between different categories within a dataset. These models are trained on labeled datasets where each data point has a predefined category label. The model learns to recognize the patterns associated with each category.

The continuous iterative process improves the quality of the data generated. The objective of this is to reach a stage where the discriminator cannot easily find the difference between the real and the synthetic data. Training GANs can be challenging as issues like gradient disappearance and poor sample data can arise. GAN models have been trained to enhance the training stability and the quality of the output generated (Goodfellow et al., 2020).

Transformers: This is in GenAI due to its strength in finding patterns in data while maintaining the context, proving to be a powerful tool for tasks that rely on text. Some applications include generating new text, structuring text, human-like chat interactions, and translating languages. This neural network helps GenAI learn the relationship between different words or sentences fed as tokens based on its use of the self-attention mechanism to allow an AI model to learn long-range and complex relationships between words (Vaswani et al., 2017; Bandi et al., 2023).

Figure 2. Different variants of Transformer Models

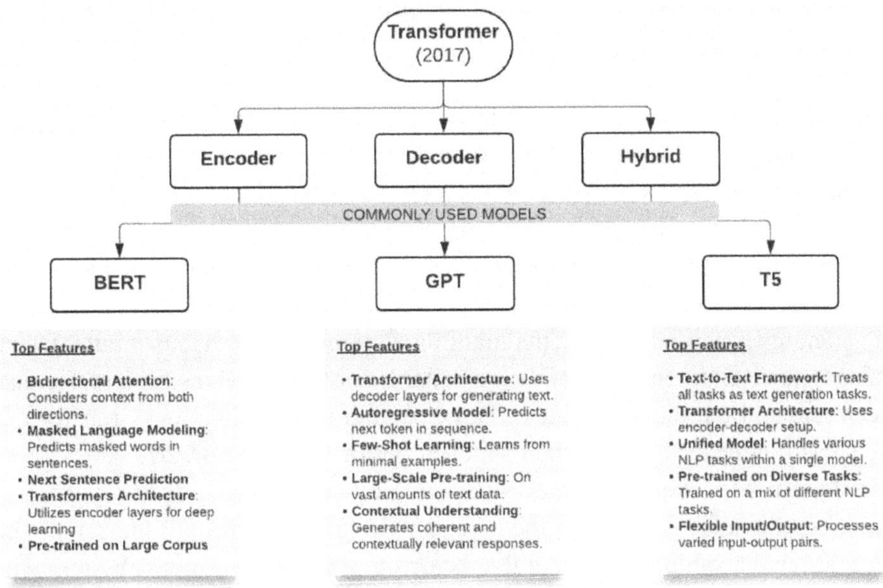

Source: Compiled by author

Figure 2 illustrates the different variants of Transformer Models (Bandi et al., 2023).

- BERT (Bidirectional Encoder Representations from Transformers) (2018)
- GPT (Generative Pre-trained Transformer) (2018)
- T5 (Text-to-Text Transfer Transformer) (2019)

Other specialized transformer models are (Bandi et al., 2023; *The Transformer Model Family*, n.d.)

- AdaSpeech, whose focus is directed towards the creation of authentic and organic-sounding synthetic speech.
- CuBERT and CodeBERT focus on code-related tasks, .eg., code comprehension, generation, and bug detection.
- MotionCLIP and Muse generate both text and visual Information. Both handle multimodal tasks.
- Grapher analyzes and comprehends visual Information, identifying interconnections and correlations among organized data.

- LaMDA (Language Models for Dialog Applications) enhances dialogue understanding and generates conversational AI applications.
- TFix and WT5 resolve concerns in transformer models related to fairness, transparency, and explainability.

Transformers have significantly impacted various fields, such as text generation, machine translation, code generation, and conversational AI. Their ability to comprehend complex interdependencies in the data enhances the quality of the contextually relevant text (Vaswani et al., 2017).

Recurrent Neural Networks (RNNs): RNNs include variants like "Long Short-Term Memory" (LSTM) networks. RNNs are designed to handle, analyze, and generate sequential data by maintaining a memory of previous inputs. Due to its ability to retain memory, RNNs are suitable for tasks such as time-series data, speech recognition, music, and text. However, in many applications, transformers have substituted RNNs due to their superior performance (Kaur, 2024).

Diffusion Models: These models operate by simulating a diffusion process. Gradually, the noise is added to the data sample to corrupt it until the sample becomes unrecognizable. The model then learns to reverse this process by removing the noise added step by step, generating realistic data (Zhang et al., 2023; C. Zhang et al., 2023). There are three types of diffusion models (M. Zhang et al., 2023):

- Denoising diffusion probabilistic models (DDPMs) use the Gaussian noise property, which is a two-step process that reverses the data through Markov chains (Ho et al., 2020).
- Score-based Generative Models (SGMs) determine the density of the log by varying the data at different noise levels (C. Zhang, Zhang, Zhang, et al., 2023).
- Stochastic Differential Equations (SDEs) reverse the output generation through stochastic equations by formulating the diffusion process with continuous noise perturbation (Cao et al., 2023).

Other specialized diffusion models are Latent Diffusion Models (LDMs) operate in latent space (Lin et al., 2022) and generate high-quality data samples by emphasizing basic data representations (L. Zhang, Qiu, Lin et al., 2023); Geometry Complete Diffusion Models (GCDMs) enhance the quality and preservation of geometric properties in the data generated by geometrically constraining the data (Morehead & Cheng, 2023); Video Diffusion Models (VDMs) expand diffusion process for creating consistency in video sequences with temporal relationships and changes (Molad et al., 2023). All these neural models generate high-quality and realistic data (image, text, and Video).

Auto-regressive (AR) Models: The term "auto-regressive" describes a variable repressed to its historical value. AR is a time series model that runs on the assumption of linear connection and predicts the future value based on the previous value. It can make predictions by employing lagged data of the relevant variable. Probabilistic correlations in sequence elements help models predict future elements. AR models use learned patterns to generate new sequences. For instance, "is" often follows "there" in English sentences according to the model (Gandhi, 2015; What Are Autoregressive Models? - AR Models Explained - AWS, n.d.; DeepAI, 2020).

Hybrid Models: As the name suggests, a hybrid model means a combination of different types of GenAI to leverage the strength of each model to create a robust and versatile system. Hybrid models may use Gans's adversarial training application and structured latent space representation of VAEs to make the output more meaningful and upgrade its quality (Z. Zhang et al., 2017; Van Den Oord et al., 2016). Hybrid models can also incorporate diffusion models to reduce the noise and data generation steps. For example, Toyota Research Institute has devised a hybrid GenAI model that merges traditional engineering limitations and modern GenAI functionalities. This model lets the designers input initial sketches and specific constraints, like aerodynamic drag and chassis dimensions, to produce the optimum vehicle designs. By combining text-to-image GenAI tools with optimization theory, the hybrid model can streamline the process design and minimize the number of iterations required (Mast, 2023).

Hybrid models are more useful when comprehending and capturing complex details, and maintaining high-quality outputs such as images or videos is essential. These types of models strive to enhance the accuracy, diversity, and quality of generated data, making them beneficial for complicated applications across diverse sectors, such as healthcare, finance, and creative fields. By incorporating different techniques, these models are capable of handling a wide range of tasks and offer flexibility to address the challenges or improve performance and generate data that resonates with real-world data.

Table 1 is a reference to GenAI model types discussed above and the key components that the models revolve around that allow them to create diverse applications for businesses to use. The best use cases included in the table can help conclude which industries would benefit from the respective GenAI model type.

Table 1. Types of GenAI: Overview

Generative Model	Key Components	Best Use Cases
GAN (Generative Adversarial Network)	Generator Network, Discriminator Network	High-fidelity images, domain-specific data generation

continued on following page

Table 1. Continued

Generative Model	Key Components	Best Use Cases
VAE (Variational Autoencoder)	Encoder, Decoder	Quick data generation with less detail required
Diffusion Models (DDPMs)	Forward Diffusion, Reverse Diffusion	High-quality image and audio generation foundation models
Transformer	Transformer Architecture	Natural language generation, text-based AI applications
RNNs	Recurrent Neurons; Long Short-Term Memory (LSTM)	Chatbots and Conversational AI; Speech-to-text
Auto-regressive (AR)	Sequential Data Input; Parameters (Coefficients)	Speech Synthesis; Time Series Forecasting; Music Generation
Hybrid	Model Integration (e.g., RNNs with CNNs or transformers)	Image Captioning; Multimodal Learning; Time Series Forecasting

Source: Compiled by author

Each model of GenAI has distinct characteristics with unique mechanisms and applications. The models leverage different techniques to generate data that is original in the form of text, image, Video, or audio. It is imperative to understand the nuances among these different types of models to select an appropriate model for specific tasks and enhance their application in both business and research settings.

FOUNDATIONAL MODEL

A foundational model is a large-scale, powerful, and deep-learning neural network. It has revolutionized the way data scientists used to work with machine learning. It is trained on massive datasets that are generalized and unlabelled and can be adjusted as per particular tasks. This model is capable of comprehending and generating enhanced data across various processes, such as images, texts, or codes, as shown in Figure 3. It is highly versatile and effective for a broad range of applications.

Figure 3. Categorization of tasks done by the Foundational Model

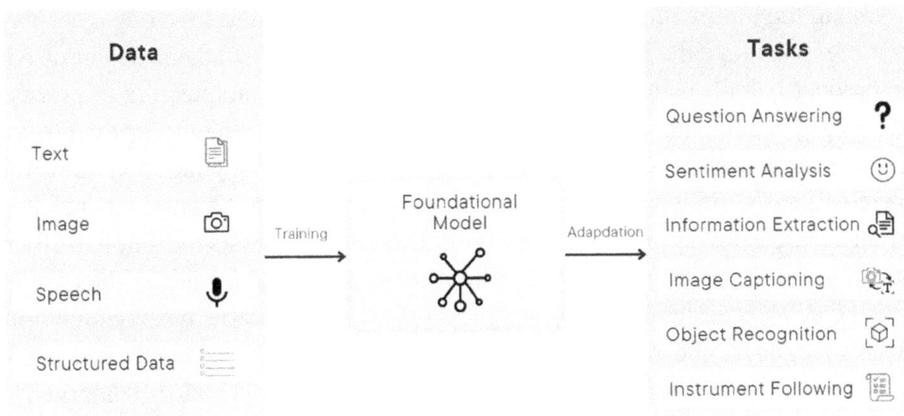

Source: Compiled by author

Due to the distinctive characteristics of foundation models of adaptability, they are capable of executing a diverse array of different kinds of assignments with high-level accuracy, contingent upon the input prompts provided. These tasks could include NLP, question answering, and image captioning. Foundational Models are unique from conventional machine learning models, which were designed to perform specific functions such as trend prediction or image classification; foundational models, on the other hand, can perform general and versatile tasks.

Foundation Models can serve as a basic model for developing more specialized applications. For example, GPT-4, a product of OpenAI, was trained using 170 trillion parameters and a 45 GB training dataset in 2023. As per OpenAI, the "computational power required for foundation modeling has doubled every 3.4 months" (Amodei & Hernandez, 2018; "*What Are Foundation Models? - Foundation Models in Generative AI Explained*" - *AWS*, n.d.). Some of the tasks the Foundational model can perform are language processing, visual comprehension, code generation, human-centered engagement, and speech-to-text (*What Are Foundation Models? - Foundation Models in Generative AI Explained* - *AWS*, n.d.; Merritt, 2024)

Examples of Foundation Model

BERT, released in 2018 and trained on Wikipedia and plain text using 3.3 billion tokens, can predict the next value in the sequence after analyzing the entire sequential data. BERT was one of the first foundation models. Apart from predicting the values, it predicts sentences, translates text, and answers questions (Buhl, 2023; "What Is Generative AI? " 2024).

Among the top trending names is ChatGPT, a product developed by a not-for-profit organization called OpenAI to spread awareness of the advancement of AI. The product is built using a 12-layer transformer decoder architecture to process text input by a human and generate text-based answers with the most possible accuracy. In simple terms, ChatGPT can produce human-like language text, however unoriginal, based on what the end user poses as a question. OpenAI has reached current accuracy levels in generating answers gradually since its inception in 2015, starting with GPT-1 (trained on 117 million parameters) and leading to the current version GPT-4, proficient in converting text to text, image, code, picking emotions in user's language, and more. The latest version of GPT-4 was launched in 2023. GPT-4 passed the "Uniform Bar Examination" and scored 297 (76%) (*"What Is GPT AI? - Generative Pre-Trained Transformers Explained" - AWS*, n.d.).

Another example to quote is Amazon Titan, which was developed by AWS. Similar to GPT-3 or GPT-4, this is also a General-Purpose Foundation model, that is, a machine learning model designed to cater to diverse needs such as question answering, summarizing text, converting speech to text, and more. It is pre-trained on massive datasets. Titan is based on two models, Generative and Embedding LLM. Having been trained using a natural language processing toolkit, Titan can be used by businesses as well as developers for language translation, generating both summarized and detailed texts and creating personalized applications such as chatbots, virtual assistants, and similar features. The second model is an embedding LLM that converts text inputs, encompassing words, phrases, and significant text units, into numerical representations (embeddings) that contain the semantic essence of the text (Amazon Titan Text Premier Demo (1:30), n.d.).

Hugging Face, the next popular example of a foundational model, offers open-source tools for the general public, organizations, researchers, and developers to build applications customized to their needs. The application makes this possible by giving the end-user access to a massive database featuring approximately 200,000 pre-trained models created by their community and 30,000 datasets, allowing high-level potential. It is now a popular tool for deploying pre-trained models and building AI applications ("Hugging Face – the AI Community Building the Future.", n.d.).

PRACTICAL APPLICATIONS OF GENAI IN THE BUSINESS WORLD

With the latest developments and improved accuracy of GenAI applications, all business functions can benefit from technological advancements. Big corporations such as Google, Microsoft, and OpenAI (Shaw, 2023) are constantly working to ascertain ways to improve business processes by finding ways to include artificial intelligence in the daily operations of any business. In this context, daily operations can refer to diverse corporate functions, especially the ones that primarily rely upon content, either verbal or written. Some examples of such major functions would be customer experience, sales, marketing, and software development (codes and scripts) ("Global CEE 2023," 2024; Deveau et al., 2023; (Communications, 2023).

GenAI is a magnificent subset of NLP (*What Is Generative AI?* n.d.) capable of generating human-quality text, having been trained on massive amounts of text-based, image-based, and similar datasets. Based on its command of NLP, GenAI possesses an understanding of different styles of writing, meaning it can impressively detect whether the end user is using formal or informal language, whether they are using sarcasm or humor in the language (Pathak, 2024). GenAI applications are capable of doing so by finding a pattern in the choice of words of the end user, having been trained on data from various sources from around the web. In addition to this potential, the tool can answer the user's questions in a thoughtful and informative way by accessing its vast internal libraries in a matter of seconds (Sajid, 2023; Brady, 2024).

To quote an example of how this ability can help a business, let's consider a small project in which a startup, recently funded, has a goal to market itself and create a brand name for itself within the next six months. To achieve this, the company can be assumed to rely heavily on marketing content to be used over social media, digital advertisements, and co-branding to quote a few methods. Given that it is a startup, the business can be assumed to have a low employee count, surviving on alternative approaches. In this scenario, GenAI is the tool that would benefit the business extensively. Using trending applications such as ChatGPT, Gemini, and Copilot, the business can raise requirements as questions asking for creatively written content as many times as required, even adjusting the parameters, including the length of the copy, providing keywords for SEO optimization, and more. The startup team can be specific to the extent that they can provide the brand's mission and vision statement for context within the prompt to personalize the final content for increased relatability (*Deep Analysis of ChatGPT, Microsoft Copilot, and Google Gemini*, 2024; Chui et al., 2022; Team, 2023; Valchanov, 2024). Figure 4 explains the application of GenAI across various industries and sectors.

Figure 4. Application of GenAI across Industries

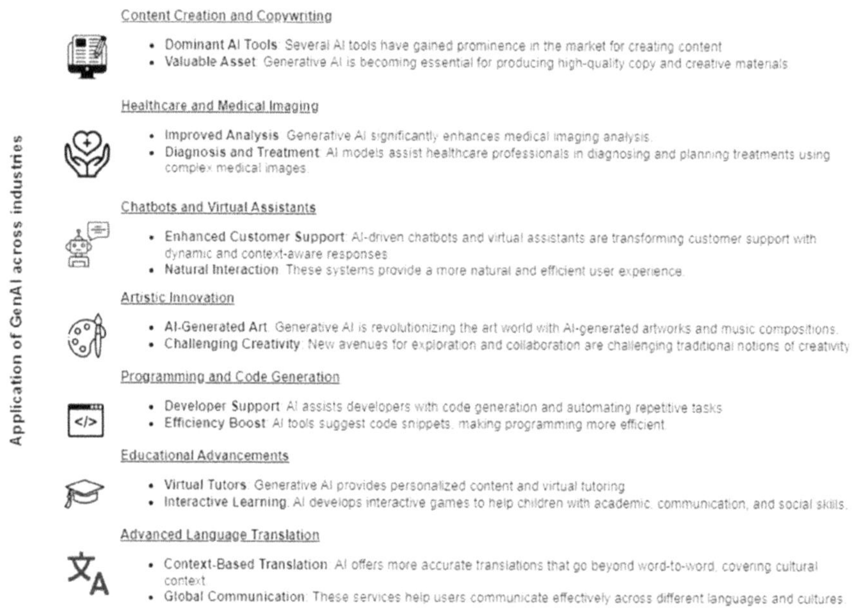

Source: Compiled by author

TRANSFORMING CUSTOMER EXPERIENCE WITH GENAI

GenAI is trained in a manner that understands human interactions, helping itself interact as a customer service employee would (Valchanov, 2024). Digital chatbots are one such example of AI-based products that are trending and are found on common brand websites such as Apple and Microsoft. These bots appear as small icons to end users, providing automated responses to common queries (Capacity, 2023). Here, common queries refer to the ones that can conveniently be resolved through step-by-step instructions such as troubleshooting. Another major feature Chatbots include is helping users connect with customer support teams in real-time or schedule appointments at service centers (Caballero, 2023). Since the internet allows them to be available 24x7, customers can benefit from Chatbot's features at their convenience.

Another commonly found use case of GenAI is Virtual Assistants, i.e., voice and text-incorporated interfaces that ease the process of using technology. The most common names in the market include Apple's Siri, Google Assistant, and Amazon's

Alexa (Schwartz, 2022; Bathia, 2021). Having been trained in NLP, these virtual assistants are even better at handling complex queries and interactions. To quote a trending scenario of businesses implementing virtual assistants, television brands such as Sony and Xiaomi have initiated partnering with Google to provide Google Assistant in their Smart TVs (Bathia, 2021; Nair & Suvarna, 2023). This business tie-up provides the TV brands' customers with a new feature to give voice-based instructions to the system, like which movie to play or which OTT app to access content on.

Morgan Stanley Wealth Management (MSWM) has introduced AI @ Morgan Stanley Debrief, an OpenAI-powered tool that generates meeting notes, identifies action items, drafts emails, and saves notes into Salesforce, enhancing efficiency for Financial Advisors (FAs). This follows the AI @ Morgan Stanley Assistant, a GenAI chatbot providing quick access to company knowledge, used by 98% of FA teams. The partnership with OpenAI, established in March 2023, highlights MSWM's commitment to using AI to streamline workflows and improve client engagement while maintaining the importance of the human touch in client relationships (Morgan Stanley, n.d.).

TRANSFORMING MARKETING & SALES WITH GENAI

GenAI has rapidly captured marketing and sales domains, given its potential in creative content writing. Trained thoroughly on Auto-Regressive Models (Auto-Regressive Models in Generative AI | Saturn Cloud, 2023), it can generate text-based responses paraphrased as many times as the end users create a prompt. The technology accommodates end users' requests, interests, and preferences based on the prompts raised, allowing it to produce highly customized responses. This skill has helped GenAI gain popularity in advertising, copywriting, and even social media influencing (Generative AI in Marketing, n.d.).

However, these responses produced could potentially lead to copyright violations or plagiarism (Mittal, 2024). This risk is caused by the fact that GenAI can yet only produce paraphrased content but also original. Here, it means that GenAI can only rephrase content infinitely from the database it is built over and not something in its original form. Hence, even with these advancements, human oversight is still significant in auditing content quality and content personalization as per business needs ("Navigate the Minefields of Generative AI," 2023).

Some major benefits in this area can be listed as (Christison, 2024):

- Content creation
- SEO Optimization

- Improved conversion rate
- Increased lead count

For example, Michaels Stores has leveraged GenAI to transform its sales and marketing by personalizing customer engagement. They use GenAI to tailor email and SMS campaigns to individual shoppers, increasing the frequency and relevance of interactions. This personalization has led to improved click-through rates and more effective marketing efforts, significantly enhancing customer engagement and driving sales (Harkness et al., 2023).

TRANSFORMING SOFTWARE ENGINEERING WITH GENAI

A major task within the software engineering domain, consuming massive amounts of resources and time, is Coding. Due to the manual efforts it requires, along with the requirement of developers, testers, and management, to name a few roles, it is a great example of how all industries would benefit from a reduction in human effort in this area (*Generative AI for Software Engineering - Capgemini USA*, 2023).

The characteristics that GenAI has received from the "Transformers" architecture allow it to effectively provide text-to-code functions to the end user (*Text Generation Models*, n.d.). Architecture models have trained GenAI well to understand the user's instructions and generate code accordingly for coding languages including but not limited to Python, SQL, Java, and VBA. One example of such an application is Codex, a product by OpenAI dedicated to code generation (Zaremba et al., 2021). While Codex is primarily associated with generating code to reduce human effort, it is capable of directly relevant additional tasks such as completing a code that is a work in progress shared by the user, sharing suggestions on code improvement, and saving time by performing complex coding tasks based on constant efforts within performance improvement of GenAI applications.

The business sector benefits from this feature by reducing its annual spending in this business function. This is possible primarily by letting GenAI take care of core activities at their initial stage, which are further verified by respective teams for accuracy. Some examples of these activities would include early-stage drafts of code, code improvement, automating repetitive code pieces, root cause analysis where necessary in case of errors, and so on.

GitHub Copilot utilizes OpenAI Codex to suggest software code and entire functions based on natural language prompts. It is affordable, with costs for individual programmers at $10 a month and $19 per month per business user. This tool demonstrates how GenAI can become an invaluable coding partner for developers, significantly enhancing productivity and reducing the time needed to deploy new software (*The Startling Power Generative AI Is Bringing to Software Development*, n.d.).

POWER OF GENAI TO GOVERN AN ENTIRE INDUSTRY

Similar to how the Large Language Models (LLM) are capable of influencing a business function, we can adjust their usage for an even deeper impact by implementing the advancements across a whole industry. The said impact can be planned by adjusting how well GenAI is used overall across different functions within the respective industry.

Some key examples that may help understand the potential GenAI holds to change the industry through changes across domains can be stated:

- Pharmaceuticals Industry - Drafting the usage instructions and precautions for consuming a medicine while strategizing the resale of the drug. This should help automate a whole subset of research and documentation, leaving a big overall impact on the industry.
- Banking Industry - A common example here that we can observe in our daily routines, due to its heavy usage by the banking sector, is how they are conducting branding of their name as well as their products (such as credit cards). At present, the marketing and sales team of any bank heavily relies on personalized advertising, such as SMS marketing, email marketing, personalized automated calls, and more. By using GenAI in these processes, banks can tailor content to each existing and potential customer's credit profile and history within the banking industry.

EXPLORING HOW GENAI CAN STRENGTHEN THE COMPETITIVE POSITION OF AN ORGANISATION IN THE MARKET

Businesses can boost customer experience and add new capabilities by integrating their systems with GenAI. If the businesses have sufficient potential to build their own GenAI model using their data and expertise, they can become proficient and outdo publicly available tools like ChatGPT. But before diving into their model building, businesses should understand the uniqueness of their data, assess the trustworthiness of customer feedback, and calculate the cost of getting credible customer feedback.

Understanding the Uniqueness of the Data

For businesses to leverage GenAI, the key factor is to have Information that is specific to the business and not easily replicated by publicly available data for the competitors. For instance, Netflix, an online streaming platform, has a plethora of user behavior data that is specific to its platform, including data on user preferences, frequency, and most searched content. Netflix can use this data and leverage GenAI to create highly personalized recommendations for users (Gomez-Uribe & Hunt, 2015). This type of data, which is idiosyncratic, is crucial to developing AI models that outperform publicly available tools because they provide personalized insights and solutions. If businesses strategically understand the value of such idiosyncratic data, they will gain a competitive advantage (Gao & Verrecchia, 2012).

Assessing the Trustworthiness of Customer Feedback Through AI

Companies that can supervise how their products or services are helping customers fulfill their needs have the advantage of leveraging GenAI technologies. For instance, Deep Brew, a predictive analytics and machine learning tool of Starbucks, helps to personalize their communications, loyalty programs, and marketing efforts and enhance customer experience by recommending drinks based on the weather, location, and past purchasing habits (Lumoa, 2024). However, not all businesses have access to direct feedback, as Starbucks and Netflix do. In such a case, businesses rely on indirect feedback metrics, like session length, bounce rate, and engagement within an app. Duolingo, a language learning platform, uses data such as lesson completion rates, daily streaks, and lag effect. This type of data helps Duolingo enhance the user experience by adjusting the difficulty level of the lessons and introducing new learning modules that are bespoke to users' needs. It also enables the company to anticipate users' propensity to retain words at a given moment, thereby enhancing learning through the spacing effect, a concept which believes that learning is more effective when distributed over time and is founded upon psycholinguistic theory and further refined through the utilization of machine learning (Settles, 2022). This type of data, while informative, may not provide straightforward insights about customer satisfaction. It is important to acknowledge that the effectiveness of data differs considerably among different product ranges. For example, Spotify incorporates a complex algorithm to evaluate listening patterns, song choices, and genres, and it saves to understand user preferences and recommend personalized music instead of depending on feedback like ratings (Stacy, 2021). Binary feedback, in this case, may only be useful to assess a customer's satisfaction with the recommended music.

It is crucial to acknowledge that the efficacy of feedback varies considerably among various product categories. Straightforward indicators of approval or disapproval (for instance, thumbs-up/thumbs-down) can be extremely effective in evaluating user satisfaction with a movie recommendation on content platforms. However, such binary feedback may not be as valuable when it comes to evaluating intricate features of enterprise software, as user contentment relies on a multitude of factors that extend beyond mere approval.

IMPACT OF GENAI ON PRODUCTIVITY AND ECONOMIC GROWTH

Technology has changed the way people interact with their work; it has enhanced their capabilities with machines and computers. GenAI has further evolved in its progression by automating tasks and transforming a range of activities and professions. In 2017, it was estimated that with the help of current technology, about half of the work tasks could be automated, but widespread adoption depends on economic feasibility and integration time (*A Future That Works: Automation, Employment, and Productivity*, 2017).

Historical shifts, such as the agricultural employment in China, which was 82 percent in 1962 and reduced to 13 percent in 2013, exemplify how work activities evolved. In the United States, millions of people leave their jobs every month. The workforce is facing the challenge of adjusting and maintaining income stability. Demographic changes, like aging populations, require increased labor efficiency to sustain economic growth (*"Job Openings and Labor Turnover Survey News Release"*, n.d.).

Integrating GenAI with existing technologies can hasten the automation and enhance workforce capabilities, impacting knowledge workers earlier than anticipated. This highlights the potential capabilities of GenAI in driving economic growth and changing work activities.

Transforming Knowledge Work

Figure 5 gives the anticipated timeline for technology to achieve human-level performance across several technical abilities. In the figure, the timeline compares estimates regarding the achievement of human-level work by GenAI. Recent developments in GenAI have increased the technological capabilities. It is now expected to match "median human performance reaches top quartile performance earlier than previously estimated across a wide range of capabilities" (Chui et al., 2023).

The timeline differentiates between the expected median and top-quartile human performance levels. Blue dots represent the estimates made in 2023, and grey dots represent estimates made in 2017. The lines that are connecting the dots indicate the predictions made by the experts. The timeline for reaching median human performance in natural-language understanding has advanced from 2027 to 2023, indicating the rapid progress in GenAI capabilities. Comparable progress can be observed in logical reasoning and creativity.

Figure 5. Breakdown of tasks typically performed by human workers in business settings

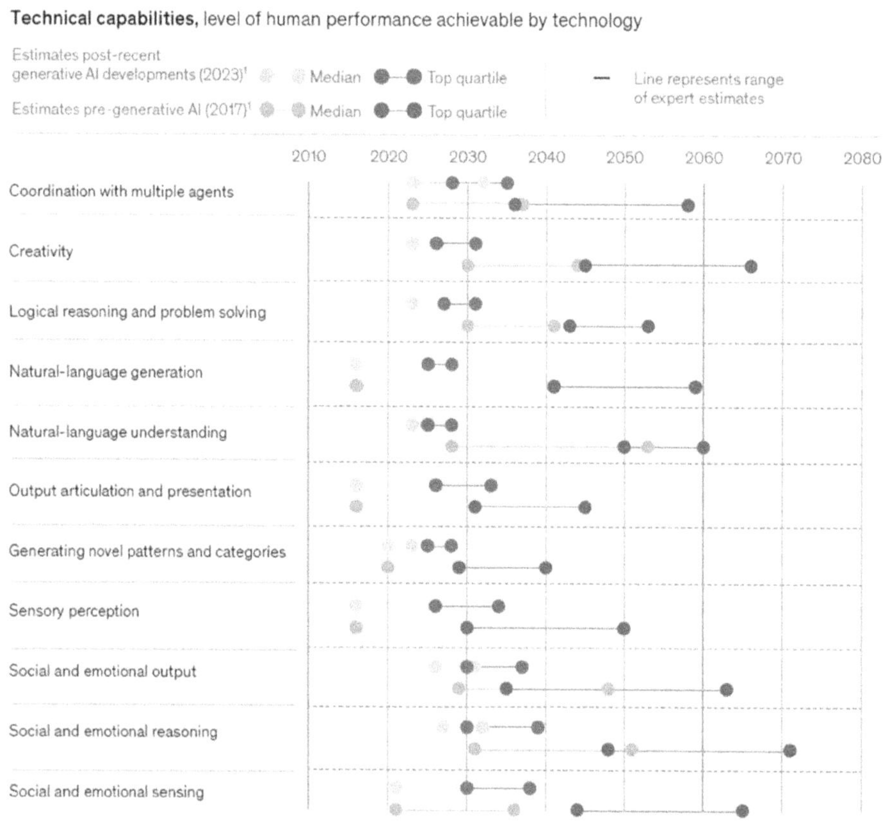

Source: "McKinsey Global Institute occupation database; McKinsey analysis Coordination"

As a result of these technological advancements, the potential for automation has significantly risen. The existing technology has made it possible to increase automation in the working environment from 50% to approximately 60%-70%. This sharp increase is attributed to the continuous improvement in the NLP capabilities of GenAI ("*The State of AI in 2023: Generative AI&Rsquo;S Breakout Year*, 2023"; Ellingrud et al., 2023).

Recent studies and reports indicate that the impact of GenAI on the automation of work activities has transformed industries (Chui et al., 2023; "*The State of AI in 2023: Generative AI&Rsquo Breakout Year*", 2023). Research conducted by Deloitte supports the findings of the study done by McKinsey & Company, where the theoretical percentage of automation increased the work hours from 50% to 60-70%. This emphasized that GenAI can boost productivity and reduce labor costs by automating tasks. This ability of the technology to efficiently process and generate high-quality output has increased the automation potential across multiple areas ("*Generative AI and the Labor Market: A Case for Techno-optimism,*" 2024; *Generative AI in Sourcing and Procurement Operations*, n.d.).

The adoption scenarios for GenAI encompass the consideration of integration time regarding technological capabilities, cost comparisons relative to human labor, and the diffusion rate of these technologies throughout the economy. The importance of strategic planning for CEOs is discussed by Boston Consulting Group (BCG) in their report. The report also highlights that GenAI can nearly disrupt every industry by providing competitive advantages, opening doors for new opportunities, and introducing new efficiencies. The businesses should focus on used cases, adapt organizational structures, and establish ethical guidelines to harness the potential of this technology effectively and efficiently (Candelon et al., 2023). KPMG, in its report, also highlights that the tasks that earlier required humans, such as preparing tests or evaluating student work, can now be done with the help of GenAI, freeing up time for more interactive teaching and learning (Heneghan & Henninger, 2023).

The cumulative perspectives collectively emphasize the transformative potential GenAI holds in reshaping the workforce, boosting productivity, and fostering economic growth. Hence, there is a need for businesses to adapt to these changes and invest in the technology, make workers learn the skills, and develop strategic approaches to leverage the capabilities of AI technology.

BUSINESS AND SOCIETAL CONCERNS

History has demonstrated that novel technologies can change societies, and AI is no exception. AI has impacted our daily lives to a great extent. The rapid advancement of GenAI could augment and enhance AI's overall influence, potentially resulting in trillions of dollars in additional value annually and revolutionizing the work culture.

Besides these advancements, these novel technologies also present significant challenges. These obstacles must be promptly addressed, given the rate at which GenAI is making advancements. Concerns have already emerged regarding the data generated by GenAI systems, including probable breaches of intellectual property due to the "plagiarism" within the training data utilized to establish foundational models. Additionally, several uncertainties continue to persist regarding the accuracy and comprehensibility of the output generated by LLMs, along with the question of whether the output is biased or unbiased, possibly mimicking detrimental stereotypes (Rash, 2024).

The economic consequences are also noteworthy because the scale at which the workforce transitions are happening is alarming. In the midpoint adoption scenario, approximately a quarter to a third of work assignments could change in the upcoming years. Table 2 shows other risk categories that GenAI is believed to pose to society. Therefore, it is crucial to manage the potential opportunities and challenges of GenAI simultaneously.

Table 2. Potential risks of GenAI

Risk Category	Description
Fairness	There is a risk of algorithmic bias due to imperfect training data or decisions made by developers.
Intellectual Property (IP)	AI training data and outputs can infringe on legally protected materials, necessitating a careful review of data origins and uses.
Privacy	Privacy concerns arise if user input data is inappropriately included in AI outputs, potentially identifying individuals.
Security	AI can be exploited by bad actors to enhance cyberattacks or manipulated to produce harmful outputs via prompt injection.
Explainability	The complexity of AI models makes it difficult to explain how outputs are generated.
Reliability	AI models may produce varying answers to the same prompts, complicating the assessment of accuracy and reliability.
Organisational Impact	AI could significantly affect the workforce with potentially disproportionate negative impacts on certain groups and communities.
Social and Environmental Impact	The training of AI models can have detrimental effects on society and the environment, such as increased carbon emissions.

Source: Compiled by author

It is possible that GenAI models, due to imperfect training data or the developer's subjective decision-making, start exhibiting algorithm bias, which could lead to unfair outcomes. The outcome could further affect marginalized groups and could even amplify those biases, resulting in discriminatory output (Heneghan & Henninger, 2023; *The State of AI in 2023: Generative AI&Rsquo;S Breakout Year*, 2023; Bockting et al., 2023).

There is also a risk to intellectual property. These include copyright violations, trademark, and patent infringement. The risk is persistent even while using third-party GenAI tools. Companies need a system that is robust enough to manage intellectual property, alleviate the risks associated, and avoid legal actions (Heneghan & Henninger, 2023).

The global deployment of AI technologies is complicated by inconsistent regulations across countries, leading to challenges for multinational companies and opportunities for regulatory arbitrage. Additionally, the complexity of assigning responsibility and liability when AI causes harm, especially in scenarios involving multiple parties across different jurisdictions, underscores the need for international cooperation and harmonization of AI regulations. Clear, consistent frameworks are essential to address both cross-border regulatory challenges and liability issues (Klein & Patrick, 2024).

Privacy is another concern that arises. This could potentially make the individuals identifiable. The risk is increased when the GenAI processes sensitive Information and could unintentionally disclose private Information. Businesses should implement strict data-handling policies to comply with data protection regulations. GenAI applications and tools can also be used by malevolent individuals to conduct cyberattacks, creating advanced phishing schemes. It can also be used to change the generated data and manipulate it to produce harmful results. Strengthening policies to develop a robust cybersecurity system could help mitigate the risk of exploiting it. When GenAI models are based on deep learning, operating as black boxes with billions of parameters, it is difficult for users to comprehend how the particular data is generated, and they lack transparency. It makes the system complicated and questions the accountability of the output *("Generative AI and Cybersecurity: Strengthening Both Defenses and Threats," 2023).*

It is observed that the models can generate different outputs to the same input, creating inconsistencies and unreliability towards the output. This difference makes it challenging for the users to judge the data for accuracy and decreases the dependability. To avoid this, regular testing and validation of the processes should be done so that performance reliability is ensured ("*The State of AI in 2023: Generative AI&Rsquo's Breakout Year*", 2023; *A Survey of Generative AI Applications*, n.d.).

GenAI has automated several tasks that earlier required manual labor, causing job displacement and economic disruption of specific groups and communities. Businesses should motivate their workforce and invest in reskilling programs to lessen the impact. However, the rapid pace of AI development can outstrip the ability of their reskilling, creating a gap. This can be resolved by creating accessible training programs through public-private partnerships and promoting learning through digital platforms and workshops. There is also a risk of this technology being concentrated among a few, exacerbating wealth inequality and threatening small businesses that struggle to compete with larger, AI-enabled companies. Promotion of human-AI collaboration, investment in creative industries, and enforcement of ethical guidelines for transparent usage might mitigate these effects (Alonso et al., 2020). An increase in surveillance by AI-driven systems could create a high-pressure work environment, leading to job dissatisfaction. This makes it crucial for AI creators to implement ethical conduct to restrict excessive surveillance and enhance technology's complex problem-solving skills (Giacosa et al., 2023). Another major risk is carbon emission. Energy consumption in the training of an LLM contributes to environmental degradation. A set of practices promoting sustainability effective immediately could be the best plausible solution (*"The State of AI in 2023: Generative AI&Rsquo's Breakout Year"*, 2023; *Stanton & Higgins, 2023; "Generative AI and Cybersecurity: Strengthening Both Defenses and Threats," 2023).*

ETHICAL CONCERNS

As GenAI continues to innovate and is poised to redefine various industries, it presents both exciting opportunities and significant ethical challenges. While its ability to automate tasks and generate creative outputs holds great promise, the ethical implications demand careful attention. To ensure that GenAI is developed and deployed in a manner that is both responsible and equitable, it is important to address some of the issues mentioned in Figure 6 (Ahuja, 2024; Dey, 2024).

Figure 6. Ethical Concerns GenAI

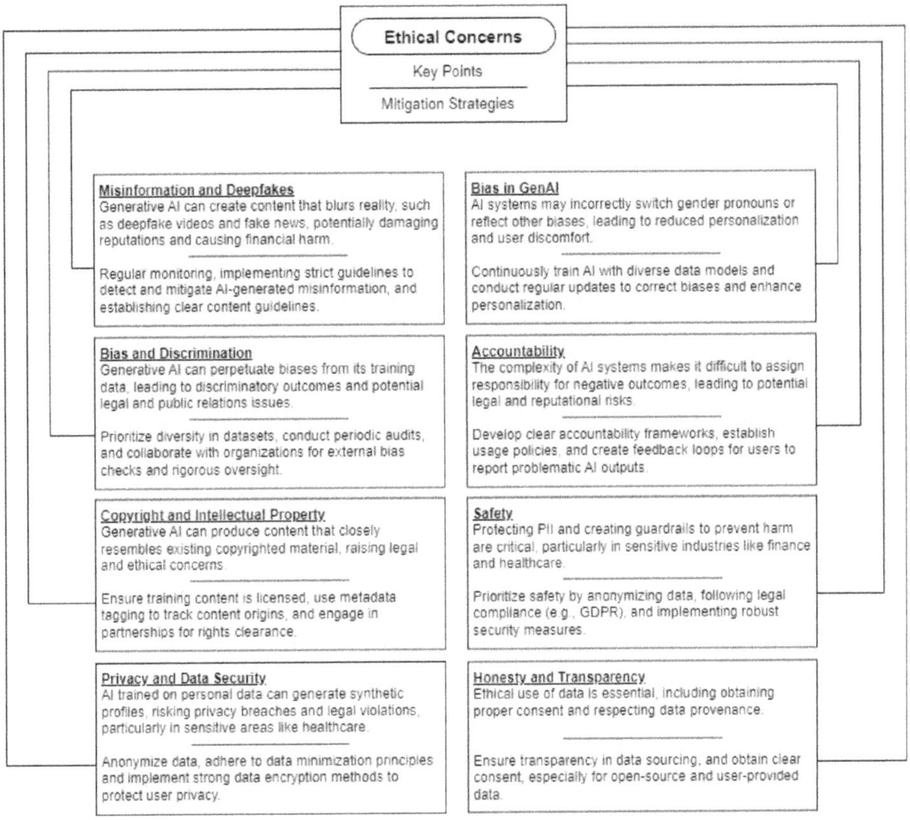

Source: Compiled by author

FUTURE TRENDS IN GENAI

Since GenAI has gained a lot of traction and attention in the last few years across all industries and sectors, it seems to have promising use cases that can be implemented in several sectors; NLP models are being improved to generate content like art, music, and storytelling, enhance emotional intelligence for better human interaction, and facilitate tasks such as summarization, translation, and efficient workflow management. Figure 7 further describes the future trends in GenAI.

Figure 7. Evolutions in GenAI

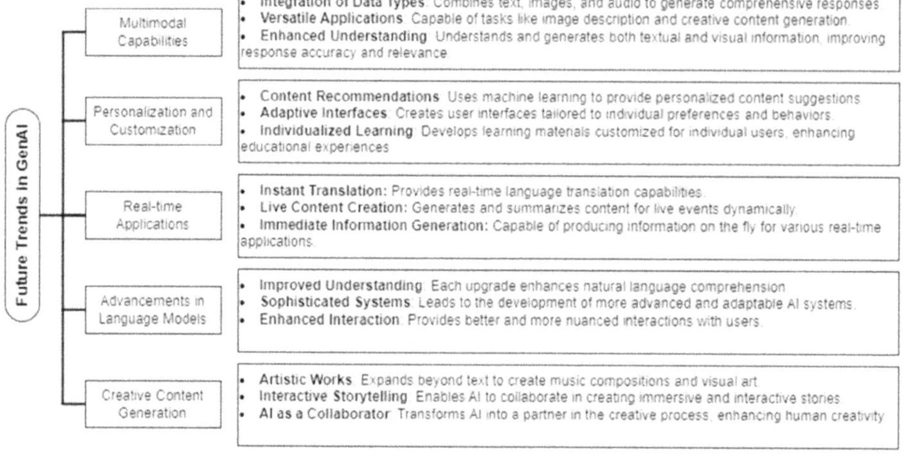

Source: Compiled by author

CONCLUSION

A technological innovation that emerges rapidly and spreads widely seemingly overnight can evoke both admiration and apprehension. GenAI, introduced in the autumn of 2022, made businesses and individuals adopt it, incorporate it, explore this technology, and become a prime example of this occurrence. As we begin this journey to comprehend the potency, scope, and capabilities of GenAI, in the coming years, these technological advancements and breakthroughs may continuously change our perception of the impact of AI on our daily lives and work.

Since the adoption of GenAI is rapid, there is a need to act to hasten the digital transformation and retain workforces by making them learn the skills that are required to survive in this dynamic environment. Because these tools contend with the expenses of adjusting to and lessening the effects of climate change, they hold great value for the global economy. However, GenAI also can be more disruptive than previous AI technologies. Its proficiency with language, which is considered to be a key aspect of knowledge and expertise, can be both advantageous and detrimental. The best way to summarize the introduction and growth of Generative AI is how it holds the potential to encourage individuals as well as markets to adapt to rapidly growing technological innovations.

REFERENCES

A future that works: Automation, employment, and productivity. (2017, January 12). McKinsey & Company. https://www.mckinsey.com/featured-insights/digital-disruption/harnessing-automation-for-a-future-that-works/de-DE

A survey of Generative AI Applications. (n.d.). Ar5iv. https://ar5iv.labs.arxiv.org/html/2306.02781

Ahuja, V. (2024, June 3). Ethical Considerations for Generative AI. Forbes. https://www.forbes.com/sites/forbestechcouncil/2024/02/14/ethical-considerations-for-generative-ai/

Alonso, C., Kothari, S., & Rehman, S. (2020, December 2). *How artificial intelligence could widen the gap between rich and poor nations*. IMF. https://www.imf.org/en/Blogs/Articles/2020/12/02/blog-how-artificial-intelligence-could-widen-the-gap-between-rich-and-poor-nations

Amazon Titan Text Premier Demo (1:30). (n.d.). [Video]. Amazon Web Services, Inc. https://aws.amazon.com/bedrock/titan/

Amodei, D., & Hernandez, D. (2018, May 16). *AI and compute*. IBM. https://openai.com/index/ai-and-compute/

Auto-Regressive models in generative AI | Saturn Cloud. (2023, July 14). https://saturncloud.io/glossary/autoregressive-models-in-generative-ai/

Bandi, A., Adapa, P. V. S. R., & Kuchi, Y. E. V. P. K. (2023). The power of Generative AI: A review of requirements, models, Input–Output formats, evaluation metrics, and challenges. *Future Internet*, 15(8), 260. DOI: 10.3390/fi15080260

Bathia, U. (2021, September 23). *Conversational AI Examples: How Siri, Alexa & Google Assistant have Human-like Conversations*. CFTE. https://blog.cfte.education/conversational-ai-examples-how-siri-alexa-google-assistant-have-human-like-conversations/

Benedikt Frey, C., & Osborne, M. (2023). *Generative AI and the Future of Work: a reappraisal – The Brown Journal of World Affairs*. The Brown Journal of World Affairs. https://bjwa.brown.edu/30-1/generative-ai-and-the-future-of-work-a-reappraisal/

Bockting, C. L., Van Dis, E. M., Van Rooij, R., Zuidema, W., & Bollen, J. (2023). Living guidelines for generative AI — Why scientists must oversee its use. *Nature*, 622(7984), 693–696. DOI: 10.1038/d41586-023-03266-1 PMID: 37857895

Brady, D. (2024, February 8). *What developers need to know about generative AI - The GitHub Blog*. The GitHub Blog. https://github.blog/2023-04-07-what-developers-need-to-know-about-generative-ai

Brown, P. F., Cocke, J., Della Pietra, S. A., Della Pietra, V. J., Jelinek, F., Lafferty, J., Mercer, R. L., & Roossin, P. S. (1990). *A statistical approach to machine translation*. ACL Anthology. https://aclanthology.org/J90-2002

Brown, T. B., Mann, B., Ryder, N., Subbiah, M., Kaplan, J., Dhariwal, P., Neelakantan, A., Shyam, P., Sastry, G., Askell, A., Agarwal, S., Herbert-Voss, A., Krueger, G., Henighan, T., Child, R., Ramesh, A., Ziegler, D. M., Wu, J., Winter, C., . . . Amodei, D. (2020). Language Models are Few-Shot Learners. arXiv (Cornell University). https://arxiv.org/pdf/2005.14165.pdf

Caballero, B. (2023, August 17). 8 best AI Chatbots for Customer Service. *Fit Small Business*. https://fitsmallbusiness.com/best-ai-chatbots-for-customer-service/

Candelon, F., Gupta, A., Krayer, L., & Zhukov, L. (2023, May 25). *The CEO's guide to the Generative AI Revolution*. BCG Global. https://www.bcg.com/publications/2023/ceo-guide-to-ai-revolution

Cao, Y., Li, S., Liu, Y., Yan, Z., Dai, Y., Yu, P. S., & Sun, L. (2023). A Comprehensive Survey of AI-Generated Content (AIGC): A History of Generative AI from GAN to ChatGPT. *arXiv (Cornell University)*. https://doi.org//arxiv.2303.04226DOI: 10.48550

Capacity, T. (2023, April 11). *Everything you need to know about Helpdesk Chatbots (2023) - capacity*. Capacity. https://capacity.com/learn/helpdesk-chatbots/

Christison, C. (2024, April 8). *The 15 best AI tools for Social Media in 2024*. Social Media Marketing & Management Dashboard. https://blog.hootsuite.com/best-ai-tools-for-social-media/

Chui, M., Roberts, R., & Yee, L. (2022). *Generative AI is here: How tools like ChatGPT could change your business*. https://www.mckinsey.com/~/media/mckinsey/business%20functions/quantumblack/our%20insights/generative%20ai%20is%20here%20how%20tools%20like%20chatgpt%20could%20change%20your%20business/generative-ai-is-here-how-tools-like-chatgpt-could-change-your-business.pdf

Clifford, C. (2018, February 1). *Google CEO: A.I. is more important than fire or electricity*. CNBC. https://www.cnbc.com/2018/02/01/google-ceo-sundar-pichai-ai-is-more-important-than-fire-electricity.html#:~:text=%E2%80%9CAI%20is%20one%20of%20the,the%20potential%20of%20artificial%20intelligence

Communications, H. (2023, September 6). HubSpot unveils HubSpot AI and new sales hub at INBOUND 2023. *HubSpot*. https://www.hubspot.com/company-news/inbound-2023-hubspot-ai-sales-hub

Davenport, T. H. (2023, August 15). *How generative AI is changing creative work*. Harvard Business Review. https://hbr.org/2022/11/how-generative-ai-is-changing-creative-work?utm_medium=paidsearch&utm_source=google&utm_campaign=intlcontent_bussoc&utm_term=Non-Brand&tpcc=intlcontent_bussoc&gad_source=1&gclid=Cj0KCQjw6auyBhDzARIsALIo6v80evxQNjT6M-JEvCAm_-InTpL3Y-1i8TE5yXus9ri1dFgphEmXHQAaAiuREALw_wcB

Deep, A. I. (2020, June 25). *Autoregessive model*. DeepAI. https://deepai.org/machine-learning-glossary-and-terms/autoregressive-model

Dey, S. (2024, February 20). Which ethical implications of generative AI should companies focus on? *Forbes*. https://www.forbes.com/sites/forbestechcouncil/2023/10/17/which-ethical-implications-of-generative-ai-should-companies-focus-on/

Ellingrud, K., Sanghvi, S., Dandona, G. S., Madgavkar, A., Chui, M., White, O., & Hasebe, P. (2023). Generative AI and the future of work in America. In *McKinsey & Company*. https://www.mckinsey.com/mgi/our-research/generative-ai-and-the-future-of-work-in-america

Explained: generative AI. (2023, November 9). MIT News | Massachusetts Institute of Technology. https://news.mit.edu/2023/explained-generative-ai-1109

Floridi, L., Cowls, J., Beltrametti, M., Chatila, R., Chazerand, P., Dignum, V., Luetge, C., Madelin, R., Pagallo, U., Rossi, F., Schäfer, B., Valcke, P., & Vayena, E. (2018). AI4People—An Ethical Framework for a Good AI Society: Opportunities, risks, principles, and recommendations. *Minds and Machines*, 28(4), 689–707. DOI: 10.1007/s11023-018-9482-5 PMID: 30930541

Gandhi, V. (2015). Interfacing brain and Machine. In *Elsevier eBooks* (pp. 7–63). DOI: 10.1016/B978-0-12-801543-8.00002-8

Gao, P., & Verrecchia, R. E. (2012). Economic consequences of idiosyncratic Information in diversified markets. *Social Science Research Network*. DOI: 10.2139/ssrn.2008364

Generative AI and cybersecurity: Strengthening both defenses and threats. (2023, September 18). Bain. https://www.bain.com/insights/generative-ai-and-cybersecurity-strengthening-both-defenses-and-threats-tech-report-2023/

Generative AI and the labor market: A case for techno-optimism. (2024, March 25). Deloitte Insights. https://www2.deloitte.com/us/en/insights/economy/generative-ai-impact-on-jobs.html

Generative AI for Software Engineering - Capgemini USA. (2023, August 11). Capgemini USA. https://www.capgemini.com/us-en/solutions/generative-ai-for-software-engineering/

Generative AI in marketing. (n.d.). https://www.marketsandmarkets.com/industry-practice/GenerativeAI/genai-marketing

Generative AI in sourcing and procurement operations. (n.d.). Deloitte United States. https://www2.deloitte.com/us/en/blog/business-operations-room-blog/2023/generative-ai-in-procurement.html

Giacosa, E., Alam, G. M., Culasso, F., & Crocco, E. (2023). Stress-inducing or performance-enhancing? Safety measure or cause of mistrust? The paradox of digital surveillance in the workplace. *Journal of Innovation & Knowledge*, 8(2), 100357. DOI: 10.1016/j.jik.2023.100357

Global, C. E. E. 2023. (2024, January 18). *KPMG*. https://kpmg.com/xx/en/home/insights/2024/01/global-cee-2023.html

Gomez-Uribe, C. A., & Hunt, N. T. (2015). The Netflix Recommender system. *ACM Transactions on Management Information Systems*, 6(4), 1–19. DOI: 10.1145/2843948

Goodfellow, I., Pouget-Abadie, J., Mirza, M., Xu, B., Warde-Farley, D., Ozair, S., Courville, A., & Bengio, Y. (2020). Generative adversarial networks. *Communications of the ACM*, 63(11), 139–144. DOI: 10.1145/3422622

Hao, K. (2022, January 10). The messy, secretive reality behind OpenAI's bid to save the world. *MIT Technology Review*. https://www.technologyreview.com/2020/02/17/844721/ai-openai-moonshot-elon-musk-sam-altman-greg-brockman-messy-secretive-reality/

Harkness, L., Robinson, K., Stein, E., & Wu, W. (2023, December 5). How generative AI can boost consumer marketing. McKinsey & Company. https://www.mckinsey.com/capabilities/growth-marketing-and-sales/our-insights/how-generative-ai-can-boost-consumer-marketing#/

Heath, D., & Heath, C. (2007). *Made to Stick: Why Some Ideas Survive and Others Die. Random House*. Random House.

Heneghan, L., & Henninger, P. (2023, April 17). Generative AI models – the risks and potential rewards in business. *KPMG*. https://kpmg.com/xx/en/home/insights/2023/04/generative-ai-models-the-risks-and-potential-rewards-in-business.html

Ho, J., Jain, A., & Abbeel, P. (2020). Denoising diffusion probabilistic models. *Neural Information Processing Systems, 33*, 6840–6851. https://proceedings.neurips.cc/paper/2020/file/4c5bcfec8584af0d967f1ab10179ca4b-Paper.pdf

Hugging Face – The AI community building the future. (n.d.). https://huggingface.co/

Job Openings and Labor Turnover Survey News Release. (n.d.). U.S. Bureau of Labor Statistics. https://www.bls.gov/news.release/jolts.htm

Kaur, H. (2024, February 2). Types of Generative AI Models and Its Applications. *Signity Solutions*. https://www.signitysolutions.com/blog/types-of-generative-ai-models

Klein, E., & Patrick, S. (2024, March 21). *Envisioning a global regime complex to govern artificial intelligence*. Carnegie Endowment for International Peace. https://carnegieendowment.org/research/2024/03/envisioning-a-global-regime-complex-to-govern-artificial-intelligence?lang=en

Lin, C., Gao, J., Tang, L., Takikawa, T., Zeng, X., Huang, X., Kreis, K., Fidler, S., Liu, M., & Lin, T. (2022). Magic3D: High-Resolution Text-to-3D content Creation. *arXiv (Cornell University)*. https://doi.org//arxiv.2211.10440DOI: 10.48550

Lo, A. W., & Ross, J. (2024). Generative AI from Theory to Practice: A Case Study of Financial Advice. *An MIT Exploration of Generative AI*. DOI: 10.21428/e4baedd9.a1f6a281

Martineau, K. (2024, May 1). *What is generative AI?* IBM Research. https://research.ibm.com/blog/what-is-generative-AI

Mast, G. (2023, June 20). *Toyota Research Institute unveils new generative AI technique for vehicle design - Toyota USA Newsroom*. Toyota USA Newsroom. https://pressroom.toyota.com/toyota-research-institute-unveils-new-generative-ai-technique-for-vehicle-design/

Merritt, R. (2024, January 24). *What are foundation models? | NVIDIA Blogs*. NVIDIA Blog. https://blogs.nvidia.com/blog/what-are-foundation-models/

Mittal, A. (2024, January 9). *The Plagiarism Problem: How generative AI models reproduce copyrighted content*. Unite.AI. https://www.unite.ai/the-plagiarism-problem-how-generative-ai-models-reproduce-copyrighted-content/

Molad, E., Horwitz, E., Valevski, D., Acha, A. R., Matias, Y., Pritch, Y., Leviathan, Y., & Hoshen, Y. (2023). Dreamix: Video Diffusion Models are General Video Editors. *arXiv (Cornell University)*. https://doi.org//arxiv.2302.01329DOI: 10.48550

Morehead, A., & Cheng, J. (2023). Geometry-Complete diffusion for 3D molecule generation and optimization. *arXiv (Cornell University)*. https://doi.org//arxiv.2302.04313DOI: 10.48550

Morgan Stanley. (n.d.). Launch of AI @ Morgan Stanley Debrief | Morgan Stanley. https://www.morganstanley.com/press-releases/ai-at-morgan-stanley-debrief-launch#:~:text=The%20new%20AI%20%40%20Morgan%20Stanley,meetings%20and%20surfaces%20action%20items

Nair, N. P., & Suvarna, S. (2023, April 13). Xiaomi Smart TV X Pro Series with Google TV, Dolby Vision IQ Launched in India: All details. *Gadgets 360*. https://www.gadgets360.com/tv/news/xiaomi-smart-tv-x-pro-43-50-55-inch-price-india-rs-32999-41999-47999-launch-sale-date-april-19-specifications-features-3945522

Navigate the minefields of generative AI. (2023, August 15). *KPMG*. https://kpmg.com/my/en/home/insights/2023/08/navigate-the-minefields-of-generative-ai.html

Novet, J. (2023, April 10). *Microsoft's $13 billion bet on OpenAI carries huge potential along with plenty of uncertainty*. CNBC. https://www.cnbc.com/2023/04/08/microsofts-complex-bet-on-openai-brings-potential-and-uncertainty.html

Open, A. I., & Pilipiszyn, A. (2021, March 25). GPT-3 powers the next generation of apps. *OpenAI*. https://openai.com/index/gpt-3-apps/

Pathak, P. (2024, January 25). *GenAI vs. LLMs vs. NLP: A Complete Guide*. Scribble Data. https://www.scribbledata.io/blog/genai-vs-llms-vs-nlp-a-complete-guide/

Rahwan, I. (2018). Society-in-the-loop: Programming the algorithmic social contract. *Ethics and Information Technology*, 20(1), 5–14. DOI: 10.1007/s10676-017-9430-8

Rash, W. (2024, May 9). Generative AI exposes users to new security risks. *Forbes*. https://www.forbes.com/sites/waynerash/2024/02/07/generative-ai-exposes-users-to-new-security-risks/?sh=452246c52dfe

Schwartz, E. H. (2022, July 1). *Alexa beats Google Assistant and Siri in voice assistant popularity as voice AI market expands - Voicebot.ai*. Voicebot.ai. https://voicebot.ai/2022/06/24/alexa-beats-google-assistant-and-siri-in-voice-assistant-popularity-as-voice-ai-market-expands/

Settles, B. (2022, July 6). *How we learn how you learn*. Duolingo Blog. https://blog.duolingo.com/how-we-learn-how-you-learn/

Shaw, F. X. (2023, November 17). Microsoft Ignite 2023: AI transformation and the technology driving change - The Official Microsoft Blog. The Official Microsoft Blog. https://blogs.microsoft.com/blog/2023/11/15/microsoft-ignite-2023-ai-transformation-and-the-technology-driving-change/

Shukla, N. (2024, April 11). *Evolution of Language models: From Rules-Based Models to LLMs*. Appy Pie. https://www.appypie.com/blog/evolution-of-language-models

Skalic, M., Jiménez, J., Sabbadin, D., & De Fabritiis, G. (2019). Shape-Based Generative Modeling for de Novo Drug Design. *Journal of Chemical Information and Modeling*, 59(3), 1205–1214. DOI: 10.1021/acs.jcim.8b00706 PMID: 30762364

Stacy. (2021, October 28). Adding that extra 'You' to your discovery: Oskar Stål, Spotify vice president of personalization, explains how it works — Spotify. Spotify. https://newsroom.spotify.com/2021-10-13/adding-that-extra-you-to-your-discovery-oskar-stal-spotify-vice-president-of-personalization-explains-how-it-works/

Stanton, C., & Higgins, M. (2023, December). *Harvard Business School*. https://www.hbs.edu/faculty/Pages/item.aspx?num=65160#:~:text=Generative%20AI%20seemed%20poised%20to,Creation%2C%20and%20the%20labor%20market. https://hbsp.harvard.edu/product/824130-PDF-ENG

Susskind, D. (2020). A world without work: technology, automation and how we should respond. https://openlibrary.org/books/OL28086183M/World_Without_Work

Szudarski, P. (2023). *Collocations, corpora and language learning*. DOI: 10.1017/9781108992602

Team, S. (2023, September 21). *Maximizing SEO Impact with ChatGPT: A Comprehensive Guide*. Semrush Blog. https://www.semrush.com/blog/chatgpt-seo/

Text generation models. (n.d.). OpenAI; OPenAI. https://platform.openai.com/docs/guides/text-generation/json-mode

The CEO's Guide to Generative AI. (2024, April). IBM Institute for Business Value. Retrieved May 18, 2024, from https://www.ibm.com/thought-leadership/institute-business-value/en-us/report/ceo-generative-ai

The startling power generative AI is bringing to software development. (n.d.). KPMG. https://kpmg.com/us/en/articles/2023/generative-artificial-intelligence.html

The state of AI in 2023: Generative AI's breakout year. (2023, August 1). McKinsey & Company. https://www.mckinsey.com/capabilities/quantumblack/our-insights/the-state-of-ai-in-2023-generative-AIs-breakout-year

The Transformer model family. (n.d.). https://huggingface.co/docs/transformers/en/model_summary

Toews, R. (2024, February 20). The next generation of large language models. *Forbes.* https://www.forbes.com/sites/robtoews/2023/02/07/the-next-generation-of-large-language-models/?sh=1e80997e18db

Valchanov, I. (2024, April 23). *35 Best ChatGPT Prompts for SEO in 2024 [Expert Sourced].* Team-GPT. https://team-gpt.com/blog/chatgpt-prompts-for-seo/

Van Den Oord, A., Dieleman, S., Zen, H., Simonyan, K., Vinyals, O., Graves, A., Kalchbrenner, N., Senior, A., & Kavukcuoglu, K. (2016). WaveNet: a generative model for raw audio. *arXiv (Cornell University).* https://doi.org//arxiv.1609.03499DOI: 10.48550

Vaswani, A., Shazeer, N., Parmar, N., Uszkoreit, J., Jones, L., Gomez, A. N., Kaiser, L., & Polosukhin, I. (2017). Attention is All you Need. *arXiv (Cornell University), 30*, 5998–6008. https://arxiv.org/pdf/1706.03762v5

Weizenbaum, J. (1966). ELIZA—A computer program for the study of natural language communication between man and Machine. *Communications of the ACM,* 9(1), 36–45. DOI: 10.1145/365153.365168

West, S. M., Whittaker, M., & Crawford, K. (2019). *Discriminating Systems: Gender, Race and Power in AI.* AI Now Institute.

What are Autoregressive Models? - AR Models Explained - AWS. (n.d.). Amazon Web Services, Inc. https://aws.amazon.com/what-is/autoregressive-models/

What are Foundation Models? - Foundation Models in Generative AI Explained - AWS. (n.d.). Amazon Web Services, Inc. https://aws.amazon.com/what-is/foundation-models/

What is generative AI? (2024, April 2). McKinsey & Company. https://www.mckinsey.com/featured-insights/mckinsey-explainers/what-is-generative-ai

What is generative AI? (2024a). In *McKinsey & Company.* McKinsey Global Publishing.

What is Generative AI? (n.d.). Darktrace. https://darktrace.com/cyber-ai-glossary/generative-ai

What is GPT AI? - Generative Pre-Trained Transformers Explained - AWS. (n.d.). Amazon Web Services, Inc. https://aws.amazon.com/what-is/gpt/

Wittenberg, C., Epstein, Z., Berinsky, A. J., & Rand, D. G. (2024). Labeling AI-Generated Content: Promises, Perils, and Future Directions. *An MIT Exploration of Generative AI.* DOI: 10.21428/e4baedd9.0319e3a6

Xiaomi Smart TV X Series 2023 Edition. (n.d.). Xiaomi. https://www.mi.com/in/product/xiaomi-smart-tv-x-series-2023/

Zaremba, W., Brockman, G., & Open, A. I. (2021, August 10). OpenAI Codex. *OpenAI.* https://openai.com/index/openai-codex/

Zhang, C., Zhang, C., Zhang, M., & Kweon, I. S. (2023). Text-to-image diffusion models in Generative AI: a survey. *arXiv (Cornell University).* https://doi.org//arxiv.2303.07909DOI: 10.48550

Zhang, C., Zhang, C., Zheng, S., Zhang, M., Qamar, M., Bae, S., & Kweon, I. S. (2023). A survey on Audio Diffusion models: text to speech synthesis and Enhancement in Generative AI. *arXiv (Cornell University).* https://doi.org//arxiv.2303.13336DOI: 10.48550

Zhang, L., Qiu, Q., Lin, H., Zhang, Q., Shi, C., Yang, W., Shi, Y., Yang, S., Xu, L., & Yu, J. (2023). DreamFace: Progressive Generation of Animatable 3D Faces under Text Guidance. *arXiv (Cornell University).* https://doi.org//arxiv.2304.03117DOI: 10.48550

Zhang, Z., Song, Y., & Qi, H. (2017). Age Progression/Regression by Conditional Adversarial Autoencoder. *arXiv (Cornell University).* /arxiv.1702.08423DOI: 10.1109/CVPR.2017.463

Chapter 2
Bridging IoT and AI for Business Intelligence:
A Case of Smart Home Solutions

Paul Philip V.
Christ University, India

Tiakumzuk
https://orcid.org/0009-0006-7836-9758
Christ University, India

Anita H. B.
Christ University, India

ABSTRACT

Home solutions that unite the internet of things (IoT) and artificial intelligence (AI) offer a ground-breaking opportunity. This chapter discloses how IoT devices and AI algorithms can change household inventory management and security. The main focus is on sensors and AI-based analytics for optimizing data collection, analysis, and decision-making processes. Predictive maintenance becomes possible when sensors allow real-time monitoring of stock levels. On the other hand, integrated artificial intelligence surveillance systems, entry controls, anomaly detection algorithms help to improve home safety measures. Also, this project examines whether these home solutions can be used in commercial environments. By embracing this cutting-edge technology, the initiative emphasizes optimized inventory management, heightened security measures, and operational efficiency. For businesses seeking to merge IoT with AI technologies in future intelligent living spaces, this chapter provides useful information about them.

DOI: 10.4018/979-8-3693-9246-1.ch002

INTRODUCTION

The development of technology has been proven to change and improve several fields, such as industry, healthcare, and education. The proposed project aims at applying new technologies – the Internet of Things (IoT) with Artificial Intelligence (AI) in the field of Business Intelligence (BI), which involves two main components. First is the smart home inventory management system, and second is the security protection system. The idea of a "smart home" has moved from being only limited to descriptions of fictional and science fiction constructs to becoming an actual realistic thing. This development in smart homes is where different devices converge with sensors; these things are integrated to create an intelligent living space (R. K. Kodali et al., 2016).

In the past, inventory management depended on manual tracking, periodic audits, and reactive responses to shortages or excesses in stock levels (Chin et al., 2023). Similarly, traditional security systems include alarm systems, cameras, and passive surveillance. Now, this has changed dramatically with IoT. Inventory monitoring systems and asset tracking software enable us to trace RFID tags/RFID sensor/ radio frequency identification tags (RFID tags), temperature and humidity sensors, as well as any other type of environmental sensor through it, becomes possible for us to monitor our inventories by using RFID tag readers or any other types of wireless scanners that can read sensory data from various points of the store floor by scanning the movement between certain points that have been fitted with detectors or transponders. Some cameras can now spot irregularities and intruders using AI-powered image recognition; this makes them much smarter than motion sensors, which send alarms, or human guards, which are augmented in shops. Even at your home, smart cameras can recognize strange actions and possibly forecast the anticipated act of burglary. In addition, they can consider the danger of opening the door to strangers.

This chapter's main aim is to investigate how artificial intelligence (AI) and Internet of Things (IoT) technologies could converge to enhance business intelligence (BI). The study will emphasize smart home security and inventory management. This aims to invent a holistic solution that can optimize inventory management, minimize losses, and harden the security position of current homes by developing an integrated system that uses IoT sensors and AI algorithms. This research also seeks to show how much these systems can be scalable,, such as in factories and industries.

The importance of this research is huge when it comes to emerging technologies and business intelligence. The combination of IoT and AI technologies offers unparalleled efficiency, security, and innovation opportunities when the world increasingly relies on interconnections between devices and data-driven decision-making processes. These technologies discussed under smart home inventory management and security are directed toward current critical issues. First, there is an urgent need for efficient household inventory management because item tracking can be challenging and lead to errors. By using IoT sensors as well as AI algorithms, the main objective of this project is to streamline this process so that homeowners can easily monitor and control their possessions. Secondly, it is crucial to emphasize security, especially with increased cases of cyber insecurity. Incorporating IoT-based security systems improves overall smart home security posture by providing real-time monitoring, intrusion detection, and remote access control. Furthermore, the outcomes of this research are applicable in many more areas other than residential ones. Similar systems have been implemented in industries and factories for improved operations, reduced losses, and better security measures.

Finally, this study adds to ongoing discourses on utilizing modern technologies to develop smarter, safer, and more efficient living and working spaces through which numerous advancements may occur within diverse sectors by applying these ideas in a domestic setting, making it possible for innovation in different industries towards intelligent, secure data-driven operations era. The system will use machine learning models on IoT sensors and cameras for inventory management to monitor the presence and position of different items within a house. They include necessities, documents, essentials, valuable belongings, and other consumables. By doing so, this AI-based system will enable users to get regular updates on the status of these items while facilitating their efficient management and replacement when necessary (Guo et al., 2019).

This work, however, has limitations. The prototype is primarily a proof-of-concept project demonstrating how IoT and AI can be utilized in a home setting. However, it will be deployed only on a smart house simulation rather than an entire residential system. However, all other edge cases and real-world scenarios may still need to be addressed even if some benefits presented in these core functionalities of this prototype refer us to the proposed system. Furthermore, for future advancements or reiteration purposes, analyzed according to its scope analysis would have considered the system's scalability and adaptability towards different user preferred-house configurations (H. Kazmi et al., 2017).

LITERATURE REVIEW

The revolutionization of traditional household management and the integration of Internet of Things (IoT) technology in smart homes has ushered in a new era of convenience, efficiency, and safety. Scholars and practitioners have shown much interest in this subject matter, which has led to much research on how these technologies can be used, their benefits, and the challenges surrounding them in smart homes. As essential components of security systems applied in smart home settings,

IoT sensors have attracted significant attention from researchers. These are active devices embedded within building materials with the main aim of continuously monitoring for any irregularity in real time. Consequently, they play a big role in detecting unusual activities or changes occurring within a particular environment, thereby boosting the general safety levels of smart houses (Dan Ding et al., 2011).

Developments in IoT sensor technologies have made it possible for advanced anomaly-detecting mechanisms to be realized within smart homes. To detect anomalous behaviors, new algorithms and frameworks have been tested. Thus, by using machine learning algorithms, such devices separate normal actions from anomalies, given that they allow security threats to be detected promptly (Samad et al., 2020). Another case study about supply chain management also reveals interest in smart homes. Smart systems with AI (Artificial intelligence) and ML (Machine Learning) are being developed to perform home product stock control automatically. They follow up consumption, replenishment, and arrangement for several other things in the house through IoT sensors and cameras (Azeem et al., 2024).

Additionally, object recognition in smart home inventory has yielded favorable results when ML algorithms are used. These algorithms can uniquely identify objects and classify them based on Internet of Things (IoT)-driven image or video data, enabling effective inventory control and retrieval (Frédéric Bergeron et al., 2018). The convergence between IoT technologies and industrial settings resulted in the development of industrial-based IoT solutions aimed at tracking assets in factories and warehouses. These solutions rely on sensor networks that interconnect different devices to follow real-time positioning, movement, and condition of assets. In this way, such systems improve overall efficiency, minimize losses, and optimize resource utilization within the business environment (B. Sai Subrahmanya Tejesh et al., 2018).

There has also been a growing interest in the use of IoT for supply chain optimization. Through the employment of IoT sensors and analytics tools along the company's supply chain, organizations can achieve unseen visibility into the flow of goods, detect bottlenecks, and streamline logistics processes to cost reduction, shorter delivery times, and better overall performance (Hiranmoy Samanta et al., 2023). However, it is still important to ensure that data security and privacy are handled robustly, especially when using IoT technologies. Recently, several studies have focused on different forms of data encryption, access control mechanisms, and authentication processes that can be used to protect IoT devices from cyber threats. This is important since user trust is required for confidence in IoT systems' reliability (Biljana L et al., 2016). Moreover, much research has been dedicated to developing effective data privacy mechanisms for IoT devices. In such investigations, there have been some efforts towards encrypting protocols, anonymizing techniques, or even secure communication protocols to reduce privacy risks related to collecting and processing sensitive information within IoT environments (Hussain Kazmi et al., 2017).

Analysis

Many strengths and weaknesses arise from the literature on integrating Internet of Things (IoT) technologies with Machine Learning (ML) models for use in smart homes. This review examines what these studies add to our knowledge, their limitations, and areas in which possible improvements can be made. Home automation has improved: IoT devices can help automate household chores. This technology offers comfort and energy-saving measures by turning off lights, changing thermostats, or scheduling daily activities according to the desires of the occupiers (Sepasgozar S et al., 2020).

Real-Time Monitoring and Alerts: Real-time monitoring for home environments is a strength of IOT sensors. These include motion detection, door or window opening, and environmental condition sensors. They are responsible for notifying security personnel when there are attempts at intrusion, security violations, or any other type of threat (U. S. Kameswari et al., 2015). Smart Security Features: Combining IoT devices with ML algorithms enhances home security through predictive analytics and anomaly detection. By recognizing that certain actions, such as unusual movements within premises or unauthorized access to your place, could signal danger, ML models spot these exceptions by learning what normal behavior in a typical household looks like.

Privacy and Security Concerns: However, privacy and security issues have arisen from IoT devices despite their numerous advantages. Even though manufacturers make their software more sophisticated every day, they do not guarantee total safety since there is always a possibility that hackers might corrupt them; for example, bugs in firmware, data leaks, and hacking into important systems may all be considered weaknesses. Interoperability Issues: For IoT devices of different manufacturers to integrate seamlessly, standardization is lacking, which results in the absence of interconnectivity (Sharma et al., 2021). This implies compatibility problems exist and intricate setups are needed to connect different gadgets in a smart dwelling network (Kumar S et al., 2019).

Complexity of User Interface: Some studies reveal that user interfaces for smart home systems can be too complicated. In the context of IoT devices, many application settings and arrangements need to be managed; this may slow down uptake across the board. Prior research highlights the strides in applying Internet of Things (IoT) technologies alongside Machine Learning (ML) models in smart homes. Nevertheless, as these strengths indicate convenience, efficiency, and safety gains, it is vital to address weaknesses and areas for improvement to be embraced by all parties and sustainability over the long term. Future research on privacy, interoperability, user experience, and predictive maintenance scalability, among others, will also offer new perspectives for building tomorrow's smarter, safer, more efficient homes, with blockchain technology being one of them and others like internet security protocol, etc.

METHODOLOGY

Research Design

Our research is experimental, employing quantitative analysis to develop and access a smart home system by incorporating Internet of Things (IoT) devices with machine learning (ML) models. The ultimate goal is to create an effective inventory control system combined with security measures that suit modern homes.

1. Sensor Placement: We strategically placed several sensors throughout the house.
2. Camera-Based Object Detection: The security cameras record Live video feeds once IR sensors detect any movement. The video streams from these cameras are run through a Convolutional Neural Network (CNN) model that has been fine-tuned for object detection purposes (Alzubaidi et al., 2021).
3. Optimizing CNN model: The CNN model used for object detection training is optimized and converted into the TensorFlow Lite model. This makes the model less bulky; hence, it can be run on edge devices such as Raspberry Pi.
4. Edge Computing Based on Raspberry Pi: This enables Raspberry Pi to act as a central processing unit (CPU) for smart home object detection by running the TensorFlow Lite Model. This approach reduces latencies and allows quick responses.
5. Secure Cloud Storage: All information acquired from sensors of IoT gadgets interpreted by the ML model is securely sent to cloud storage. It ensures that the data is always there and cannot be corrupted since it resides in the cloud.
6. Mobile Application Interface: A mobile phone application program interface (API) lets users control their smart houses. Possession of this instrument helps homeowners who want to monitor their belongings, see things in real-time through the cameras, get security alarms, or decide how some systems will function.

Evaluation Metrics and Testing

1. Object Detection Accuracy: The model is then tested for home-based object detection accuracy to determine how well it can identify objects. Consequently, precision, recall, and f1-score metrics are used during the evaluation.
2. Response Time to Security Events: In this case, the speed at which system users get informed about an incident occurring within their residence through mobile app screens from the instances where motion triggers through to alerts being generated has been looked into.

MATERIALS

The primary materials and components used encompass a variety of sensors, devices, and technologies that collectively enable the system to function efficiently. Here is an overview of the main components involved:

IoT Sensors

- Motion Sensors (IR): They can sense any movement occurring in each of the houses, which triggers such activities as object tracking by the camera system. The motion sensor is given in Figure 1.

Figure 1. Motion Sensor

- Flame Sensors: To detect possible fire hazards or abnormal heat sources inside a home. The diagrammatic representation of the Flame Sensor is given in Figure 2.

Figure 2. Flame Sensor

- Temperature & Humidity Sensors: To keep track of ambient conditions in a house to suit comfort and safety purposes. The Temperature and Humidity Sensor (DTH11) is given in Figure 3.

Figure 3. Temperature and Humidity Sensor (DTH11)

- Tilt Sensors: Used to note changes in the position/orientation of objects for motion detection or disturbance identification. The general diagram of the Tilt Sensor is given in Figure 4.

Figure 4. Tilt Sensor

- Sound Sensors: Help pick up any strange noise or sound, thus facilitating security monitoring. Figure 5 illustrates the Microphone Sensor.

Figure 5. Microphone Sensor

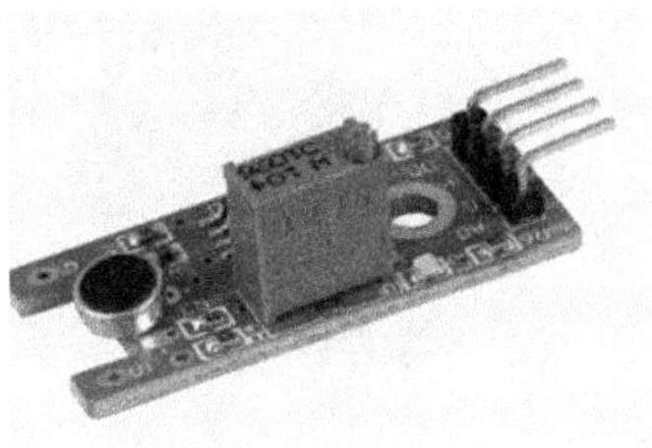

- Knock Sensors: These sensors detect vibrations or knocks on doors/windows, enhancing security measures. The overall diagram of the Knock Sensor is given in Figure 6.

Figure 6. Knock Sensor

1. Camera System:

Convolutional Neural Network (CNN) Model: This identifies and follows actual objects. It is especially useful where homes have inventories.

TensorFlow Lite Model: The converter transforms the CNN model to a TensorFlow Lite model optimized for Raspberry Pi and other edge devices.

2. Raspberry Pi:

It is the device responsible for processing sensor data and the CNN model at the edge. In Figure 7, you can see the general design of the Raspberry Pi 4B.

Figure 7. Raspberry Pi 4B

3. Mobile Application (UI):

The system's user interface can be found through a mobile app to make it easy to monitor and control.

4. Data Storage and Communication:

Cloud Storage: All collected information, including sensor readings, camera footage, and inventory logs, is securely stored in the cloud.

Wi-Fi Connectivity: This allows information to be transmitted seamlessly between Raspberry Pi, sensors, and cloud storage.

These elements combined work together to develop an all-inclusive smart home system that performs intelligent inventory management and strong surveillance. On the other hand, it has instantaneous facts about diverse environmental aspects of one's house that allow its user to escape this problem. According to this design, communication between sensors and machine learning computer systems such as Raspberry Pi with TensorFlow Lite or Cloud storage will always go through the main CPU unit during execution.

Procedure

1. Sensor Installations and Setups:

First, home sensors are strategically placed in it. While motion sensors are put on strategic points to detect movements that might be happening in those areas around your house, flame sensors are put near potential fire hazards. Simultaneously, temperature and humidity sensors observe objects' surroundings; tilt detectors detect an object's movements while sound detectors give out sounds. Doors and windows also have knock sensors installed. Every sensor is connected to Raspberry Pi, which acts as its CPU through suitable connections and wiring. The schematic diagram of the Wiring design is given in Figure 8.

Figure 8. Wiring Diagram

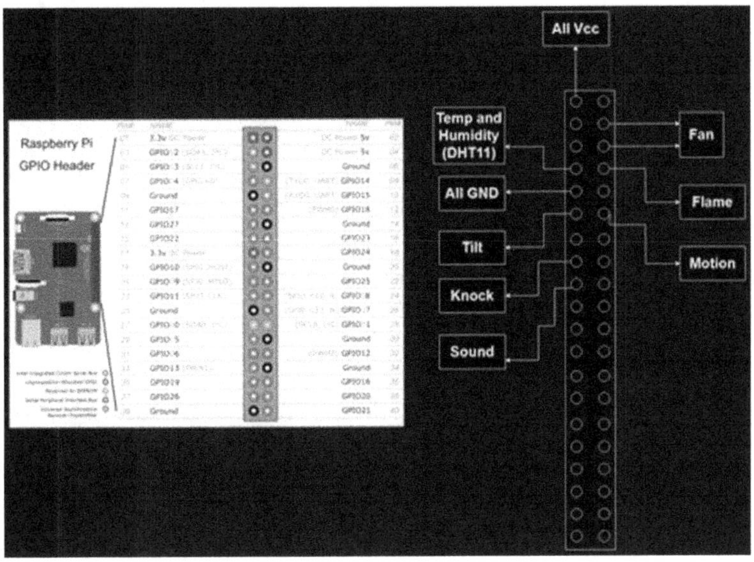

2. CNN Model Development

This project involves the development of a Convolutional Neural Network (CNN) model for efficient object detection and tracking via TensorFlow. For example, this specific model has been trained using pictures of usual household items like groceries, electrical gadgets, and private effects that identify and classify them. This model is taught by feeding labeled images to help it accurately differentiate between different objects based on learned patterns.

3. Conversion to TensorFlow Lite

After training and optimization processes are completed, the CNN model is converted to TensorFlow Lite format. Designed for mobile and edge devices such as Raspberry Pi, TensorFlow Lite ensures that the model runs efficiently without affecting its accuracy level. This conversion process enables seamless integration of the model into the Raspberry Pi's architecture.

4. Incorporation with sensors

After that, Raspberry Pi is merged with the TensorFlow Lite model and peripheral devices. When the sensor triggers motion, heat, or sound, the CNN model will run on Raspberry Pi. The detection of objects occurs as the model processes live camera feed from the camera module for Raspberry Pie and tracks what has been identified in it respectively.

5. Object detection and Tracking

This happens when the CNN model analyzes each frame in the video camera. Based on trained categories, it identifies objects in a frame, such as a water bottle or cup. Thus, the locations of these items within a home are updated continuously as they move across frames in real-time by this model.

6. Alerts And Notifications

This system sends Alarms out when there are abnormal activities like unauthorized movement, fire outbreaks, and strange noises. Instant messages to users through a mobile app allow immediate response to future events. With these alerts, they can see all their current movements at home or in their room; an alarm could be used if needed.

7. Data Cloud storage and logging

We store all sensor data safely on a cloud camera feed and track a log of things moving using objects. This particular kind of storage in the cloud supports historical analysis of home events, inventory control, and monitoring security accidents. Access this information through a mobile app to be informed about your house status and past.

8. Constant Monitoring and Updating

This system operates unceasingly, checking for changes or occurrences within one's homestead. Studies have shown that the CNN model must be regularly upgraded and maintained to make it more accurate in object detection. The purpose is for the users to personalize their settings to suit their needs via the mobile application system, thus enabling them to receive alerts on their phone whenever anything happens; hence, they can customize according to their needs and preferences. In this comprehensive smart home outline, we are going to discuss how IoT sensors are used with a CNN Model based on TensorFlow Lite for real-time object detection using Raspberry Pi as a central hub, which makes it an extensive platform that provides support for home monitoring, security, and inventory management. Figure 9 depicts the general structure of the workflow of the system.

Figure 9. Workflow of the System

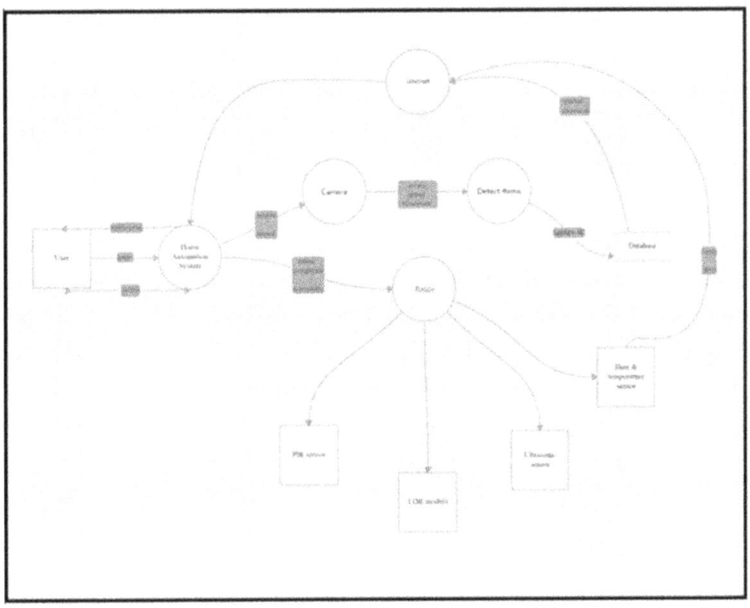

Data Collection

Data collection is a remarkable feature embodied in the smart home inventory and security system, enabling it to obtain information about a home's environment, user activities, and security events. The procedure for gathering and storing these is given below;

1. Sensor Data Acquisition

Data relevant to the system comes from IoT sensors spread over several points in the house. Upon sensing an event such as motion, temperature change, sound, or vibration, the sensor sends a signal to the central processing unit, Raspberry Pi. For instance, when activity is detected by motion sensors in some areas of our home, data collection begins. These sensors have been designed to send live updates to Raspberry Pi so that they can stream sensor measurements continuously.

2. Camera Feed Analysis

Using a Raspberry Pi camera module, this device can take live videos of what happens inside our homes. These video feeds are processed in real time with an integrated CNN model for object detection and tracking. While being recognized and tracked by a CNN model within frames from videos, positioning coordinates and movement produce valuable data about them. Surveillance monitoring depends on this information for inventory control, including detecting abnormal happenings.

3. Cloud Storage in Firebase

The system has been developed with Firebase, a Google cloud-based platform for storing and managing collected data safely. For instance, Firebase has real-time database features that make it perfect for storing sensor readings, camera footage, and object-tracking data. It is sensor events and object detection by the CNN model. Then, the data will be sent to Firebase instantly to store it. Firebase's NoSQL database structure facilitates flexible and scalable data storage that caters to the dynamic nature of smart home systems.

4. Data Structure and Organization

Within Firebase, motion events like temperature readings or sound detections are categorized into separate datasets. Each type of sensor has its collection; hence, specific data points can be easily accessed or retrieved from them. The set of information, for example, includes timestamps, locations, and durations of detected movements on the motion sensor dataset. Information acquired from camera feed analysis, such as event timestamps, tracking paths, and object classifications, are stored in Firebase collections.

5. Real-time Updates and Notifications

Since database storage is in Firebase, the system can give users real-time updates/ notifications. Mobile devices owned by users receive instant notifications when sensor events occur or when the CNN model detects any abnormal activity. These notifications include details such as the type of event, location within the home, and timestamp, allowing users to respond promptly.

6. Robust and efficient data processing

Firebase guarantees the safety of users' data using strong encryption protocols and authentication mechanisms. Inactive or idle user information like personal details and home logs of activities being stored are securely kept. It is hard for those who want to get it by protecting it in a way no one can understand.

7. Continuous data monitoring and maintenance

This ensures the sensor's flow of real-time camera data is up-to-date and accurate whenever it checks them. These include the reliability and efficiency of Smart Home inventory and its security systems ensured through proactive data surveillance during routine maintenance that can verify any abnormalities in automated validation processes. Firebase saves this for a safe, scalable, real-time infrastructure for smart home systems; this contains different forms of monitoring, analytics, and notification capabilities that can help them effectively manage their homes with confidence.

DATA ANALYSIS

In conjunction with information from the camera, it feeds on large amounts of sensor data extraction, which is an important part of the smart home inventory and security system.

Here are ways to analyze the data

1. Anomaly Detection and Pattern Recognition

Such systems use pattern recognition algorithms to recognize normal behavior patterns in a house. Typical activities like habitual routines, preferred activity time, or normal motion patterns for occupants have been learned from historical sensors' collected data and camera feeds. These anomalies deviating from known patterns can signify safety threats or strange happenings. For instance, if the system detects an unusual movement when no one is supposed to be inside, it raises an alert for further investigation. For instance, if the system detects an unusual movement when no one is supposed to be inside, it raises an alert for further investigation.

2. Occupancy Trend Analysis

The system traces and appraises developments in occupancy concerning different areas within the house over time. By capturing how often and how long motion events occur along with sound detections, occupancy profiles are created for each room by the system. This study facilitates optimum energy consumption by adjusting heating, cooling, and lighting according to room occupation. Similarly, it helps identify unexpected trends in occupancy, like prolonged absence or sudden appearance within specific locations.

3. Health Monitoring and Well-being Analysis

The system does health trend analysis for room occupants using data sourced from health sensors such as heart rate monitors and temperature sensors. It is possible to gain insight into how a person feels by continuously monitoring their vital signs alongside environmental variables. It monitors sleep trends, stress levels discerned through audio analysis, and the environment (temperature and humidity). This will ensure that occupants or caregivers are notified about any major deviations from normal health patterns in time for timely medical response and attention.

4. Object Tracking and Inventory Management

CNN model, which is used for object detection and tracking, provides valuable information on inventory management in the home. Every object detected/tracked by the system is logged with its location, motion, and usage frequency, among others. This data creates an inventory profile of the house, indicating the presence of various items, their locations, and their status. Through real-time updating of inventories, users receive notifications when stocks run low, while tracking objects within the house can be done conveniently.

5. Insights into Behavior Patterns:

The system can also gain insight into what a person likes or dislikes, how he behaves, and so forth by going through his sensor data, camera surveillance videos, or even interacting with him through the system. For instance, the system learns about preferred light intensities depending on historical usage trends, entertainment choices, and room temperatures. This results in personalized experiences for inhabitants, such as automatic adjustments of room settings based on individual preferences. Through behavioral analysis, the system can also predict future actions or needs of its users, thus enhancing their proactive behavior. In Figure 10, a Bar Chart that compares the usage of each sensor is shown.

Figure 10. Bar Chart that Compares the Usage of Each Sensor

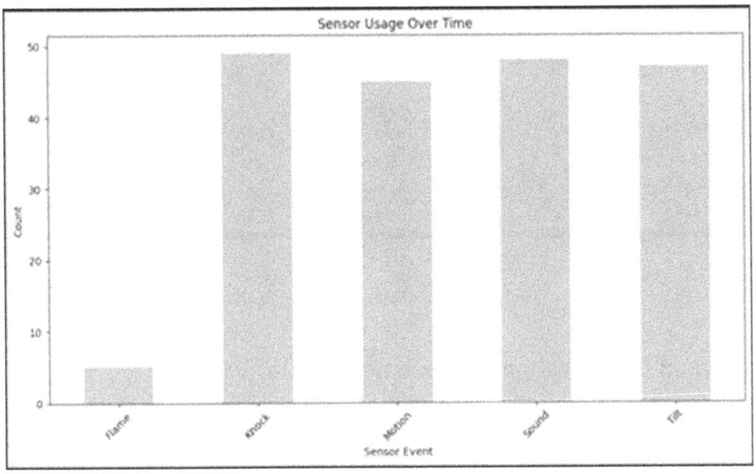

6. Visualization of Historical Data and Reporting

The user has visual reports that explain previous analytics made by the system. However, homeowners can understand occupancy trends, health indicators, energy utilization levels, and inventory stocks, amongst others, using easy-to-understand graphs and charts. These figures enable people to identify trends in managing homes that may act as inputs in the decision-making process. These findings could look at different aspects of the house environment, meaning it would be possible to see what happens daily within a homestead. The visualization of Temperature and Humidity across several days is featured in Figure 11.

Figure 11. Visualization of Temperature and Humidity Across Several Days

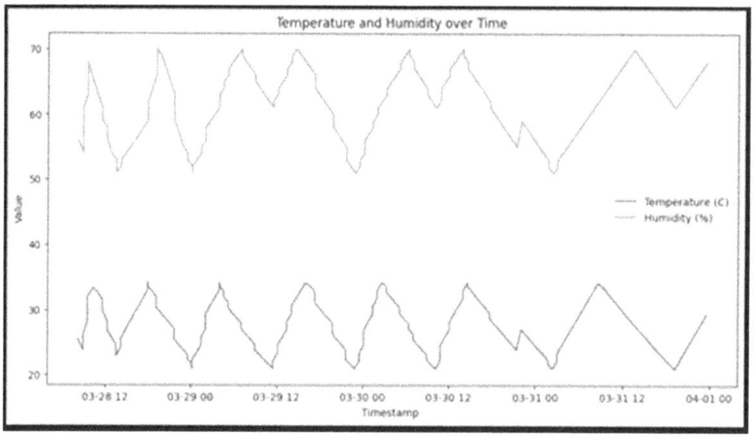

7. Training and Optimization of Machine Learning Models

New data inputs, as well as feedback from end users, constantly refine machine learning models. Training algorithms analyze historical data to improve the accuracy of object detection, anomaly detection, and health trend predictions. Model optimization techniques, such as hyperparameter tuning and feature selection, are applied to enhance the system's performance. The goal is to create a self-learning system that adapts to the changing needs and dynamics of the home environment.

8. Real-Time Alerts and Recommendations

When the analysis is done, the system generates real-time alerts and recommendations. Whenever abnormalities are identified, messages are sent as alerts through a mobile app, making it possible for users to take instant steps. The other aspect is also giving suggestions on how to save energy, improve health, or manage inventory. These notifications enable users to make prompt decisions to boost safety, economy, and welfare at home.

Smart home inventory and security systems convert raw sensor data and camera feeds into actionable insights through advanced analytic techniques. It gives a view of the entire home environment, including occupancy patterns, health trends, and user preferences. With these facts, users can manage their homes properly so they do not fall victim to insecurity risks while enhancing both quality of life and health standards, making homes secure.

RESULTS

There were some key findings from developing the smart home inventory and security system that incorporated Internet of Things (IoT) technologies with AI-powered business intelligence:

1. Efficient Inventory Management

The system's accuracy that used convoluted neural network (CNN) models in detecting objects was high, thus ensuring that household items were well managed and recorded. The locations of the products were monitored in real-time to give insights into how often a given product is used and their preferred areas. This information enabled optimized inventory placement, ensuring that frequently used items were easily accessible while reducing clutter.

2. Behavioral Insights

Household activities were analyzed using motion sensor data to identify different patterns. More activity during working hours revealed a weekly usage pattern, indicating regular daily routines. On weekends, there was a shift towards more relaxed and sundry home activities, indicating leisurely home dynamics. These insights allowed customization of the system's responses based on the time of day and detected activity levels.

3. Abnormality detection

This system could well notice such anomalies as unexpected movement or temperature changes within one's house. Users received instant alerts via mobile phones regarding security breach possibilities or any unusual activities. This feature gave homeowners peace of mind by proactively addressing safety and security concerns.

4. Inventory Optimization

The system produced Placement optimization suggestions, which analyzed the usage of different items over time. To enhance accessibility and efficiency, users needed advice on rearranging their stuff. Daily routines became streamlined, and search for frequently used items was reduced, making it useful to busy households.

5. Remote Monitoring and Control

Users could access their home environment from any part of the world through a mobile application interface that allowed them to connect remotely with the system. This enhanced security and peace of mind by visualizing household activities using real-time camera feeds. Apart from this, users could control IoT devices such as thermostat settings or lights on/off even when far away, thus making it easier and energy efficient.

6. Scalability and Adaptability

The system's initial design was intended for domestic purposes, but its scalability made its eventual operation possible in commercial or industrial areas. With modular architecture, more sensors or various other devices could be integrated without difficulties, allowing for evolving user needs and preferences.

7. Safety and Security Alerts

In integrating IoT sensors for security, safety was significantly boosted at home. The alerts randomly triggered by the system include fluctuations in temperature, increased decibels, or strange movements. For example, one night, a certain sensor started to detect knocks on the door of an individual's place, and this made the system send a message to his mobile phone through an installed application. In addition, when there was an unusual increase in the temperature recorded from readings of thermometers in the kitchen section where they are kept, as shown below, it raised the alarm, hence calling for an investigation. This gave room for the personal requirements of users during expansion so that different environments/and uses could be well-suited by the system.

In conclusion, smart home inventory and security systems effectively enhanced household organization, security, and efficiency. The system employed IoT technologies with AI-driven insights that provided valuable information regarding usage patterns of inventory items, behavioral insights into abnormalities detection, and remote monitoring methodologies. These significant findings point to further improvements in smart homes, thereby giving individuals a comprehensive way of managing their living spaces, which is informed by intelligence.

INTERPRETATION

This study on combining Internet of Things (IoT) technology into our AI-controlled business intelligence for smart home inventories and security systems has shown possibilities and consequences. This section seeks to interpret results in depth, discussing the relevance of integrating IoT in changing home management and security.

Enhanced Home Monitoring and Safety

This research revolved around monitoring home activities to keep occupants safe through real-time alerts and intelligent data analysis. Findings illuminate how transformative IOT sensors can create a proactive home environment. The system allows homeowners to observe motion patterns, identify objects, analyze user behaviors, etc., offering a comprehensive snapshot of the state inside their houses and predicting prospective safety hazards. For instance, this is evident when identifying unusual sounds or movements, such as someone knocking on the door at night or a sudden temperature spike that would enable homeowners to respond quickly before anything happens. In line with the broader theme of leveraging emerging technologies to build safer habitats, this proactive approach towards securing homes characterizes modern smart living environments.

Maximum Efficiency and Resource Management

However, the research has significant implications related to home resource optimization. Herein, integrating IoT sensors leads to intelligent energy management, which may involve adjusting lighting, heating, and appliance usage depending on occupancy patterns. This will improve energy efficiency and help homeowners save money. The system can recommend personalized suggestions for optimizing consumption by analyzing past energy consumption trends. For example, changing thermostat settings could be suggested during low occupancy periods or appliance utilization could be scheduled outside peak hours. Also in line with this broad theme is the role of IoT in business intelligence as far as environmental sustainability is concerned with efficient resource management.

Customized Home-Based Experiences

This study shows how adopting internet-of-things technology can create tailor-made home-based experiences. Through user behavior analysis, it can predict preferences and adjust the physical environment accordingly. A welcoming and comfortable ambiance is achieved when such technologies can automatically change lighting and temperature settings upon detecting the presence of homeowners. Moreover, the system's object detection and tracking capabilities pave the way for intelligent inventory management. Homeowners can easily track the location of items, receive alerts for misplaced belongings, and even generate shopping lists based on inventory levels. This personalized approach to home management not only enhances convenience but also improves overall quality of life.

Scalability in Industrial and Commercial Applications

Even though the main focus is on the implications of this for a smart home, its scope is much broader than that. Combining Internet of Things (IoT) technologies with artificial intelligence (AI) based analytics successfully shows the possibility of scalability in industrial and commercial apps. Some examples include inventory tracking systems in manufacturing industries and asset management systems in healthcare sectors. One of its potential applications can be observed in a manufacturing setting where AI algorithms and IoT sensors are used to optimize inventory levels, predict maintenance needs for machinery, and streamline production processes. Also, IoT-enabled devices are suitable for healthcare facilities since they help monitor patients' conditions and keep track of medical equipment.

Ethical and Privacy Considerations

Nonetheless, there should be moral concerns regarding combining the Internet of Things. According to the report, it is imperative to have robust data security mechanisms, user consent methodologies, or transparent ways to handle data. As identified by the survey results, there must be enforcement and adherence to strict guidelines on matters related to ethical data practice. This will lay a solid foundation upon which IoT-enabled product trust can be built.

THE FUTURE OF SMART HOME MANAGEMENT

This research confirms that this is a definitive moment in home management and security with AI merging with IoT technologies. This means we can make life safer, more efficient, and personalized by viewing findings from various perspectives. This study is about new technology trends that would work well in AI-powered Business Intelligence, such as IoT; therefore, it suggests that these technologies can change home management aspects and shape future industries and businesses that are currently entering this era. As pioneers of smart home management in the future, ethical concerns and privacy issues are critical to ensuring that innovation comes with accountability.

Through our research, we imagine a future where homes are ecosystems that enhance and secure life rather than just shelters. This can only be achieved if houses can be transformed into smart ecosystems by integrating IoT technologies, such as those discussed in this chapter. In starting on this transformative path, there is an unlimited scope of possibilities marking a new age of intelligent living catalyzed by innovation and intelligence. In starting on this transformative path, there is an unlimited scope of possibilities marking a new age of intelligent living catalyzed by innovation and intelligence.

LIMITATIONS

Like any other emerging idea, the proposed smart home inventory and security system is limited in some ways that should be considered. These limits provide insights that are valuable about areas where there could be further refinements and advancements:

Single-Camera Limitation

The current stage of our system uses only one camera for object detection and tracking. While it has been able to offer an affordable and easily implemented solution for residential use, this kind of setup comes with coverage shortcomings. In homes that are bigger or have complex structures, there may exist blind spots that adversely affect the overall effectiveness of monitoring all parts.

Limited Sensor Integration

The array consists of IR, fire, temperature, humidity, tilt, sound, and knock sensors; however, the integration is restricted by device compatibility. This means not all available sensors may fit into the system seamlessly, hindering its potential for extensive data collection and analysis.

Dependency on Cloud Connectivity

We depend on cloud-based storage for data management and analysis. This ensures remote access and scalability but also subjects us to reliance on online connection. When internet coverage might prove unstable or non-existent, the system may lose its real-time status.

Performance Constraints of Raspberry Pi

Raspberry Pi, though a flexible and cost-effective computing platform, has limitations that stem from its processing power and heat dissipation. In this regard, system performance may be affected when running computationally expensive tasks such as real-time object detection using Convolutional Neural Networks (CNNs). The Raspberry Pi can get hot under extended heavy loads, which might disturb the stability of the whole system.

Privacy and Data Security

The combination of IoT devices with AI-based analytics raises concerns regarding data privacy. As much as we have implemented encryption protocols and secure data handling practices, the possibility of security breaches or unauthorized access is still an issue. We continue to improve our cybersecurity efforts to match the ever-changing threats.

Limited Scalability for Industrial Applications

While the system proves effective in homes, it may not scale up to larger industrial or commercial environments. The extensive inventory management and robust security required in such setups need complex and scalable solutions beyond what our current project covers.

User Learning Curve

When introducing a new technology system into their households, some users who may not be used to IoTs may experience difficulties adapting. The learning curve associated with system setup, configuration, and troubleshooting could impact user adoption and satisfaction.

FUTURE IMPROVEMENTS

Anticipating these will be solved in future versions of our smart home system, it is aimed at solving the challenges above and further improving its abilities:

Multiple Camera Integration

This would improve coverage while also ensuring a broader view of the space being monitored due to expanding the scope. This enhancement necessitates synchronized data processing as well as object-tracking algorithms.

Diverse Sensors and Compatibility

Integrating more environmental sensors into different sensor technologies has to be facilitated by further research. These include looking at ways RFID can be used for inventory tracking and increasing the number of monitoring capabilities for various environmental parameters within this system.

Offline Functionality

Creating a hybrid model that can work offline while still connected to the cloud would help minimize this system's reliance on continuous internet connectivity. Thus, if there are any interruptions in internet connectivity, it will not affect its operation.

Better-Data Encryption

Moreover, ensuring data encryption protocols are strengthened and advanced cybersecurity measures are taken will help increase resilience against potential security breaches. Furthermore, security audits should occur frequently, and updates should be made regularly to align with industry best practices.

Scalability for Industrial Application

Development of designed versions applied to industrial and commercial use cases will include scalability testing, work with existing inventory management systems, and customization for different use cases.

User-Friendly Interface

One way is to add a user-friendly intuitive interface in the mobile application as well as help buttons that would guide users on how to go about each step in the process and, at last, provide the user guide so that it can be easily adopted by people who are not technology savvy thus reducing its learning curve. Therefore, the smart home inventory and security system aims to be a versatile, reliable, and user-friendly solution for homes and offices of residential or industrial use by addressing these limitations and pursuing future enhancements.

CONCLUSION

This project on smart home inventory and security is a unique amalgamation of advanced technologies into one complete intelligent system. We developed an integrated solution by combining Internet of Things (IoT) devices with complex machine learning algorithms that enhance safety and security in living spaces and promote efficient stock management within smart home environments. The constructional details and operations of our intelligent home have been examined thoroughly in this chapter. For instance, IoT sensors were used for detecting anomalies or strange activities while we developed a customized Convolutional Neural Network (CNN) model for object recognition and tracking purposes. All these components have been carefully created to play their part in delivering a safe and effective house. This chapter is a hands-on and thought-provoking look into how AI-powered BI can be used in smart home environments. Highlighting how IoT and AI integration can improve inventory management, predict future maintenance requirements efficiently, and provide a real-time security management system to check all impending threats, we offer a model for businesses that want to leverage technological advances. The fusion of Artificial Intelligence (AI) and the Internet of Things (IoT) has an excellent potential to transform our homes and workplaces as we enter the era of connected things and intelligent systems. This project exhibits how technology can revolutionize our living spaces by improving safety standards, energy efficiency levels, and eco-friendliness.

Furthermore, this chapter delves deeper into the book's theme on IoT and AI advances being channeled into data-driven decision-making processes. We contribute towards this debate by providing insights that can be acted upon and practical examples from business spheres where emerging technologies have been deployed for strategic gains in the fast-paced current commercial settings. This chapter also showcases various ideas and innovations that form the basis of future efforts in the field of AI intersecting with Business Intelligence as we move towards a more networked and smarter world.

Finally, this chapter displayed how these technologies work and shed light on their huge potential to reshape our interaction with and manage residential premises. In the future, this merger between IoT and AI will guide innovation, thus improving efficiency and making us live in smarter societies.

REFERENCES

Alzubaidi, L., Zhang, J., Humaidi, A. J., Al-Dujaili, A., Duan, Y., Al-Shamma, O., Santamaría, J., Fadhel, M. A., Al-Amidie, M., & Farhan, L. (2021). Review of deep learning: Concepts, CNN architectures, challenges, applications, future directions. *Journal of Big Data*, 8(1), 53. DOI:10.1186/s40537-021-00444-8 PMID:33816053

Azeem, M. (2024). Internet of Things (IoT) impact on inventory management. IGI Global. https://www.igi-global.com/chapter/internet-of-things-iot-impact-on-inventory-management/339252

Bergeron, F., Giroux, S., Lachiver, G., & Bouchard, B. (2018). Tracking objects within a smart home. ResearchGate. https://www.researchgate.net/publication/326162154_Tracking_objects_within_a_smart_home

Chin, C., Ramiah, S., & Razali, N. F. (2023). Inventory Management Systems (IMS). 2600-7304. https://www.researchgate.net/publication/372653862_Inventory_Management_Systems_IMS

Din, D. (2010). Sensor technology for smart homes. ResearchGate. https://www.researchgate.net/publication/51087784_Sensor_technology_for_smart_homes

Guo, X., Shen, Z., Zhang, Y., & Wu, T. (2019). Review on the application of artificial intelligence in smart homes. Smart Cities, 2(3), 402-420. https://www.mdpi.com/2624-6511/2/3/25

Kameswari, U. S., & Babu, I. R. (2015). Sensor data analysis and anomaly detection using predictive analytics for process industries. 2015 IEEE Workshop on Computational Intelligence: Theories, Applications and Future Directions (WCI), Kanpur, India. https://ieeexplore.ieee.org/document/7495528

Kazmi, H., Mehmood, F., & Amayri, M. (2017). Smart home futures: Algorithmic challenges and opportunities. 2017 14th International Symposium on Pervasive Systems, Algorithms and Networks & 2017 11th International Conference on Frontier of Computer Science and Technology & 2017 Third International Symposium of Creative Computing. https://ieeexplore.ieee.org/abstract/document/8121811

Kodali, R. K., Jain, V., Bose, S., & Boppana, L. (2016). IoT based smart security and home automation system. International Conference on Computing, Communication and Automation, IEEE, 1286-1289, April 29-30. https://ieeexplore.ieee.org/abstract/document/7813916

Kumar, S., Tiwari, P., & Zymbler, M. (2019). Internet of Things is a revolutionary approach for future technology enhancement: A review. *Journal of Big Data*, 6(1), 111. DOI:10.1186/s40537-019-0268-2

Risteska Stojkoska, B. L., & Trivodaliev, K. V. (2017). A review of Internet of Things for smart homes: Challenges and solutions. ScienceDirect. https://www.sciencedirect.com/science/article/abs/pii/S095965261631589X

Samad, T. (2020). A review of IoT devices and applications. ScienceDirect. https://www.sciencedirect.com/science/article/pii/S1389128620312068

Samanta, H., & Golui, K. (2023). Smart inventory control: Proposed framework on basis of IoT, RFID, and supply chain management. ResearchGate. https://www.researchgate.net/publication/370272042_Smart_Inventory_Control_Proposed_Framework_on_Basis_of_IoT_RFID_and_Supply_Chain_Management

Sepasgozar, S., Karimi, R., Farahzadi, L., Moezzi, F., Shirowzhan, S., Ebrahimzadeh, M. S., Hui, F., & Aye, L. (2020). A systematic content review of artificial intelligence and the Internet of Things applications in smart home. *Applied Sciences (Basel, Switzerland)*, 10(9), 3074. DOI:10.3390/app10093074

Sharma, A., Sharma, S., & Gupta, D. (2021). A review of sensors and their application in Internet of Things (IOT). *International Journal of Computer Applications*, 174(24), 27–34. DOI:10.5120/ijca2021921148

Tejesh, B. S. S., & Kumar, T. R. (2018). Warehouse inventory management system using IoT and open-source framework. ResearchGate. https://www.researchgate.net/publication/329230209_Warehouse_inventory_management_system_using_IoT_and_open_source_framework

Chapter 3
The Social Media Influence Journey:
A Comprehensive Study of Techniques and Strategies for Effective Diffusion

Minnu Elizabeth Ittan
https://orcid.org/0009-0000-1992-9416
Division of Computer Science, School of Engineering, Cochin University of Science and Technology, India

Sudheep M. Elayidom
Division of Computer Science, School of Engineering, Cochin University of Science and Technology, India

Midhun P. Mathew
Division of Computer Science, School of Engineering, Cochin University of Science and Technology, India

ABSTRACT

The expanding use of social networks has opened up a wealth of possibilities for practical applications, with influence analysis emerging as a crucial tool for understanding real-life dynamics. This study aims to offer a comprehensive survey of the state-of-the-art in influence analysis, focusing on strategies to address these challenges. The authors discuss various methods, including greedy algorithms, heuristics, and machine learning approaches, and also looks at how these algorithms are used in public health, marketing, and information sharing in the real world, giving a thorough review of how they might be used to effectively utilize the influence of social media. The main goal is to compare the advantages and disadvantages of

DOI: 10.4018/979-8-3693-9246-1.ch003

the research methodologies and procedures used within these categories through a critical assessment. Additionally, the authors provide an in-depth overview of existing methods that address the key issue of privacy protection in influence maximization. This study also analyzes the latest applications where social influence is being used.

INTRODUCTION

Social media has entirely changed the way we engage, communicate, and consume information in the digital age. It has significantly influenced various aspects of our lives, from shaping public opinion and influencing consumer behavior to driving social change and impacting individual identities. Through their content and engagement with their audience, social media influencers can influence opinions, promote products, and even influence electoral outcomes. Social media connects people globally, cultivating community and belonging. It quickly shares information, especially during crises, and serves as a powerful tool for businesses in networking, customer service, marketing, and advertising. It also offers access to educational content and communities, supporting personal branding, emergency response, and public opinion and behavior insights.

Motivation

Need for diffusion Techniques, Influence Maximization Algorithms, and Artificial Intelligence techniques in social network analysis. Diffusion techniques in social networks predict content popularity, identify influencers, and understand information spread, aiding in influencer engagement, campaign planning, and crisis management. They help predict trends, enhance marketing and communication, and support informed decision-making. These methods allow organizations to identify opinion leaders, test strategies, and evaluate campaigns, maximizing influence in public health, politics, and marketing while promoting ethical influence maximization.

Figure 1. Strategy of Information Diffusion

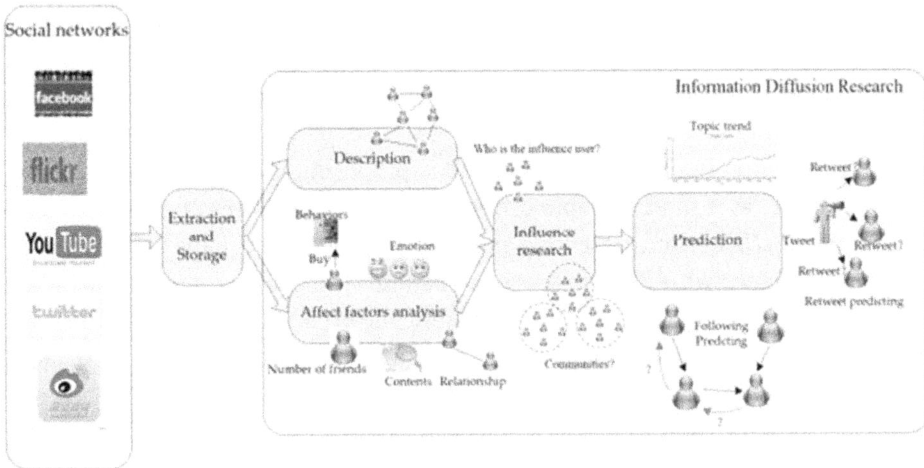

Figure 1 in the work of Li et al., (2017) describes the research strategy on information diffusion. After extracting and storing data, the initial task involves describing the diffusion process and assessing the influential factors that impact information spread. This examination enables the prediction of future diffusion patterns. Influence maximization algorithms identify key individuals or groups in social networks to optimize political advertising, public health messaging, and marketing by improving resource allocation and understanding of information spread. AI enhances this process by standardizing influential entity identification and increasing message reach, providing a competitive advantage to businesses and marketers.

RELATED WORKS

Shakeel et al., (2022) present a comparative analysis of influence maximization techniques in social networks, evaluating the performance of four algorithms on various datasets and highlighting the efficiency of the Discrete Shuffled Frog-Leaping Algorithm (DSFLA) in terms of computational time. Uthayasuriyan et al., (2023) compare the performance of the Greedy, CELF, and CELF++ algorithms for influence maximization in social networks, highlighting the efficiency and effectiveness of CELF++ in solving the problem more effectively than the other two algorithms. Lin et al. (2019) present a Hybrid Distribution Value Accumulation Algorithm for Influence Maximization in social networks, which outperforms traditional Hill

Climbing and Greedy Algorithms, as demonstrated through experiments on two datasets, and addresses the need for further improvements in time complexity.

D'Angelo et al., (2023) present improved upper bounds on the adaptivity gap for influence maximization under full-adoption feedback, focusing on the independent cascade model and general graphs and introducing new techniques to relate adaptive policies with non-adaptive ones. Qin et al., (2017) explore the challenges of influence maximization in online social networks, proposing the Three Steps Cascade Model (TSCM) as a more suitable diffusion model compared to the Independent Cascade Model (ICM) and presenting an efficient algorithm scalable for large social networks. Kimura et al., (2006) propose two natural special cases of the Independent Cascade Model (ICM) for efficiently estimating the expected number of influenced nodes in large-scale social networks, demonstrating their effectiveness in extracting influential nodes and providing good approximations to the ICM, especially when the propagation probabilities through links are small.

Gursoy et al., (2018) discuss the problem of Influence Maximization in social networks under the Deterministic Linear Threshold Model and propose a novel algorithm, the TArgeted and BUdgeted Potential Greedy (TABU-PG) algorithm, to address this problem, along with real-world mimicking techniques for generating influence weights, thresholds, profits, and costs, and empirical evaluations of the proposed heuristics. Tang et al., (2019) introduce a second-order diffusion model for influence maximization in social networks, proposing that the second-order IC model considers previous activations and demonstrates its effectiveness in accurately simulating influence diffusions through extensive experimental evaluations. Lotf et al., (2022) propose a novel approach using a dynamic generalized genetic algorithm to address the influence maximization problem in dynamic social networks, aiming to identify influential individuals and optimize influence spread while considering changes in network structures over time.

Guo et al., (2024) present a novel algorithm for influence maximization in heterogeneous networks, utilizing the propagation field's multi-scale propagation strength and repulsive force and outperforming baseline algorithms in experimental evaluations. Liang et al., (2023) discuss the Customized Influence Maximization (CIM) problem in attributed social networks, addressing the need to consider users' attributes and external queries in identifying influential nodes and propose two algorithms, DD-CIM and AACS-CIM, to solve this problem. Dizaji et al., (2024) discuss the application of reinforcement learning and deep Q-learning methods to address the influence maximization problem in dynamic networks, focusing on optimizing information propagation and influence spread.

Contributions

Influence maximization algorithms play a crucial role across various industries, from enhancing social network analysis to identifying key influencers for marketing and supporting public health initiatives. These algorithms are important in optimizing political campaign strategies, increasing recommendation systems, predicting viral social media content, and detecting fraudulent activities to support security. Essentially, they increase the effectiveness of campaigns and deepen our understanding of network dynamics. Current research focuses on enhancing interdisciplinary approaches, efficiency, robustness, and ethical considerations. This involves creating faster algorithms for large-scale networks, making strategies for addressing ethical implications, and integrating insights from multiple disciplines to create more comprehensive models for increasing influence in social networks.

The Framework of Social Network Analysis

Figure 2 of Peng et al., (2018) describes the social network analysis framework. Analyzing social influence involves collecting big data from various sources, preprocessing to filter irrelevant and sensitive information, and using metrics like node centrality and interaction frequency to identify top influencers. Influence maximization algorithms, often based on improvements to the greedy algorithm, aim to pinpoint influential nodes efficiently. Performance is validated through simulations, ensuring algorithms achieve broad influence with minimal computational complexity.

Figure 2. Framework of Social Network Analysis

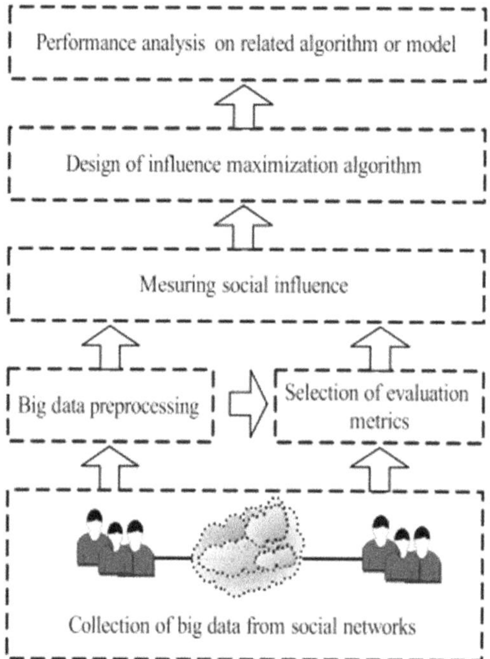

METHODOLOGY

Diffusion Methods

In the early 2000s, research on influence maximization in social networks focused on how information, behaviors, and innovations spread using models like the Linear Threshold Model (LTM), Independent Cascade Model (ICM), and Heat Diffusion Model (HDM). The LTM determines node activation based on neighbors' influence thresholds, while the ICM involves nodes attempting to activate neighbors with a probability. Epidemic models like the SIR model from epidemiology were also adapted to describe adoption processes in social networks.

Figure 3. Classification of Information Diffusion Models

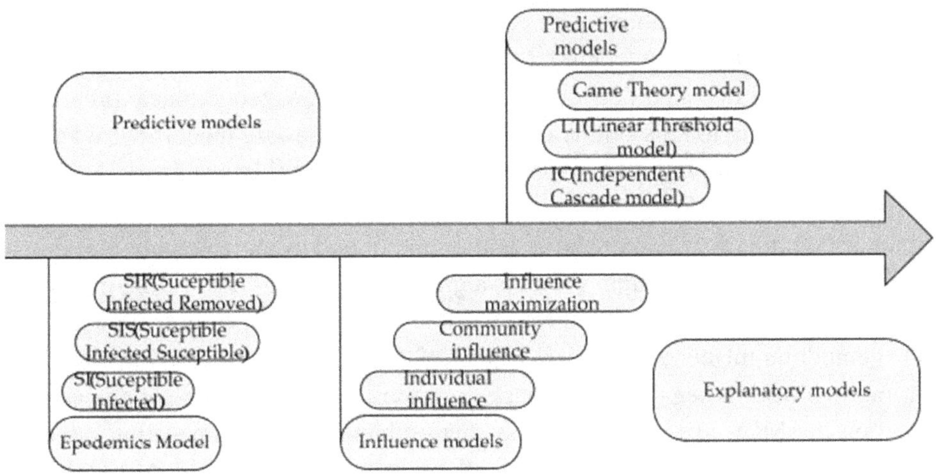

Figure 3 depicts the information diffusion models based on the view in Li et al.'s (2017) work. In social network analysis, explanatory models seek to understand the dynamics of social connections and information spread. In contrast, predictive models forecast future interactions or trends based on past behavior and network structure. Context-aware diffusion models incorporate situational factors and individual attributes to predict information diffusion.

Figure 4. Comparison of the Properties of Context-Aware Diffusion Models

Comparison of the properties of context-aware diffusion models.

Context-aware model	Categories	Diffusion model	Social network		Submodularity		Activation	
			Directed	Weighted	Diminishing return	Monotone	Time constraint	Multiple activation
Location-aware (LAM)	Online location-aware	ICM	√	√	√	√	×	×
		LTM	√	√	√	√	×	×
	Offline Billboards	OI	√	-	×	×	×	×
		MI	√	-	-	-	×	×
		LI	√	-	-	-	×	×
Time-aware (TAM)	Discrete-time	IC-M	√	√	√	√	√	×
		CTIC	√	√	-	-	√	×
		LAIC	√	√	√	√	√	×
	Continuous-time	TV-IC	√	√	√	√	√	×
		TV-LT	√	√	√	√	√	×
		DynaDiffuse	√	√	√	√	√	×

Figure 4 of Jaouadi et al., (2019) compares the properties of context-aware diffusion models. The table compares the properties of various context-aware diffusion models. It categorizes these models into Location-aware models (LAM) and Time-aware models (TAM). Each model is further evaluated based on multiple criteria: whether the diffusion is directed, weighted, exhibits diminishing returns, is monotone, has time constraints, and allows multiple activations. The table details several specific models under each category and indicates which properties each model possesses with a checkmark (√) or lacks with a cross (×). Location-aware diffusion models in social network analysis focus on how geographical proximity influences information dissemination. In contrast, time-aware models consider the temporal aspects that impact the spread of information within a network.

Kempe et al., (2005) introduce a decreasing cascade model for influence propagation, accounting for the reduced activation likelihood as more neighbors attempt to influence a node. The paper presents a greedy algorithm for selecting initial influencers, proving its efficiency through mathematical demonstration of monotone and submodular influence spread. Wu et al., (2017) focus on minimizing misinformation by blocking negative influences. It proposes scalable heuristic algorithms based on the MIA structure to address this problem in two propagation models. Wang et al., (2018) propose a multiple-path asynchronous threshold (MAT) model for viral marketing to capture influence accumulation in three stages. It introduces the IV-Greedy heuristic to address the influence maximization problem. Hosseini et al., (2018) assess influence maximization in signed social networks, proposing models incorporating distrust relationships and demonstrating their effectiveness through experiments on real-world datasets.

Li et al., (2017) analyze information diffusion research, categorizing models as explanatory or predictive, comparing them, and discussing future challenges and directions relevant to public opinion and marketing. Gu et al., (2023) propose the Com-LT model for influence spread across multiple non-competitive products. It introduces approximation algorithms to optimize lower bounds of objective functions and demonstrates their effectiveness through real-world experiments. Saxena et al. (2019) discuss influence maximization, focusing on user connectivity and activity-based models. It introduces activity-based models (AbIC and AbLT) and presents experimental results to enhance understanding of influence spread. Teng et al., (2018) introduce the MuG-IC model for revenue maximization in multi-grade products and propose the PS (Pricing-Seeding) algorithm to provide pricing suggestions and iterative customer seeding. Rezvanian et al., (2023) discuss a new stochastic diffusion model for influence maximization, proposing an algorithm based on robust optimization for finding seed nodes with a performance comparison of the LA-IP algorithm. Tian et al., (2020) propose a machine-learning framework for designing greedy heuristics for topic-aware influence maximization, introducing diffusion models for advertisement spread, and demonstrating effectiveness through experiments.

Li et al., (2021) propose an agent-based influence-diffusion model considering multiple influences and contexts, introducing new algorithms for influence minimization and evaluating their performance, highlighting the impact of influence context. Xie et al., (2015) discuss influence maximization in dynamic networks with the DYNADIFFUSE model, allowing network evolution over time and proposing the FASTMARGIN algorithm to address continuous time-constrained influence maximization. Juul et al., (2021) analyze information diffusion mechanisms by comparing cascade structural properties like reach and speed, emphasizing size control in studying differences between cascade types and explaining diffusion variations for different content types. Jaouadi et al., (2019) comprehensively review influence maximization, covering theoretical and algorithmic aspects, context-aware extensions, deep learning applications, stability, submodularity challenges, and future research directions.

Nguyen et al., (2017) propose a probability-based multi-hop diffusion algorithm for influence maximization, considering diversity in diffusion probabilities and two-hop neighbors, showing performance improvements over existing algorithms by 6-16%. Chen et al., (2021) introduce the Independent Cascade model with Meeting events (IC-M) for time-critical influence maximization, capturing time delays in influence diffusion. It presents efficient heuristics MIA-M and MIA-C based on Maximum Influence Arborescence. Hosseini-Pozveh et al., (2018) discuss influence propagation through distrust relationships, proposing new distrust propagation schemes, evaluating models in cascade-based and threshold-based classes, and assessing influence function non-monotonicity, non-submodularity, and NP-hardness.

Case Study: Promoting an Eco-Friendly Product

A business wants to increase the use of a new sustainable household product it has created, like reusable water bottles, on social media. To increase awareness and encourage adoption, they wish to use the dynamics of networks and social influence.

Models of Diffusion:

Model of Independent Cascades (ICM):

Model Description: In the Independent Cascade Model (ICM), each node in the network has an independent probability of adopting a product based on its neighbors' usage.

Implementation: A business identifies influential nodes within the network and sends them free samples to initiate product adoption. These key nodes may then influence their neighbors to adopt the product as well.

Linear Threshold Model (LTM):

A node in the Linear Threshold Model (LTM) has a threshold that indicates the minimal percentage of its neighbors that must adopt the product for the node to adopt it.

Implementation: The business analyzes user data or conducts surveys to establish the adoption thresholds of different nodes in the network. They target nodes with lower thresholds to initiate a chain reaction of product adoptions.

Application and Outcomes:

The business employs ICM and LTM techniques to maximize the product's effect on the social network. It measures the rates of adoption and changes in product awareness over time.

Figure 5 by Jaouadi et al., (2019) summarizes the comparison of the properties of classical diffusion models.

Figure 5. Comparison of Classical Diffusion Models

Classical diffusion model	Categories	Social network		Submodularity		Activation	
		Directed	Weighted	Diminishing return	Monotone	Time constraint	Multiple activation
Threshold Model (TM)	Linear TM (LTM)	√	√	√	√	×	×
	Majority TM	√	√	√	√	√	×
	Small TM	√	√	√	√	×	×
	Unanimous TM	√	√	×	×	×	×
	Separated TM	√	√	×	×	×	×
	LT-C	√	√	√	√	×	×
Cascading Model (CM)	Independent CM (ICM)	√	√	√	√	×	×
	Weight CM	√	√	√	√	×	×
	DICM	√	×	√	×	√	×
	Opinion CM	√	√	√	×	√	√
	ICM-NO	√	√	√	√	×	×
Triggering Model (TRM)	Decreasing CM	√	√	√	√	×	×
Epidemic Model (EM)	SIR	√	√	×	√	×	×
	SIS	√	√	×	√	×	√
	SIRS	√	√	×	√	×	√

Figure 5 provides a comparison of various classical diffusion models across several properties. The models are categorized into Threshold Models (TM), Cascading Models (CM), Triggering Models (TRM), and Epidemic Models (EM). Each model evaluates whether the diffusion process is directed, weighted, exhibits diminishing returns, is monotone, has time constraints, and allows for multiple activations. Checkmarks (√) indicate the presence of a property, while crosses (×) indicate its absence. This detailed comparison helps us understand each diffusion model's distinct characteristics and capabilities. Traditional diffusion models, categorized as threshold, cascade, trigger, and epidemic models, operate stochastically. They propagate information among nodes based on probabilistic rules, sequentially transferring information to neighboring nodes.

Influence Maximization Algorithms

Figure 6. Taxonomy of Influence Maximization Techniques

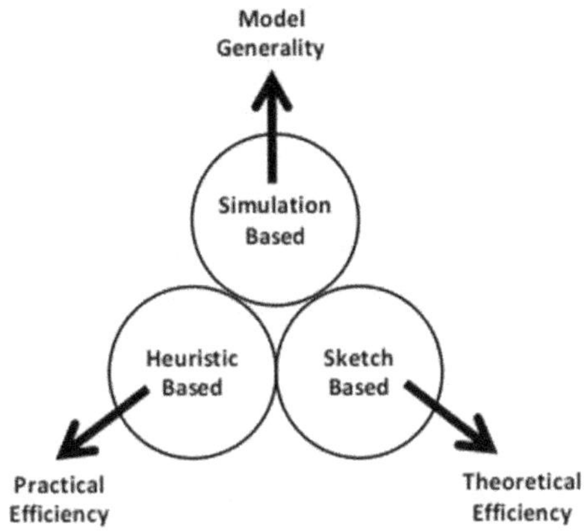

The existing IM algorithms are classified into three categories: the simulation-based approach, the proxy-based approach, and the sketch-based approach, as shown in Figure 6 of Li et al., (2018). In the taxonomy of influence maximization algorithms, the simulation-based approach involves running multiple simulations to estimate the influence spread in a network. The proxy-based approach utilizes proxies or representative nodes to approximate influence spread efficiently. In contrast, the sketch-based approach influences compact data structures (sketches) to estimate influence spread accurately with reduced computational complexity. Singh et al.. (2019) propose an efficient algorithm, C2IM, to address the challenges of influence maximization in social networks. It combines a community-based framework with contextual features, such as user interests, to improve the effectiveness of seed selection. The proposed algorithm outperforms existing methods regarding influence spread and running time, making it a promising approach for maximizing social influence in networks.

Talukder et al., (2017) discuss the Reverse Influence Maximization problem, introducing two models (R-RIM and RLT-RIM) to estimate the minimum nodes needed to activate targets in a social network, and evaluating their performance on Facebook and Twitter datasets, showing cost and time trade-offs between the models. Shang et al., (2016) discuss influence maximization in large-scale networks, introducing a community-based framework (CoFIM) that optimizes influence spread through seed expansion and intra-community propagation, demonstrating competitive results and efficiency improvements over state-of-the-art algorithms. Kim et al., (2017) discuss influence maximization in social networks, proposing the RSRS algorithm that outperforms TIM+ and SKIM in running time and influence spread and addresses the challenge of dynamic graphs with a framework for processing them using RSRS.

Li et al., (2017) discuss the location-aware influence maximization (LAIM) problem, which aims to find a seed set to maximize influence spread on targeted users with geographical preferences. It proposes a PR-tree index structure to efficiently identify targeted users and compute their preferences for given queries. It also proposes a community-based seed selection algorithm to address the challenge of efficient seed selection. The proposed algorithms are evaluated on real-world datasets and shown to be effective and efficient compared to state-of-the-art algorithms. Yang et al., (2018) discuss the problem of influence maximization in social networks and introduce a new model called IM-CM, which considers individual cost and influence. It proposes a multiobjective discrete particle swarm optimization algorithm for IM-CM and presents experimental results on three real-world networks to demonstrate the effectiveness and efficiency of the algorithm. The document also outlines future research directions, including extending the model to multiple and dynamic networks.

Gu et al., (2023) discuss the significance of user roles in social networks for information dissemination, introducing the RbNE network embedding algorithm to capture users' global roles and the RbneIM heuristic algorithm for selecting the most influential seed set, aiming to enhance understanding and future developments in social networks and information dissemination. Goyal et al., (2011) discuss CELF++, an optimized algorithm for influence maximization in social networks, improving upon CELF by exploiting submodularity to enhance efficiency, leading to significant improvements in running time and node lookups. Guille et al., 2013 describe the survey methods for information diffusion in online social networks, including popular topic detection, diffusion modeling, and identifying influential spreaders, discussing challenges and approaches for detecting topic bursts and influential individuals and presenting a taxonomy of methods with a comprehensive analysis. Lv J et al., (2014) discuss influence maximization in social networks for viral marketing, introducing CELF and CELF++ algorithms and proposing improved versions (Lv_CELF and Lv_CELF++) to enhance efficiency, evaluating their performance on Epinions data with Monte-Carlo simulations to measure influence spread.

Guo, et al., (2013) discuss personalized influence maximization on social networks, focusing on finding the top-k influential nodes for a target user, presenting a new approach that emphasizes local optima for personalized services, including a low-variance random function, efficient algorithms with approximation guarantees, and a scalable approximate algorithm for online social network applications. Lin et al. (2014) discuss the General Competitive Independent Cascade model and Competitive Influence Maximization in online social networks, addressing challenges in single source and competitive influence maximization, and propose the TCIM algorithmic framework to solve the CIM problem under the GCIC model, offering a (1 − 1/e −)-approximate solution with high probability and efficient time complexity, marking it as the first general framework with performance guarantee and practical efficiency for CIM.

Borgs et al., (2014) discuss the problem of maximizing social influence in a network by targeting a set of individuals to initiate a cascade. It presents an algorithm for influence maximization on directed graphs, providing a nearly optimal approximation to the problem. The document also references previous work on improving the algorithm's runtime, and alternative formulations of influence spread as a submodular process. Rao et al., (2020) discuss the significance of matching relationships in viral markets, particularly for online group-buying and social network propagation, introducing diffusion-matching processes and proposing algorithms to maximize matching influence in social networks, with experimental results on real-world datasets showing improved matching influence and reduced ineffective influence.

Kim et al., (2017) discuss the problem of influence maximization in social networks, focusing on selecting influential nodes for viral marketing. It introduces a new algorithm, RSRS, which outperforms existing algorithms TIM+ and SKIM regarding running time and influence spread. Additionally, the document addresses the challenges of applying influence maximization algorithms to dynamic graphs and proposes a framework for processing dynamic graphs using the new RSRS algorithm. Huang et al., (2017) discuss influence maximization in social networks, providing a rigorous analysis of the SSA and D-SSA algorithms, comparing them with previous methods, identifying inaccuracies in reported results, and suggesting opportunities for scaling up influence maximization with approximation guarantees while refuting some theoretical claims about SSA and D-SSA. Li et al., (2018) comprehensively overview social influence analysis (SIA) models, methods, and evaluation. It discusses microscopic and macroscopic models and various methods for influence maximization, minimization, flow, and individual influence. The document also highlights the strengths and weaknesses of existing models and methods and identifies future research directions and potential applications in social influence analysis.

Ko et al., (2018) discuss Influence Maximization in social networks, proposing a hybrid-IM approach that combines PB-IM and CB-IM, introducing the PB-CD and G-CELF strategies to enhance performance. Experimental results demonstrate that hybrid-IM significantly outperforms existing methods and achieves influence spread close to state-of-the-art methods. Li et al., (2018) is a survey on influence maximization (IM) algorithms, focusing on the computational complexity and approximation algorithms for IM. It also discusses context-aware IM problems, such as topic-aware, time-aware, and location-aware influence maximization, and how existing IM algorithms are applied in designing efficient context-aware IM algorithms. The survey comprehensively reviews state-of-the-art algorithmic methods for IM, including a fine-grained classification and rigorous theoretical analysis of existing IM algorithms.

Sun et al., (2018) discuss the Multi-Round Influence Maximization (MRIM) problem, which involves selecting influential nodes in multiple rounds to maximize the expected number of activated nodes in a social network. It introduces both non-adaptive and adaptive versions of MRIM and algorithms that exhibit trade-offs between efficiency and effectiveness. The study also presents scalable algorithms based on the reverse influence sampling approach and demonstrates their effectiveness through experiments on real-world social networks.

Peng et al., (2018) is a comprehensive survey on social influence analysis in social networks, covering topics such as evaluation metrics, influence maximization algorithms, and future trends in the field. It discusses the challenges of analyzing social influence, such as distinguishing positive, negative, and controversial influences, as well as collecting big social networking data. The document also highlights the need for effective evaluation metrics and integrating knowledge from cross-disciplines to advance research in social influence analysis. Kim et al., (2013) discuss using social network services (SNS) for viral marketing and the Influence Maximization Problem, aiming to identify top-k seed users for maximizing information spread, introducing Markov Clustering and leveraging community structures to enhance heuristic approaches' performance, demonstrating scalability beyond greedy methods and superior influence spread compared to heuristic approaches.

Wang et al., (2018) discusses the problem of influence maximization in social networks and introduces a new algorithm called ITÖ-IM to solve this problem. The algorithm uses a new index called Node Key Degree (NKD) to calculate a given node's influence diffusion range accurately. Experimental results show that the proposed ITÖ-IM algorithm outperforms other state-of-the-art algorithms regarding solution quality on six real-world datasets.

He X et al., (2018) discuss the evaluation of algorithms and heuristics for robust influence maximization in social networks. It focuses on real-world datasets, including Twitter and MemeTracker, and considers different diffusion models and network structures. The goal is to understand the performance of these algorithms and heuristics in optimizing influence functions under noise and uncertainty. Hafiene et al., (2019) survey the current status and future directions of influential node detection in social networks, categorizing literature into static, snapshot, and dynamic network models and identifying open issues like generalizing algorithms for different diffusion models and scalability challenges in dynamic networks.

Banerjee et al., (2020) is a comprehensive survey on the Social Influence Maximization (SIM) Problem, focusing on its variants, solution methodologies, and current research trends. It provides an overview of the problem, its background, and its associated challenges while presenting a taxonomy for classifying existing solution methodologies. The survey aims to understand the SIM Problem and its future research directions thoroughly. Tang et al., (2020) discuss the problem of influence maximization in social networks under a partial-feedback model, where seeds can be selected at intermediate stages of the diffusion process. It introduces a novel greedy adaptive policy, called the β-Greedy policy, which aims to balance the tradeoff between delay and performance. The paper also highlights the limitations of existing results on adaptive submodular maximization and proposes a new approach to address these challenges.

Sheshar et al., (2021) focus on information diffusion models and influence maximization (IM) algorithms for both single and multiple networks. It discusses challenges and future directions for IM problems, categorizing them based on different frameworks and providing insights into stability, submodularity, and group norms. The paper also compares existing IM algorithms regarding performance metrics and lists open problems that need further exploration. Purba et al., (2021) review recent studies on Influence Maximization (IM) in social media, discussing graph-based and non-graph-based approaches, categorization of IM, diffusion models, limitations, and the need for improvements in areas like noise, relation impacts, and general optimizations, highlighting challenges in validating IM algorithms in real-life scenarios, suggesting future research directions, and focusing on maximizing positive influence in competitive social networks with a trust-based solution, introducing the TrCID model for competitive influence spread simulation, estimating trust values, and developing the TrCIM algorithm for efficient calculation of positive marginal gains for non-seed users, with comparisons on synthetic and real-world datasets showing effectiveness and efficiency against other baselines.

Kumar et al., (2022) propose a novel algorithm called Communities-based Spreader Ranking (CSR) for influence maximization in social networks. It identifies influential bridge nodes based on community diversity, modularity, and density. It utilizes two information diffusion models, independent cascade (IC) and susceptible-infected-recovered (SIR), to measure the spread of information. The algorithm is evaluated on eight real-life networks, demonstrating commendable results in effectively identifying influential nodes and spreading information. Li et al., (2022) investigate the impact of crowd emotion on information propagation in social networks, proposing an attributed influence maximization method that considers emotional content and structure. It introduces a two-factor propagation model and demonstrates the algorithm's effectiveness through improved time performance and superior results to heuristic methods in real scenarios.

Ye et al., (2022) discuss using deep learning techniques to solve social network influence maximization (IM) problems. It comprehensively reviews existing advanced IM algorithms, including traditional diffusion models and those using deep learning techniques. The paper also identifies theoretical difficulties and feasible solutions for the IM problem and offers future research directions and challenges. Chen et al., (2023) present a three-stage iterative framework for influence maximization (TSIFIM) in complex networks, aiming to find a set of seed spreaders with high influence. The proposed algorithm selects initial candidate seed nodes based on global communicability and local resource allocation similarity index. It employs an adaptive search strategy to find the optimal solution among these candidates. Experimental results show that TSIFIM outperforms popular influence maximization algorithms regarding influence spreading, sensitivity analysis, seed dispersion, and statistical tests.

Jaouadi et al., (2019) focus on the influence maximization problem in social networks, addressing problem formulation, diffusion models, and influential node selection, noting the absence of comprehensive surveys and aiming to provide a detailed overview of the field, discussing extensions like distributed and dynamic networks, and proposing future research directions for dynamic and large-scale networks. Gong et al., (2024) focus on influence maximization on hypergraphs under the Susceptible–Infected spreading model with Contact Process dynamics, introducing the Multi-hop Influence Estimation (MIE) and Adaptive Neighbourhood Coefficient Algorithm (Adeff) to accurately estimate node influence ranks and solve the maximization problem, significantly outperforming existing methods by up to 450% and highlighting the importance of studying influence maximization on hypergraphs for social network marketing.

Guo et al., (2023) propose an influence maximization algorithm based on group trust and local topology structure (IMGTLT) to address the influence maximization problem in social networks. It introduces the concept of group trust to measure the interaction ability of groups and presents an influence propagation algorithm based on the local topological structure of groups. The experimental results demonstrate the effectiveness and efficiency of the proposed algorithm in selecting seed nodes and optimizing influence propagation in social network data sets. Li et al., (2024) present the HCCKshell algorithm, a novel approach to Influence Maximization that incorporates users' historical content and topological structure, using heterogeneous similarity and information entropy to measure influence and identify key seeds, showing superior performance in experiments on six real datasets and highlighting practical implications for viral propagation, social promotion, and product marketing, while noting the need for further research on time efficiency and large-scale datasets.

Li et al., (2023) is a comprehensive survey of the influence maximization (IM) research landscape, focusing on the diffusion of innovations and ideas in social networks. It reviews information diffusion models, analyzes classical IM algorithms, and proposes a taxonomy for key techniques and challenges. Additionally, the paper discusses extended IM techniques, application-oriented studies, and open questions for future research in influence maximization. Teng et al., (2024) discuss the problem of topic-aware most influential community search in social networks (TAMICS) and propose novel models and algorithms to address this problem. It introduces the concept of topic-based interaction graphs, influential communities, and lightweight indexing techniques to facilitate efficient processing of the TAMICS problem. The proposed models and algorithms significantly outperform previous works in solving the TAMICS problem by considering community interest, structural cohesion, and high influence within a given topic. Yuan et al., (2024) introduce "Influence Max," an innovative algorithm for efficiently identifying influential network nodes by considering influence propagation range, demonstrating superior performance in influence maximization across diverse network datasets, and outperforming existing methods in influence spread, computational time, and scalability.

Figure 7. Technical Comparison of Influence Maximization Approaches

Category		Methods	IC	LT	TR	WC	SIR	Techniques	Data sets
Static	Greedy	Wang and Yang (2009)	✓	✓	×	✓	×	RanCas(S) function	NetHEPT and NetPHY
		Leskovec et al. (2007)	✓	✓	✓	×	×	Submodular function	Blog networks and water networks
		Zhou et al. (2013)	✓	×	×	×	×	Upper bound function	ca-GrQc, ca-HepPh, Digger and email-Enron
	Heuristic	Narayanam and Narahari (2010)	×	✓	×	×	×	Shapley value function	sparse random graph, scale-free graph, Political Books, Jazz, Celegans, NIPS, Netscience and HEP
		Bae and Kim (2014)	×	×	×	×	✓	k-shell decomposition method	Twelve real networks and artificial networks
		Wang et al. (2015)	×	×	×	✓	×	Price-Performance Ratio and Influential Power	Political blogs and Neural network
	Hybrid	Cao et al. (2011)	✓	✓	×	×	×	Growth function	NetHEPT and NetPHY
		Ren et al. (2016)	×	×	×	×	✓	Evidential k-Shell Centrality	Zachary, Jazz, Dolphin and Email
		Pozveh et al. (2017)	×	✓	×	×	×	The community structure and influence based-closeness centrality measure	Political Books, Celegans, Netscience co-authorships, Zachary's Karate Club and Adjacency of nouns and Delicious
		Kitsak et al. (2010)	×	×	×	×	✓	K-shell decomposition and the imprecision function	Email, LiveJournal, Adult IMDB and CNI networks
		Song et al. (2014)	✓	×	×	×	×	Divide-and-conquer strategy with parallel computing mechanism	Four Mobile Social Networks
Snapshots	Incremental	Liu et al. (2012)	✓	×	×	×	×	Monte Carlo estimation strategy	NetHEPT, NetPHY, Enron, Epinions and Amazon
		Zhuang et al. (2013)	✓	×	×	×	×	Probing strategy	Synthetic, Twitter and Coauthor
		Chen et al. (2015)	×	×	×	✓	×	Interchange Heuristic strategy	Mobile, HepPh and HepTh
		Liu et al. (2017)	✓	×	×	×	×	Pruning strategy	Facebook, NetHEPT, and Flickr
		Wang et al. (2017)	×	✓	×	×	×	Pruning strategy	NetHEPT, Facebook, Flixster and Flickr
	Sliding window	Wei and Carley (2015)	Temporal aggregation models					Persistence and Emergence metrics	Newcomb, Enron, DBLP and Arab Spring dataset
		Wang et al. (2016)	DCBM model					Independent clustering, the extended Jaccard's coefficient and centrality metrics	Real and synthetic networks and LFR benchmarks graphs
		Wang et al. (2019)	Dynamic evolution model					The strategy of Local Detection and Updating	The Barabasi-Albert (BA) scale-free network, the Watts-Strogatz (WS) small-world network, and the Erdos-Rényi (ER) random network.
Dynamic	Interval	Aggarwal et al. (2012)	Influence pattern model					Globally optimized forward trace approach and a locally optimized backward approach	DBLP and Arnetminer Citation
		Bhowmick et al. (2019)	Temporal retweet model					Twitter Specific Metrics	Algeria, Egypt, and Nepal
	A contact sequence	Ohsaka et al. (2016)	✓	×	✓	✓	×	Reverse reachable sketch	Digg, Enron, Epinions, Facebook, DBLP, YouTube and Flickr
		Tong et al. (2017)	DIC model					Adaptive Seeding Strategy	Hep, Wiki and PL
		Yalavarthi and Khan (2018)	MIA model and IC					Localized batch updates technique	Digg, Slashdot, Epinions and Flickr

Figure 7 of Hafiene et al., (2019) describes the influence of Maximization approaches classified based on network models: static, snapshot, and dynamic, detailing their techniques and applicable datasets. Static networks model real-world scenarios but overlook changing social interactions over time. Snapshot models capture network evolution at specific moments, while dynamic networks, with precise temporal data, better represent continuously evolving social relationships.

Case Study: Maximizing Sales on an E-Commerce Platform

Background: An e-commerce platform wants to increase sales by identifying influential customers and optimizing targeted marketing campaigns. To increase sales and revenue, they try distributing advertisements as widely as possible within the client group.

Influence Maximization Algorithms

A greedy algorithm continuously choosing a seed set of powerful clients, the Greedy Algorithm considers each one's capacity to maximize the spread of promotional offers throughout the network.

Implementation: The e-commerce platform uses the greedy algorithm to select a seed group of very influential clients, such as those who have a history of making regular purchases or a significant social media following. Personalized offers are made especially to these clients to get them to share with their network.

Relevance: By taking a realistic approach to seed selection, the Greedy Algorithm allows the platform to widen the reach of advertisements throughout the consumer base.

Algorithm for PageRank: The PageRank Algorithm calculates a user's influence based on their activities on the platform, including purchases, reviews, and suggestions.

Implementation: The e-commerce platform calculates each customer's PageRank score using data gathered from their engagement history. Consumers with high PageRank scores are considered influential and seek further influence in the network through specialized campaigns.

Heuristic for Degree Discount:

Algorithm Description: The Degree Discount Heuristic accounts for diminishing rewards as a customer gains more connections by adjusting a customer's impact based on their degree centrality in the network.

Application: The e-commerce platform employs the Degree Discount Heuristic to prioritize customers with high influence potential based on their network connectivity, ensuring that it targets users most likely to impact sales through their influence.

Results and Implementation: The e-commerce platform uses these influence maximization algorithms in focused marketing campaigns to maximize sales and track revenue indicators.

Greedy Algorithm: The algorithm finds influential clients for specialized marketing campaigns, which raises the platform's sales. However, because of its computational complexity, scaling the technique to broader consumer networks might be difficult.

PageRank Algorithm: Marketing initiatives specifically targeted at clients with significant effect rankings yield better sales results based on PageRank scores. This technique allows for the location of significant clients who might not be easily visible from transaction history alone.

Degree Discount Heuristic: The platform avoids over-targeting highly connected but less influential clients by utilizing the Degree Discount Heuristic to create a more balanced distribution of marketing efforts. As a result, sales increase regularly and resources are distributed more effectively.

AI Techniques in Social Media

Gong et al., (2016) discuss optimizing influence maximization in social networks using a discrete particle swarm optimization (DPSO) algorithm to find a small group of individuals with maximal influence cascades, showing that the DPSO algorithm performs comparably to state-of-the-art methods like the CELF++ greedy method but with significantly lower running time in both the independent cascade (IC) and weighted cascade (WC) spread models. Abbas et al., (2021) provide an overview of research on text classification, sentiment analysis, and recommender systems in online social networks using deep learning methods, discussing the application of techniques like CNN, RNN, LSTM, and GRU for sentiment analysis of textual and visual data, classification of user interests, and recommendation of people and multimedia content, and addressing challenges including sentiment analysis in short texts, balancing information exposure, and reducing opinion polarization in social media.

Bhattacharya et al., (2023) discuss the development of a deep learning model called DeepInfNode, which uses Graph Convolutional Networks (GCN) to identify influential nodes in large graph-based datasets, outperforming traditional centrality-based and machine learning-based approaches in accuracy, precision, and F1 scores, especially on social media datasets like Facebook, demonstrating potential for improved prediction accuracy and effectiveness for undirected graph data. Tsai et al., (2019) discuss the influence maximization problem (IMP) in online social networks (OSN) and introduce a novel metaheuristic algorithm called Search Economics for Influence Maximization (SEIM), designed to identify users with maximum influence, beneficial for marketing strategies, understanding user relationships, and detecting abnormal behavior, with experimental results demonstrating SEIM's superior performance over other state-of-the-art algorithms, particularly for large-scale datasets, achieving better results with fewer evaluations.

Lingam et al., (2019) discuss a Deep Q-Learning algorithm proposed for detecting social bots and identifying influential users in online social networks. It also presents experimental results and compares the performance of the proposed algorithm with other existing algorithms. Fan et al., (2023) discuss the influence maximization problem in social networks, proposing a new objective function and algorithm, introducing the Harris Hawks Optimization (HHO) algorithm for potential application to influence maximization, and concluding with future research directions and acknowledgments for support. Keikha et al., (2019) discuss the development of DeepIM, a novel algorithm for influence maximization across interconnected networks, using deep learning and network embedding to extract optimal structural features of nodes on various networks, addressing challenges such as heterogeneous features, cross-links, and bridge nodes, and presenting empirical evaluations that demonstrate DeepIM's scalability and efficiency compared to other IM approaches.

Tian et al., (2020) discuss the development of a machine learning framework for topic-aware influence maximization (TIM) problems, proposing a data-driven approach to assign probabilities based on both IC and LT models, incorporating user similarity, user benefit, and inherent contact frequency, demonstrating superior performance and comparable quality in experiments on generated graph instances and real-world social networks. Camacho et al., (2020) is a compilation of research papers on applying advanced data science and artificial intelligence techniques to social network analysis (SNA), covering topics like sentiment analysis, community detection, natural language processing, and machine learning, with applications to social media data from platforms such as Twitter, Instagram, and Sina Weibo, showcasing the latest advances and developments in these areas through innovative models and algorithms to gain insights from social media data.

Zhang et al., (2021) discuss the Grain framework, a new data selection criterion for Graph Neural Networks (GNNs) combining active learning and social influence maximization. It proposes a novel propagation model and diversity functions based on both direct and indirect feature influence of GNNs over the graph and feature space. The framework significantly outperforms state-of-the-art baselines on performance and efficiency in real-world graph experiments. Singh et al., (2021) discuss a method for maximizing information spread in online social networks by predicting the upcoming snapshot of the graph and detecting the seed set that maximizes information spread. It introduces the concept of link prediction, the Independent Cascade model for information diffusion, and the Influence Maximization problem for social networks. The proposed LPINT framework uses link prediction to predict the upcoming snapshot of the graph and computes the seed set for influence maximization.

Agarwal et al., (2022) discuss the GNPA algorithm proposed for maximizing social influence in dynamic networks, evaluating its performance with real-world and synthetic datasets, detailing the algorithm's modules and experimental results, addressing limitations of existing models, and emphasizing the importance of considering user characteristics in influence maximization. Shakeel et al., (2022) discuss the issue of influence maximization in social networks, particularly in the context of viral marketing and spreading products through influential nodes. It explores using centrality measures and correlation coefficients to identify influential nodes and their impact on network propagation. The proposed approach aims to evaluate the performance of models and the size of the network, as well as to select seed nodes for maximizing influence spread.

Karoui et al., (2022) discuss a machine learning-based approach for predicting influential nodes in a social network by analyzing articles' structural and semantic features. It outlines the steps for data cleaning, modeling, and output, as well as machine learning models to analyze and predict influential articles. The approach integrates structure-based and topic-based models to calculate the influence score of an article based on its research subject and location in social networks. Ma et al., (2023) discuss influence maximization in complex networks, aiming to activate seed nodes to maximize information or influence spread, presenting various algorithms like approximate greedy and meta-heuristic algorithms, validating their effectiveness through experiments on real-world networks, and showing that an evolutionary deep reinforcement learning algorithm outperforms other baselines in influence spread and overall performance.

Li et al., (2023) discuss the PIANO framework, a deep reinforcement learning approach to influence maximization in social networks, focusing on model training, pretraining strategies, and experimental results, comparing PIANO-S, which trains from scratch and a trained model pool, concluding both achieve similar effectiveness, highlighting the use of machine learning for influence maximization, and integrating learning models for practical applications without training from scratch. Yang et al., (2024) discuss the development of BIM-DRL, a novel framework for balanced influence maximization in multi-entity social networks, featuring an entity correlation evaluation module and a balanced seed node selection module, leveraging historical behavior sequences and deep reinforcement learning to accurately assess entity correlation and select seed nodes for maximizing influence diffusion.

Liu et al., (2024) discuss the influence maximization (IM) problem and the adoption maximization problem in social networks with competitive influences, reviewing algorithms and models such as the independent cascade (IC) model, conformity-aware independent cascade (C-IC) model, and competitive independent cascade model, emphasizing the C-IC model's representation of multiple competitive influences' diffusion and the impact of conformity on user decisions, and the objective of the adoption maximization problem to find a set of users that maximizes adoption under the C-IC model. Zareie et al., (2024) discuss Fuzzy Influence Maximization (FIM) in social networks, aiming to identify a small set of users to initiate message spread and maximize influence, introducing a framework that models social networks as fuzzy graphs with users' activation levels and influence as fuzzy variables, proposing a fuzzy diffusion model to simulate spreading and evaluating methods for finding influential users.

Chen et al., (2023) discuss influence maximization (IM) on social networks and the application of deep reinforcement learning (DRL) to solve this problem, introducing the ToupleGDD model for efficient and effective seed set selection to maximize influence spread, comparing it with other DRL-based models and traditional algorithms, and showing its superior performance in influence spread and running time on various datasets. Aghaee et al., (2021) provide an in-depth review of strategies for identifying influential users in social networks to maximize impact, highlighting the importance of user interactions and discussing the Influence Maximization Problem (IMP) as an NP-hard problem aiming to find key influential nodes, classifying and comparing IMP algorithms into greedy, heuristic, community-based, influence-path based, and meta-heuristic categories, and highlighting meta-heuristic algorithms as a potential solution for balancing influence spread and computational efficiency.

RESULTS

Surveys on information diffusion models reveal how user engagement patterns and network designs affect information spread and evaluate the effectiveness of diffusion strategies on various platforms. Influence maximization techniques identify key nodes, while diffusion models and AI, particularly deep learning, enhance understanding and prediction of information propagation. Deep learning models are trending due to their ability to automatically learn features, predict diffusion, detect influential nodes, and provide real-time analysis, making them highly effective in social network research.

Graph Neural Networks (GNNs) are effective for tasks like graph and node classification, making them valuable for social network analysis, influence maximization, fake news detection, and community detection. Researchers are enhancing influence maximization with GNNs, deep learning, and reinforcement learning, exploring multi-modal data integration and Reinforcement Learning (RL) advancements. Multi-agent RL addresses simultaneous influence maximization challenges, while future research may focus on hierarchical and temporal GNNs for more accurate influence propagation predictions.

DISCUSSION

Different diffusion models offer unique advantages for influence maximization in social networks:

1. Linear Threshold Model (LTM): Reflects realistic influence dynamics but may oversimplify nonlinear behaviors.
2. Independent Cascade Model (ICM): Effective for identifying key nodes but assumes independent influence spread.
3. Network Contagion Model: Helps understand influence within network structures but may overlook diverse influence mechanisms.
4. Threshold Models with Time Delays: Reflect real-world adoption patterns but increase computational complexity.
5. Dynamic Diffusion Models: Suitable for evolving networks but may struggle with real-time changes.
6. Community-Based Models: Customize strategies for specific communities but may oversimplify inter-community relationships.

Advantages of Using Different Influence Maximization Algorithms

Influence maximization in social networks is complex, and while various diffusion models offer valuable insights, their limitations highlight the need for continued research to capture influence propagation dynamics better. Different models are essential for understanding various aspects of influence within networks.

- Greedy Algorithm: Simple and efficient, suitable for large networks.
- PageRank Algorithm: Effective for ranking nodes based on global network structure.
- Independent Cascade Model (ICM): Simulates influence spread, useful for identifying influential nodes.
- Linear Threshold Model (LTM): Captures complex dynamics and identifies nodes that can surpass adoption thresholds.
- Maximizing Spread with Limited Budget: Optimizes influence within budget constraints.
- Community Detection Algorithms: Tailors influence strategies to specific subgroups.
- Betweenness Centrality: Identifies influential nodes that bridge network areas.

- Degree Centrality: Targets well-connected nodes for wider audience reach.
- K-Core Decomposition: Focuses on tightly connected groups for effective influence.
- SIR Models: Simulates influence spread like a contagion, aiding in strategy optimization.

Disadvantages

- Greedy Algorithm: May not always find the optimal solution and is sensitive to initial node selection.
- PageRank Algorithm: Focuses on global network structure, potentially missing fine influence patterns.
- Independent Cascade Model (ICM): Assumes node independence, which may not reflect real-world influence spread.
- Linear Threshold Model (LTM): May not capture nonlinear influence dynamics and the variety of factors affecting behavior.
- Maximizing Spread with Limited Budget: Finding cost-effective methods can be challenging with insufficient funding.
- Community Detection Algorithms: May struggle with dynamic social networks and complex community structures.
- Betweenness Centrality: May neglect other influential factors and miss nodes that are not bridges.
- Degree Centrality: May prioritize well-connected nodes without considering connection quality or relevance.
- K-Core Decomposition: May miss influential nodes that connect different communities.
- SIR Models: Simplify influence dynamics, potentially ignoring real-world complexities.

Advantages of Using Different AI Techniques

AI methods, including machine learning algorithms, are crucial for maximizing influence in social networks. They analyze data to predict outcomes and identify influential nodes, which is essential for influence maximization strategies. AI can personalize approaches by understanding individual user preferences and behaviors, enhancing the effectiveness of influence campaigns. Adaptive learning through AI enables diffusion models to evolve based on real-time feedback, ensuring the continued relevance of influence maximization efforts. Additionally, AI facilitates community detection and network analysis, providing insights into information flow patterns and improving the precision of influence maximization techniques. AI enhances influence maximization in social networks by enabling real-time adaptation, scalability, and predictive capabilities. However, it also faces challenges such as data bias, lack of transparency, and difficulty in adapting to new trends. These limitations can affect the effectiveness of AI in social network influence maximization.

CONCLUSION

Different Diffusion models are vital for understanding behavior spread, information flow, and increasing influence in social networks. Influence maximization algorithms determine the powerful nodes, optimize influence, and leverage community structures. Artificial intelligence techniques like machine learning, natural language processing, and network analysis algorithms improve social network analysis. They allow for more sophisticated data processing, pattern recognition, and predictive modeling, which helps with insights processing, trend detection, and influence strategy improvement. By using a combination of diffusion models, influence maximization algorithms, and AI techniques, social network analysis continues to evolve, offering innovative approaches to understanding complex network dynamics, predicting influence outcomes, and optimizing strategies for effective communication, marketing, and community engagement in the digital age.

Future Scope of Diffusion Models in Social Media Analysis

Future diffusion models will integrate complex network structures like multiplex and temporal networks to enhance information propagation analysis. Hybrid models combining different diffusion approaches may offer deeper insights into social media dynamics. Behavioral theories could be incorporated to simulate individual interactions with information, while personalized models may improve prediction accuracy. Additionally, real-time adaptive models and predictive tools using live data streams could enable more timely responses and better forecast trends and viral content.

Future diffusion techniques in social media analysis will focus on optimizing content dissemination through personalized recommendations, identifying key influencers, and leveraging cross-platform strategies. Advanced network models, real-time analytics, and multi-objective algorithms will enhance content distribution and engagement. Multidisciplinary approaches and AI advancements will enhance influence methods, ethical considerations, sentiment analysis, trend prediction, and targeted advertising, increasing the effectiveness and trustworthiness of social media campaigns.

REFERENCES

Abbas, A. (2021). Social network analysis using deep learning: Applications and schemes. *Social Network Analysis and Mining*, 11(1), 106. Advance online publication. DOI: 10.1007/s13278-021-00799-z

Agarwal, S., & Mehta, S. (2022). GNPA: A hybrid model for social influence maximization in dynamic networks. *Multimedia Tools and Applications*, 83(1), 1–28. DOI: 10.1007/s11042-021-11606-4

Aghaee, Z., Ghasemi, M., Ahmadi Beni, H., Bouyer, A., & Fatemi, A. (2021). A survey on meta-heuristic algorithms for the influence maximization problem in social networks. *Computing*, 103(11), 2437–2477. Advance online publication. DOI: 10.1007/s00607-021-00945-7

Banerjee, S., Jenamani, M., & Pratihar, D. (2020). A survey on influence maximization in a social network. *Knowledge and Information Systems*, 62(9), 3417–3455. Advance online publication. DOI: 10.1007/s10115-020-01461-4

Bhattacharya, R., Nagwani, N. K., & Tripathi, S. (2023). Detecting influential nodes with topological structure via Graph Neural Network approach in social networks. *International Journal of Information Technology : an Official Journal of Bharati Vidyapeeth's Institute of Computer Applications and Management*, 15(4), 2233–2246. DOI: 10.1007/s41870-023-01271-1 PMID: 37256031

Borgs, C., Brautbar, M., Chayes, J., Lucier, B., & Sridhar, V. (2014). Maximizing social influence in nearly optimal time. In *Proceedings of the twenty-fifth annual ACM-SIAM symposium on Discrete algorithms* (pp. 966-977). Society for Industrial and Applied Mathematics. https://doi.org/DOI: 10.1137/1.9781611973402.70

Camacho, D., Luzon, M., & Cambria, E. (2020). New research methods & algorithms in social network analysis. *Future Generation Computer Systems*, 114, 290–293. Advance online publication. DOI: 10.1016/j.future.2020.08.006

Chen, D., Xu, G., Yang, P., & Meng, L. (2023). TSIFIM: A three-stage iterative framework for influence maximization in complex networks. *Expert Systems with Applications*, 212, 118702. DOI: 10.1016/j.eswa.2022.118702

Chen, T., Yan, S., Guo, J., & Wu, W. (2023). ToupleGDD: A fine-designed solution of influence maximization by deep reinforcement learning. *IEEE Transactions on Computational Social Systems*.

Chen, W., Lu, W., & Zhang, N. (2021). Time-Critical Influence Maximization in Social Networks with Time-Delayed Diffusion Process. In *Proceedings of the AAAI Conference on Artificial Intelligence* (Vol. 26, No. 1, pp. 591-598). https://doi.org/ DOI: 10.1609/aaai.v26i1.8204

D'Angelo, G., Poddar, D., & Vinci, C. (2023). Better bounds on the adaptivity gap of influence maximization under full-adoption feedback. *. *Artificial Intelligence*, 318, 103895. DOI: 10.1016/j.artint.2023.103895

Dizaji, S., Patil, K., & Avrachenkov, K. (2024). Influence maximization in dynamic networks using reinforcement learning. *SN Computer Science*, 5(1), 169. Advance online publication. DOI: 10.1007/s42979-023-02453-1

Fan, C., Wang, Z., Zhang, J., Zhao, J., & Rui, X. (2023). Influence maximization in social networks based on discrete harris hawks optimization algorithm. *Computing*, 106(2), 1–25. DOI: 10.1007/s00607-023-01207-4

Gong, M., Yan, J., Shen, B., Ma, L., & Cai, Q. (2016). Influence Maximization in Social Networks Based on Discrete Particle Swarm Optimization. *Information Sciences*, 367, 600–614. Advance online publication. DOI: 10.1016/j.ins.2016.07.012

Gong, X., Wang, H., Wang, X., Chen, C., Zhang, W., & Zhang, Y. (2024). Influence maximization on hypergraphs via multi-hop influence estimation. *Information Processing & Management*, 61(3), 103683. https://doi.org/10. DOI: 10.1016/j.ipm.2024.103683

Goyal, A., Lu, W., & Lakshmanan, L. V. S. (2011). CELF++: optimizing the greedy algorithm for influence maximization in social networks. In *Proceedings of the 20th international conference companion on World wide web (WWW '11)* (pp. 47-48). Association for Computing Machinery. https://doi.org/DOI: 10.1145/1963192.1963217

Gu, X., Wang, Z., Chen, X., Lu, P., Du, Y., & Tang, M. (2023). Influence maximization in social networks using role-based embedding. *Networks and Heterogeneous Media*, 18(4), 1539–1574. DOI: 10.3934/nhm.2023068

Guille, A., Hacid, H., Favre, C., & Zighed, D. (2013). Information Diffusion in Online Social Networks: A Survey. *SIGMOD Record*, 42(2), 17–28. DOI: 10.1145/2503792.2503797

Guo, C., Li, W., Liu, F., Zhong, K., Wu, X., Zhao, Y., & Jin, Q. (2023). Influence maximization algorithm based on group trust and local topology structure. *Neurocomputing*, 564, 126936. DOI: 10.1016/j.neucom.2023.126936

Guo, C., Li, W., Wang, J., Yu, X., Liu, X., Luvembe, A., Wang, C., & Jin, Q. (2024). Heterogeneous network influence maximization algorithm based on multi-scale propagation strength and repulsive force of propagation field. *. *Knowledge-Based Systems*, 291, 111580. DOI: 10.1016/j.knosys.2024.111580

Guo, J., Zhang, P., Zhou, C., Cao, Y.-N., & Guo, L. (2013). Personalized influence maximization on social networks. In *International Conference on Information and Knowledge Management, Proceedings* (pp. 199-208). DOI: 10.1145/2505515.2505571

Gursoy, F., & Gunnec, D. (2018). Influence maximization in social networks under Deterministic Linear Threshold Model. *Knowledge-Based Systems*, 161, 111–123. DOI: 10.1016/j.knosys.2018.07.040

Hafiene, N., Karoui, W., & Romdhane, L. (2019). Influential Nodes Detection in Dynamic Social Networks. In Lecture Notes in Computer Science (pp. 62-73). DOI: 10.1007/978-3-030-20482-2_6

He, X., & Kempe, D. (2018). Stability and Robustness in Influence Maximization. *ACM Transactions on Knowledge Discovery from Data*, 12(6), 66. Advance online publication. DOI: 10.1145/3233227

Hosseini-Pozveh, M., Zamanifar, K., & Naghsh-Nilchi, A. (2018). Assessing information diffusion models for influence maximization in signed social networks. *Expert Systems with Applications*, 119, 476–490. Advance online publication. DOI: 10.1016/j.eswa.2018.07.064

Huang, K., Wang, S., Bevilacqua, G., Xiao, X., & Lakshmanan, L. (2017). Revisiting the stop-and-stare algorithms for influence maximization. *Proceedings of the VLDB Endowment International Conference on Very Large Data Bases*, 10(9), 913–924. DOI: 10.14778/3099622.3099623

Jalayer, M., Azheian, M., & Agha Mohammad Ali Kermani, M. (2018). A hybrid algorithm based on community detection and multi attribute decision making for influence maximization. *Computers & Industrial Engineering*, 120, 234–250. DOI: 10.1016/j.cie.2018.04.049

Jaouadi, M., & Romdhane, L. (2019). Influence Maximization Problem in Social Networks: An Overview. In 2019 IEEE/ACS 16th International Conference on Computer Systems and Applications (AICCSA) (pp. 1-8). DOI: 10.1109/AICCSA47632.2019.9035366

Juul, J. L., & Ugander, J. (2021). Comparing information diffusion mechanisms by matching on cascade size. *Proceedings of the National Academy of Sciences of the United States of America*, 118(46), e2100786118. DOI: 10.1073/pnas.2100786118 PMID: 34750252

Karoui, W., Hafiene, N., & Romdhane, L. (2022). Machine learning-based method to predict influential nodes in dynamic social networks. *Social Network Analysis and Mining*, 12(1), 108. Advance online publication. DOI: 10.1007/s13278-022-00942-4

Keikha, M. M., Rahgozar, M., Asadpour, M., & Faghih Abdollahi, M. (2019). Influence Maximization across Heterogeneous Interconnected Networks based on Deep Learning. *Expert Systems with Applications*, 140, 112905. DOI: 10.1016/j.eswa.2019.112905

Kempe, D., Kleinberg, J., & Tardos, É. (2005). Influential nodes in a diffusion model for social networks. In Lecture Notes in Computer Science (pp. 1127-1138). DOI: 10.1007/11523468_91

Kim, C., Lee, S., Park, S., & Lee, S.-g. (2013). Influence Maximization Algorithm Using Markov Clustering. In Lecture Notes in Computer Science (pp. 112-126). DOI: 10.1007/978-3-642-40270-8_10

Kim, D., Hyeon, D., Oh, J., Han, W., & Yu, H. (2017). Influence maximization based on reachability sketches in dynamic graphs. *Information Sciences*, 394, 217–231. Advance online publication. DOI: 10.1016/j.ins.2017.02.023

Kimura, M., & Saito, K. (2006). Tractable models for information diffusion in social networks. *Lecture Notes in Computer Science*, 4213, 259–271. DOI: 10.1007/11871637_27

Ko, Y., Cho, K.-J., & Kim, S.-W. (2018). Efficient and Effective Influence Maximization in Social Networks: A Hybrid-Approach. *Information Sciences*, 465, 144–161. Advance online publication. DOI: 10.1016/j.ins.2018.07.003

Kumar, S., Gupta, A., & Khatri, I. (2022). CSR: A community based spreaders ranking algorithm for influence maximization in social networks. *World Wide Web (Bussum)*, 25(6), 2303–2322. Advance online publication. DOI: 10.1007/s11280-021-00996-y

Li, H., Xu, M., Bhowmick, S. S., Rayhan, J. S., Sun, C., & Cui, J. (2023). PIANO: Influence Maximization Meets Deep Reinforcement Learning. *IEEE Transactions on Computational Social Systems*, 10(3), 1288–1300. DOI: 10.1109/TCSS.2022.3164667

Li, H., Yang, S., Xu, M., Bhowmick, S. S., & Cui, J. (2023). Influence Maximization in Social Networks. *Survey (London, England)*.

Li, K., Zhang, L., & Huang, H. (2018). Social Influence Analysis: Models, Methods, and Evaluation. *Engineering (Beijing)*, 4(1), 40–46. Advance online publication. DOI: 10.1016/j.eng.2018.02.004

Li, M., Wang, X., Gao, K., & Zhang, S. (2017). A Survey on Information Diffusion in Online Social Networks: Models and Methods. *Information (Basel)*, 8(4), 118. DOI: 10.3390/info8040118

Li, W., Bai, Q., Liang, L., Yang, Y., Hu, Y., & Zhang, M. (2021). Social influence minimization based on context-aware multiple influences diffusion model. *Knowledge-Based Systems*, 227, 107233. DOI: 10.1016/j.knosys.2021.107233

Li, W., Li, Y., Liu, W., & Wang, C. (2022). An influence maximization method based on crowd emotion under an emotion-based attribute social network. *Information Processing & Management*, 59(2), 102818. DOI: 10.1016/j.ipm.2021.102818

Li, X., Cheng, X., Su, S., & Sun, C. (2017). Community-Based Seeds Selection Algorithm for Location Aware Influence Maximization. *Neurocomputing*, 275, 1601–1613. Advance online publication. DOI: 10.1016/j.neucom.2017.10.007

Li, Y., Fan, J., Wang, Y., & Tan, K.-L. (2018). Influence Maximization on Social Graphs: A Survey. *IEEE Transactions on Knowledge and Data Engineering*, 30(10), 1852–1872. DOI: 10.1109/TKDE.2018.2807843

Li, Y., Lu, T., Li, W., & Zhang, P. (2024). HCCKshell: A heterogeneous cross-comparison improved Kshell algorithm for Influence Maximization. *Information Processing & Management*, 61(3), 103681. DOI: 10.1016/j.ipm.2024.103681

Liang, J., Gong, Y., Wu, X., & Li, Y. (2023). Customized influence maximization in attributed social networks: Heuristic and meta-heuristic algorithms. *Complex & Intelligent Systems*, 10(1), 1409–1424. Advance online publication. DOI: 10.1007/s40747-023-01220-2

Lin, Y., & Lui, J. (2014). Algorithmic design for competitive influence maximization problems. arXiv preprint arXiv:1410.8664.

Lin, Y., Zhang, X., Xia, L., Ren, Y., & Li, W. (2019). A hybrid algorithm for influence maximization of social networks. In 2019 IEEE Intl Conf on Dependable, Autonomic and Secure Computing, Intl Conf on Pervasive Intelligence and Computing, Intl Conf on Cloud and Big Data Computing, Intl Conf on Cyber Science and Technology Congress (DASC/PiCom/CBDCom/CyberSciTech) (pp. 427-431). Fukuoka, Japan. DOI: 10.1109/DASC/PiCom/CBDCom/CyberSciTech.2019.00087

Lingam, G., Rout, R., & Somayajulu, D. (2019). Adaptive deep Q-learning model for detecting social bots and influential users in online social networks. *Applied Intelligence*, 49(11), 3947–3964. Advance online publication. DOI: 10.1007/s10489-019-01488-3

Liu, Y., Hu, Y., Yu, S., Zhou, X., & Li, K. (2024). Conformity-aware adoption maximization in competitive social networks. *Neurocomputing*, 573, 127224. DOI: 10.1016/j.neucom.2023.127224

Lotf, J. J., Azgomi, M. A., & Ebrahimi Dishabi, M. R. (2022). An improved influence maximization method for social networks based on genetic algorithm. *. *Physica A*, 586, 126480. DOI: 10.1016/j.physa.2021.126480

Lv, J., Guo, J., Yang, Z., Zhang, W., & Jocshi, A. (2014). Improved algorithms OF CELF and CELF++ for influence maximization. *Journal of Engineering Science and Technology Review*, 7(3), 32–38. DOI: 10.25103/jestr.073.05

Ma, L., Shao, Z., Li, X., Lin, Q., Li, J., Leung, V. C. M., & Nandi, A. K. (2023). Influence Maximization in Complex Networks by Using Evolutionary Deep Reinforcement Learning. *IEEE Transactions on Emerging Topics in Computational Intelligence*, 7(4), 995–1009. DOI: 10.1109/TETCI.2021.3136643

Nguyen, D.-L., Nguyen, T.-H., Do, T.-H., & Yoo, M. (2017). Probability-Based Multi-hop Diffusion Method for Influence Maximization in Social Networks. *Wireless Personal Communications*, 93(4), 903–916. DOI: 10.1007/s11277-016-3939-8

Peng, S., Zhou, Y., Cao, L., Yu, S., Niu, J., & Jia, W. (2018). Influence analysis in social networks: A survey. *Journal of Network and Computer Applications*, 106, 17–32. DOI: 10.1016/j.jnca.2018.01.005

Purba, K. R., Asirvatham, D., & Murugesan, R. K. (2021). Influence maximization algorithm: Review on current approaches and limitations. *Engineering and Applied Science Research*, 48(2), 221–229.

Qin, Y., Ma, J., & Gao, S. (2017). Efficient influence maximization under TSCM: A suitable diffusion model in online social networks. *. *Soft Computing*, 21(4), 827–838. Advance online publication. DOI: 10.1007/s00500-016-2068-3

Rao, G., Wang, Y., Chen, W., Li, D., & Wu, W. (2020). Matching Influence Maximization in Social Networks. *Theoretical Computer Science*, 857, 71–86. Advance online publication. DOI: 10.1016/j.tcs.2020.12.040

Rezvanian, A., Vahidipour, S. M., & Meybodi, M. R. (2023). A new stochastic diffusion model for influence maximization in social networks. *Scientific Reports*, 13(1), 6122. DOI: 10.1038/s41598-023-33010-8 PMID: 37059847

Saxena, B., & Kumar, P. (2019). A node activity and connectivity-based model for influence maximization in social networks. *Social Network Analysis and Mining*, 9(1), 40. Advance online publication. DOI: 10.1007/s13278-019-0586-6

Shakeel, N., & Dwivedi, R. K. (2022a). A comparative analysis of the techniques for influence maximization in social networks. In 2022 9th International Conference on Computing for Sustainable Global Development (INDIACom) (pp. 674-678). New Delhi, India. DOI: 10.23919/INDIACom54597.2022.9763226

Shakeel, N., & Dwivedi, R. K. (2022b). A learning based influence maximization across multiple social networks. In 2022 12th International Conference on Cloud Computing, Data Science & Engineering (Confluence) (pp. 53-58). Noida, India. DOI: 10.1109/Confluence52989.2022.9734145

Shang, J., Zhou, S., Li, X., Liu, L., & Wu, H. (2016). CoFIM: A community-based framework for influence maximization on large-scale networks. *Knowledge-Based Systems*, 10. Advance online publication. DOI: 10.1016/j.knosys.2016.09.029

Sheshar, S., Srivastva, S., Verma, M., & Kumar, R. (2021). Influence Maximization Frameworks, Performance, Challenges and Directions on Social Network: A Theoretical Study. *Journal of King Saud University. Computer and Information Sciences*, 34(9), 7570–7603. Advance online publication. DOI: 10.1016/j.jksuci.2021.08.009

Singh, A. K., & Kailasam, L. (2021). Link prediction-based influence maximization in online social networks. *Neurocomputing*, 453, 151–163. DOI: 10.1016/j.neucom.2021.04.084

Singh, S. S., Kumar, A., Singh, K., & Biswas, B. (2019). C2IM: Community based context-aware influence maximization in social networks. *Physica A*, 514, 796–818. DOI: 10.1016/j.physa.2018.09.142

Sun, L., Huang, W., Yu, P., & Chen, W. (2018). Multi-Round Influence Maximization. In *Proceedings of the 27th ACM International Conference on Information and Knowledge Management* (pp. 2249-2258). https://doi.org/DOI: 10.1145/3219819.3220101

Talukder, A. (2017). An approach of cost optimized influence maximization in social networks. In 2017 19th Asia-Pacific Network Operations and Management Symposium (APNOMS) (pp. 354-357). Seoul, Korea (South). DOI: 10.1109/APNOMS.2017.8094146

Tang, S., & Yuan, J. (2020). Influence maximization with partial feedback. *Operations Research Letters*, 48(1), 24–28. DOI: 10.1016/j.orl.2019.10.013

Tang, W., Luo, G., Wu, Y., Tian, L., Zheng, X., & Cai, Z. (2019). A second-order diffusion model for influence maximization in social networks. *. *IEEE Transactions on Computational Social Systems*, 6(4), 702–714. DOI: 10.1109/TCSS.2019.2921422

Teng, L., & Wang, Y. (2024). Topic-aware Most Influential Community Search in Social Networks. ArXiv, abs/2402.07601.

Teng, Y.-W., Tai, C.-H., Yu, P., & Chen, M.-S. (2018). Revenue Maximization on the Multi-grade Product. .DOI: 10.1137/1.9781611975321.65

Tian, S., Mo, S., Wang, L., & Peng, Z. (2020). Deep Reinforcement Learning-Based Approach to Tackle Topic-Aware Influence Maximization. *Data Science and Engineering*, 5(1), 5. DOI: 10.1007/s41019-020-00117-1

Tsai, C.-W., & Liu, S.-J. (2019). SEIM: Search economics for influence maximization in online social networks. *Future Generation Computer Systems*, 93, 1055–1064. DOI: 10.1016/j.future.2018.08.033

Uthayasuriyan, A., Hema Chandran, G., Kavvin, U., Mahitha, S. H., & Jeyakumar, G. (2023). A comparative analysis on influence maximization models in social networks. In *2023 International Conference on Advancement in Computation & Computer Technologies (InCACCT)* (pp. 1-6). Gharuan, India. DOI: 10.1109/InCACCT57535.2023.10141772

Wang, W., & Street, W. N. (2018). Modeling and maximizing influence diffusion in social networks for viral marketing. *Applied Network Science*, 3(6), 6. DOI: 10.1007/s41109-018-0062-7 PMID: 30839789

Wang, Y., Dong, W., & Dong, X. (2018). A novel ITÖ Algorithm for influence maximization in the large-scale social networks. *Future Generation Computer Systems*, 88, 755–763. DOI: 10.1016/j.future.2018.04.026

Wu, P., & Pan, L. (2017). Scalable influence blocking maximization in social networks under competitive independent cascade models. *Computer Networks*, 123, 38–50. Advance online publication. DOI: 10.1016/j.comnet.2017.05.004

Xie, M., Yang, Q., Wang, Q., Cong, G., & Melo, G. (2015). DynaDiffuse: A Dynamic Diffusion Model for Continuous Time Constrained Influence Maximization. *Proceedings of the AAAI Conference on Artificial Intelligence*, 29(1). Advance online publication. DOI: 10.1609/aaai.v29i1.9203

Yang, J., & Liu, J. (2018). Influence Maximization-Cost Minimization in Social Networks Based on a Multiobjective Discrete Particle Swarm Optimization Algorithm. *IEEE Access : Practical Innovations, Open Solutions*, 6, 2320–2329. DOI: 10.1109/ACCESS.2017.2782814

Yang, S., Du, Q., Zhu, G., Cao, J., Chen, L., Qin, W., & Wang, Y. (2024). Balanced influence maximization in social networks based on deep reinforcement learning. *Neural Networks*, 169, 334–351. DOI: 10.1016/j.neunet.2023.10.030 PMID: 37922716

Ye, Y., Chen, Y., & Han, W. (2022). Influence maximization in social networks: Theories, methods and challenges. *Array (New York, N.Y.)*, 16, 100264. DOI: 10.1016/j.array.2022.100264

Yuan, B. (2024). Influence max: An efficient algorithm for maximizing influence in networks. *International Journal of IoT*, 4(1), 1–4.

Zareie, A., & Sakellariou, R. (2024). Fuzzy Influence Maximization in Social Networks. *ACM Transactions on the Web*, 18(3), 32. Advance online publication. DOI: 10.1145/3650179

Zhang, W., Yang, Z., Wang, Y., Shen, Y., Li, Y., Wang, L., & Cui, B. (2021). Grain: Improving data efficiency of graph neural networks via diversified influence maximization. arXiv preprint arXiv:2108.00219.

Chapter 4

Harnessing Social Media for Trend Analysis and Behavioral Understanding: Unlocking Consumer Insights

Ambeshwar Kumar
GITAM University, India

Mahanthi Bangaru Lakshmi
https://orcid.org/0009-0005-6872-8385
GITAM University, India

Manikandan Ramachandran
https://orcid.org/0000-0001-6116-2132
SASTRA University, India

ABSTRACT

Businesses need to analyse customer behaviour and identify patterns in order to stay competitive in the rapidly evolving commercial landscape of today. Businesses now use social media as a goldmine, accessing a wealth of information that can reveal valuable patterns and provide a deeper understanding of client behaviour. Through the utilization of user-generated content on social media platforms like Facebook, Instagram, TikTok, and Twitter, businesses can obtain up-to-date insights into the assumptions and preferences of consumers regarding items. Advanced analytical techniques like machine learning, normal dialect preparation, and estimate research help firms sort through massive volumes of social media data to uncover patterns, identify designs, and predict future consumer behaviour. Additionally, these social

DOI: 10.4018/979-8-3693-9246-1.ch004

media platforms provide a coordinated way for brands and businesses to engage with their audience, fostering meaningful relationships, soliciting feedback, and establishing brand credibility.

INTRODUCTION

Hickman et al. (2022) discuss sentiment analysis, which is regularly called opinion research, which uses the control of deep learning frameworks to recognize between people's sentiments and subjective sentiments around a wide range of points. As such, it has ended up a critical instrument for distinguishing and e-rating sentences that contain enthusiastic imperfections or propensities. By methodically looking at the information behind feelings, analysts can reveal critical experiences that not only reflect current viewpoints but also anticipate future behaviors and patterns. This handle includes complex computer models that distinguish and decipher the enthusiastic tones inserted within the content. (Yang L et al., 2023; Kumar Singh et al., 2024) Whether these undercurrents are positive, negative, or impartial, with the assistance of estimation examination, we will dig into the nuances of interpersonal communication. This capacity to "get it" and express sentiments is particularly vital in our present-day computerized age, where suppositions and surveys shared through social and online stages can have a noteworthy effect on open states of mind and shopper behavior.

Estimation analysis reduces the strict implications of words and sentences and can give a more profound understanding of people's genuine sentiments (Wankhade, M. et al 2022). It is valuable in numerous areas, such as promoting legislative issues, social sciences, and indeed the improvement of fake insights, since it has the potential to degree open supposition and foresee social patterns. The reason for this article is to fortify the final few a long time of investigation within the field of assumption examination. In expansion, it points to anticipating future patterns and impacts in this quickly creating field. We center on the part of profound learning and its transformative effect on the opinion investigation approach and the possibilities, expecting how it'll proceed to refine this complex preparation of sentiments -discovery and elucidation.

As society advances, an expanding sum of data in our world has ended up with so-called "enormous information." This term implies a gigantic knowledge base in which the proportion of valuable and valuable data is moderately small. Hence, memorizing and recovering this sort of information utilizing as it where human brainpower appears unsustainable given current social developmental patterns. In differentiation, machines prepared by people have focal points such as quick information handling and keeping up a tall and steady degree of accuracy when processing huge amounts of information. This can be how the most advanced objective within the field of common dialect handling developed - to empower machines to get the human dialect. The objective is for these machines to get the same human dialect beneath distinctive conditions and make exact choices based on their understanding.

In any case, when machines experience human dialects, they have to overcome issues that people once in a while experience. The most difficult part is dealing with multimodal feelings that are regularly implanted in pictures or recordings, making machines completely competent in acknowledging data-challenged feelings. In the event that the data includes vague or vague content. The same words can have distinctive implications in diverse settings; for illustration, the word "book" is utilized to "save a room" or "examine books." Due to the need for information processing flexibility, machines are inclined to errors when managing such texts, which is particularly discernible on social media stages. In reaction to these challenges, all analysts have proposed different models and strategies aimed at preparing machines with way better dialect understanding and passionate investigation aptitudes. (Mao, Y. et al. 2024) This centers on moving forward. Machine elucidation of human feelings isn't only a major challenge but also an exciting front within the field of common dialect handling and opinion examination.

Understanding the Control of Social Media

Recognizing Social Media's Effect on Buyer Information. These days, social media is more than fundamentally a means to share engaging cat recordings and pictures from trips. It's a lively, ever-changing environment that gives an exceptional see into consumers' contemplations. (Varghese S. et. al., 2021) The effect of social media on understanding buyer data is significant and multifaceted. Gone are the days when social media stages were simply channels for sharing entertaining cat recordings and excursion photographs; they have changed into energetic, ever-evolving scenes offering unparalleled customer behavior experiences. Central to this transformative part is the unfiltered nature of client conclusions and estimations that penetrate social media. Unlike conventional showcase investigative strategies that depend on organized studies or center bunches, social media stages give people a crude and bonafide stage to express their suppositions, sentiments, and encounters unreservedly. In this way, businesses get to an affluent, weaved work of art of real-time customer bits of information, promoting a more nuanced understanding of things, brands, and exhibit designs. The speed of social media talks progresses its utility as a gadget for understanding buyer behavior. Companies can track brand takes notes, screen conclusion designs, and rapidly address client concerns, developing spry and responsive exhibiting methods. Additionally, the granular center on the capabilities of social media stages empowers businesses to reach significantly specific social occasions of individual parcels based on interface, financial, and psycho illustrations. Companies can increment engagement and change rates by displaying messages and substance to resonate with these claim-to-fame bunches of spectators, driving exchange improvement and efficiency. Social media also energizes ordinary community building and develops more significant affiliations between individuals, brands, and communities.

Ausat et al. (2023) discussed that through energetic engagement and talking with their gathering of individuals, businesses can create reliable client communities, develop user-generated substance, and accumulate invaluable bits of information in customer slants and torment centers. In addition, social media serves as a productive ground for float recognizable confirmation, with creating concepts, challenges, and thing slants regularly picking up balance. Companies that recognize and capitalize on these designs can choose a competitive edge by altering their thing offerings and fittingly exhibiting techniques. In pith, social media's effect on buyer information rises over insignificant communication channels; it has become a significant gadget for businesses looking to urge it, bolt in with it, and alter to the ever-changing scene of buyer behavior and exhibit flow.

The Value of Client Insights Within the Display Advertise

Even though important, traditional market inquiries about approaches can have limited reach and convenience. On the other hand, social media catalytic has become a game-changer, utilizing assumption investigation and information mining to get to a riches of real-time client experiences. Businesses can get invaluable data by closely watching online bunches and closely examining user-generated substances, such as:

Consumer Behavior

Social media gives a detailed insight into how clients lock in with businesses, merchandise, and each other through the utilization of engagement measurements and substance examination. Buy triggers, changing inclinations, and critical client torment spots are uncovered by this advanced ethnography.

Trend investigation

Modern patterns are continuously surfacing on social media, which could be a dynamic and ever-changing ecology. Businesses may discover the most current hashtags, item categories that are developing, and watchwords by utilizing progressed slant investigation devices. This kind of understanding empowers companies to proactively seize modern openings and predict changes within the industry.

Brand Checking

Companies can use estimation investigation to pick up knowledge into how individuals see their brand by carrying out exhaustive brand checking. This proactive methodology makes course rectification and notoriety administration possible if needed.

Audience Division

It could be a basic technique for making customized communications, and social media information makes it simpler to do so. Through groups of onlooker's divisions based on demographics, leisure activities, and online conduct, companies can make any doubt that the proper client groups are seeing their content.

Evolution of Social-Media Analytics

Customer Information in the Computerized Period Social media is utilized as a way to associate with companions, but presently, it's an energetic environment full of valuable client information. The development of social media analytics, a field that employs assumption investigation, information mining, and other strategies to uncover privileged insights into interior online communities, has paralleled this trend. Early on, the Basic Measurements and Brand Monitoring

Analytics in the early days of social media was generally concerned with brand following. Businesses utilized unrefined devices to screen notices of their brand and calculate fundamental measurements like likes, offers, and follower numbers. These measures were valuable in measuring brand awareness, but they advertised small data on the attitudes and practices of shoppers.

The Machine Learning and AI Time

The energetic spaces of artificial insights (AI) and machine learning (ML) fuel social media analytics nowadays. These cutting-edge methods allow for:

Trend Examination

Fake insights frameworks are competent in sorting through gigantic volumes of information to discover modern patterns and conversational subjects. This enables companies to take advantage of developing showcase conceivable outcomes and remain ahead of the curve.

Disposition Investigation With Granularity

Machine learning models have the capacity to examine the inconspicuous enthusiastic contrasts in a dialog in expansion to its common disposition. This empowers companies to comprehend the mental strengths impacting client behavior and adjust their methodology in like manner.

AI-Powered User-Generated Substance Bits of knowledge

AI is now able to examine a ton of user-generated content, such as pictures and recordings, to extricate data on almost how individuals see brands and utilize their items. This gives businesses the capacity to move forward with their item improvement and promotion plans.

Future Directions for Social Media Analytics

Large-Scale Computerized Ethnography: Looking ahead, advanced ethnography will likely find its way into social media analytics because it will continue to develop. Businesses will get an improved handle on the cultural quirks, social flow, and online behaviors that impact buyer choices by doing large-scale online community investigations. This all-encompassing viewpoint will empower businesses to construct encounters that are genuinely focused on the requirements of their customers instead of reacting to their wants, as shown in Figure 1.

Figure 1. Evolution of Social Media Analytics

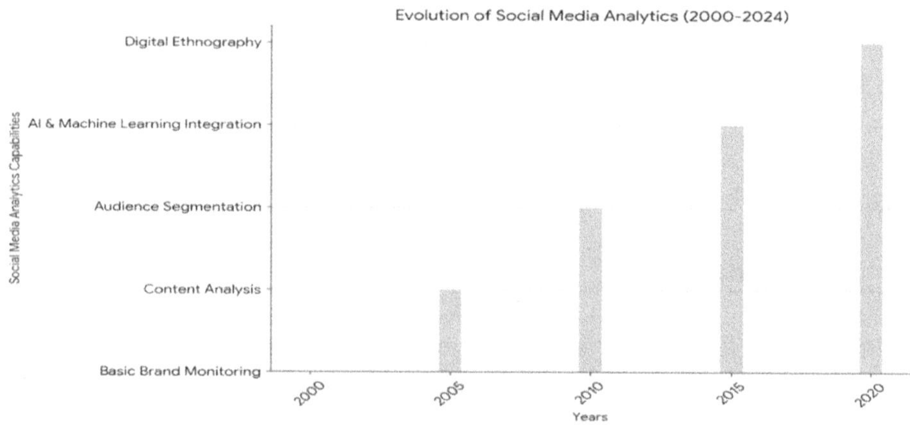

Ethics and Privacy Considerations

Acquisti et al. (2006) and Boyd et al. (2014) discuss the privacy Concerns. Social media stages regularly collect tremendous sums of individual information from clients, counting their area, browsing propensities, and individual inclinations. This raises concerns about how this information is utilized and whether it is satisfactorily secured. Clients may not continuously be mindful of the degree of information collection or the ways in which their data is shared with third parties, driving to protection breaches and potential abuse of individual data. Research Ethics: Researchers studying social media information must follow moral rules to ensure client security and guarantee the dependable utilization of information. Markham et al. (2012) and Zimmer (2020) discussed that educated consent, anonymization of information, and straightforwardness in information collection and investigation are fundamental standards in social media inquiries about morals. Moreover, moral considerations emerge when considering including helpless populaces, such as children or marginalized communities, where extra shields are vital to avoid hurt and guarantee impartial cooperation.

Analysing Consumer Trends

It is fundamental, not luxury, to comprehend customer designs inside the fast-paced showcase of these days. With its endless whole of data, social media explanations have become a solid instrument for companies looking to spot cutting-edge designs, expect changes in client conduct, and, in the long run, get a competitive advantage. Utilizing a run of procedures and social media explanations gives a comprehensive approach to analyzing client plans.

Brand Monitoring and Opinion Examination Businesses can understand how customer recognition changes by closely observing the brand, taking notes, and surveying the supposition passed on. This enables them to spot openings for progression and seize winning plans. The user-generated texture is the spine of social media stages, allowing for substance examination and trend-spotting. Finding predominant subjects, hashtags, and substance sorts to share is made less requesting by utilizing substance examination calculations. This will enable companies to keep ahead of the twist and modify their promoting plans as fundamental. Online Communities for Digital Ethnography Social media energizes the advancement of online communities centered around businesses and brands. Companies can look at these communities utilizing computerized ethnography approaches to memorize the shared experiences, social characteristics, and online hones that shape customer plans in a particular claim to fame. In showing disdain toward the truth that social media analytics can resign a wealth of unrefined data, turning it into valuable bits of knowledge requires a calculated approach. Machine Learning and Drift Forecast: Machine learning calculations are able to more absolutely distinguish rising designs and appraise future customer conduct by analyzing colossal wholes of social media data. This enables companies to take advantage of modern openings and plan ahead.

Understanding and Monitoring Pattern

Remaining ahead of the bend is basic in today's fast-paced corporate environment when buyer slants can change radically overnight. With its huge user-generated data store, social media expository has become essential for proficient incline examination and taking after. T Kietzmann et al., (2011) analyze the distinctive perspective social media expository gives, enabling organizations to comprehend changing designs and change that understanding into encounters that can be put into hone. Goh et al., (2013) emphasize that brand-watching and presumption examination businesses can overview changing buyer acumen by closely following brand notes and looking at the estimation passed on in online talks. This makes it conceivable to proactively recognize regions that require progression, take advantage of engaging designs, and address any rising negative state of intellect.

Substance examination and slant spotting of user-generated substances are ceaselessly buzzing on social media. Businesses can find predominant subjects, hashtags, and textures shared by utilizing substance examination strategies. This makes it conceivable for them to recognize cutting-edge designs early on and modify their advancing plans suitably. Machine learning and float desire Machine learning calculations can more accurately distinguish rising designs and determine future buyer conduct by analyzing sweeping entire social media data. This enables companies to form forward-thinking plans and seize modern openings a few times as late rivals do. Real-time reaction and social tuning in real-time tuning are made conceivable by social media analytics. Businesses that successfully screen web comments can recognize trending subjects and quickly address them. Expanding brand flexibility shows a commitment to client interaction and may, without a doubt, change the course of the float.

Sentiment Analysis

Client input is no longer constrained to proposal boxes or talks at the water cooler inside the progressed age. Online social occasions, social media stages, and survey destinations have made it into an energetic reverberate chamber where clients share their conclusions, experiences, and ideas about companies and stocks in genuine time. With suspicion examination, a procedure fueled by common tongue dealing with (NLP), companies can effectively investigate the tremendous ocean of online discussion and assemble speedy information from enthusiastic undercurrents. It goes past only classifying a customer's feeling as positive or negative, giving a modern comprehension of the run of sentiments they express. Conclusion estimation as positive or negative. They test progress, identifying nuanced energetic viewpoints such as desire, bother, disappointment, or certainty. Businesses can modify their reactions reasonably due to this detailed understanding. For outline, taking care of a client who is disturbed over a flaw calls for a particular method rather than managing a client who is spouting approximately.

Moral concerns around data security and straightforwardness are fundamental, as they are with any strong technology. Companies must form past any questions they are gathering and evaluate client suspicion data to understand all pertinent laws and with customers' express consent. Supposition examination includes a promising future. More complicated estimation investigation systems, such as those that can handle complex express structures, mockery discovery, and multilingual examination, are becoming conceivable due to movements in joint lingo planning. Other than that, coordination counterfeit insights (AI) gives, without a doubt, more bits of information into client states of intellect, engaging businesses to anticipate future conduct in extension to understanding customers' assumptions. Inside the continuously changing progressed environment, organizations may utilize presumption examination to fortify shopper nature, develop brand constancy, and fulfill long-term triumph by getting a handle on these improvements and putting a tall requirement on secure data administration.

Discovering Unused Designs in Buyer Patterns

Shopper patterns are always changing groups of stars of inclinations, habits, and assumptions or maybe more static, unchanging things. With its tremendous collection of user-generated information, social media analytics has ended up a strong apparatus for companies looking to recognize these modern patterns and turn them into valuable experiences. This paper analyses the diverse viewpoints that social media analytics gives, empowering companies to see past the self-evident and discover unfamiliar treasures in the continuously changing scene of buyer patterns. Mittelstadt et al., (2016), in this article, present an unmistakable perspective given by social media analytics that surpasses the impediments of routine showcase inquiry about methods. Social media stages are a hive of movement, full of postings, photographs, and videos. Content examination and point displaying are critical perspectives of this movement. Goldberg (2016), using topic modeling calculations and substance examination strategies, we are able to pinpoint not only the most prevalent subjects but also the fundamental topics and storylines that are offered to readers. This enables us to recognize diminutive designs in client interface, inclinations, and torment focuses that seem to go unnoticed or something else.

Behavioural Understanding of Consumer

Ali et al., (2024), in this study of individuals' psychological, generator, and social behaviors as they purchase, utilize, and discard goods, services, concepts, and customs, are known as buyer behavior. Declared in various ways that the study of consumer behavior focuses on the elements that influence or support the decisions that consumers make when making purchases. To become successful in the market, firms must have an adequate grasp of consumer behavior. It's also essential for producing goods and services that satisfy customers and add value. Without understanding what customers think, how they make decisions, and what inspires them to act, no firm can remain relevant for an extended period of time. Customer behavior is significant for a variety of reasons, including Improved Customer Loyalty, Competitive Benefit, Effective Marketing Techniques, and Customer Retention.

One illustration of an upsurge in consumer behavior is the way that eating habits have changed, leading to an immediate increase in the market for gluten-free goods. Companies that have recognized this market specialty have created gluten-free goods and capitalized on it. Conversely, businesses that were unable to keep an eye on customer behavior were left behind and unable to close the discrepancy in the market. Knowing how customers behave enables proactive businesses to grow their market share by foreseeing changes in consumer preferences.

Psychographic Segmentation of customer

In marketing, psychographic Segmentation is a type of market segmentation where you split up your customer base into groups according to certain parameters. Such represent the personalities, desires, and social standings of the customers. A marketing campaign might benefit from psychographic Segmentation since it enables you to highlight the value of your product to various target audiences. Because of the product's many applications, a diverse group of people are drawn to it, extending its base of customers. Blömker et al. (2024) discussed when making financial decisions, the individual's risk attitude indicates how willing they are to compromise between the expected benefits and hazards of a certain course of action. However, there can be significant individual differences in the desired level of risk-taking. As shown, previous research has greatly improved our understanding of how individual characteristics affect multichannel customer behavior.

There are two main areas in need of exploration. First off, there isn't yet a thorough and organized framework for investigating individual variations in multichannel customer behaviors. This begs the following research question: (I) How do individual differences in decision-making style, motivation, cognitive ability, risk attitudes, and personality affect multichannel customer behavior? Secondly, the effect sizes of the channel choice influencing elements are not examined from an integrative and comparative perspective. This raises the following research question: (II) With regard to channel attributes, marketing initiatives, social effects, contextual considerations, and channel experiences, what part do individual differences play in multichannel customer behavior? It is described in Figure 2.

Figure 2. Psychographic Market Segmentation

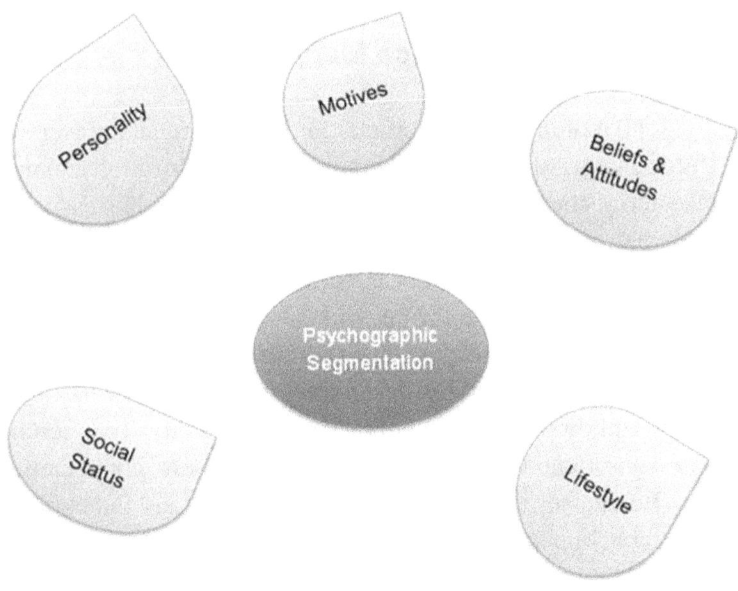

Customers are divided into groups based on psychological variables through psychographic market segmentation. These consist of customer personality, lifestyle, motives, social status, beliefs, and attitude. Handler et al., (2023) ensure that customers can be found with a variety of personalities, including those who are outspoken, friendly, creative, emotional, introverted, or outgoing. These characteristics can be used by businesses to systematically filter their customer base. Marketers of life insurance and baby products play on consumers' emotional needs, particularly their desire to provide for their families. Beliefs held by customers are an innate measure that can provide insight into their underlying actions. Every individual has their own way of thinking and being. When pursuing consumers in the market, have this idea in mind. Beliefs held by customers are an innate measure that can provide insight into their underlying actions. Every individual has their own way of thinking and being. When pursuing consumers in the market, have this idea in mind.

The following benefits of psychographic market segmentation are as:

- Effective Personalization
- Effective utilization of resources
- Efficient customer knowledge
- Determine the Reasons Behind Your Purchases

As per the survey, a few companies, including Aritzia, Indigo, and Nike, are using these strategies to judge the customer.

Network Analysis and Influence Mapping

Finding the source of influence is crucial for identifying and preventing misinformation. Unfortunately, the majority of multiple source location techniques now in use only take into account the nodes' local structures or the shortest path between them. Online social networks (OSNs) have made it easier to share information on a convenient platform. On the other hand, OSN also provides avenues for the spread of false information within the network. (Ju, W. et. al., 2024) interacts with computer infections, rumors, and fake news, which are examples of this kind of misinformation. Disseminating false information in social networks undermines society's stability and diminishes the public's trust in society. To a certain extent, rumors can hurt and offend the people concerned. Fake news could cause panic in the community. Businesses and marketers need to understand the complexities of customer shopping habits in the Omni channel era, especially how Omni channel, showrooming, and web rooming behaviors interact. Digital channels, social media platforms, and mobile devices are some of the new technologies that have changed retail business models from multichannel to omnichannel commerce. Showrooming refers to the practice of customers researching products in-store before making an online purchase from the same merchant.

In 2010, the Columbia Business School Centre released a paper characterizing Best Buy as an Amazon showroom, which gave rise to the term “showrooming.” Shao et al. (2024) talk about how retailers may suffer from showrooming, even though it is helpful for customers. The availability of channel switching and free-riding habits makes showrooming detrimental. Salespeople become disengaged due to concerns about their performance and self-efficacy, a decrease in sales efforts, and a detrimental effect on a store's earnings. Web rooming, often called Research Internet Purchase Offline (ROPO), is a practice where customers use an Internet channel to look up product information before making a brick-and-mortar purchase. Web rooming facilitates online-to-offline (O2O) retail sales or Buy Online, Pickup in Store (BOPIS or BOPS), in which customers use mobile apps and websites to search, buy, and pay, then pick up their items from specific physical establishments. It is shown in Figure 3.

Customers who engage in competitive web rooming look for information online yet buy the product from a rival seller's physical store. Competitive webrooming can have a detrimental effect on a company's perception of a product's performance, consumers' propensity to make a purchase, and the earnings of online retailers. Developing a stronger multichannel retailing strategy requires an understanding of the factors that influence consumers' channel choices and switches. Similarly, combining online and offline media can enhance company performance.

Figure 3. Three Major Factors of Influencing Customer Behavior

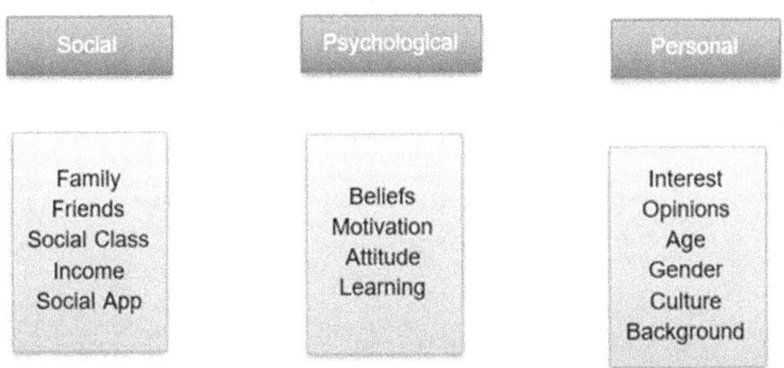

Influence Mapping is the best marketing framework; it massively improves work's effectiveness and dramatically cuts down on poor decision-making; there are two natural responses:

"Wow, this is very great! It will spare me a great deal of misery."

"What is wrong with me?" Why wasn't this mentioned to me earlier?

This is how the customer feels when they stumbled across the behavioral design of influence mapping. It is at the core of any marketing strategy to think about how the audience interacts with design to make a smart decision when purchasing any product. A decade ago, the people's actions were driven by forces around them. They always modify the strategy of how to behave in statistically measurable and influenceable ways. Marketing will be significantly more successful if it understands how people find and interact with your place. All organization needs answers about how to change customer minds on the issue, encourage them to think about a new product, alter their behavior, or educate them about fresh information. Marketers are buried without understanding how prospective customers find out about your industry, which channels they utilize at key stages of their journey, and which channels and sources persuade their decisions. Excellent marketing frameworks can assist us in making more strategic decisions, just as an organized communication process or to-do list technique can improve our talks. Influence maps can assist with all of these issues: where to test, how to influence the appropriate stages of the buyer experience, and what messages to provide. It is shown in Figure 4.

Figure 4. Influence Map Framework

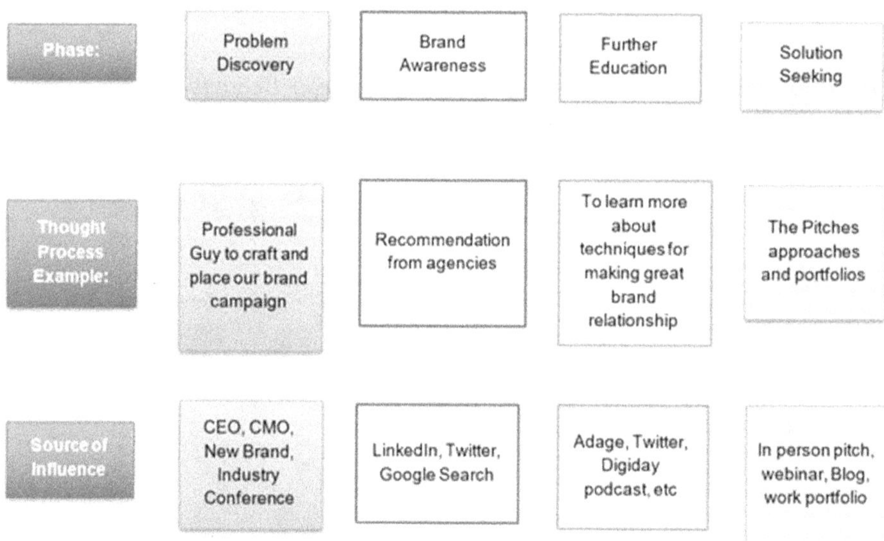

Understanding Consumer Decision-Making

The channels a consumer will use to make a purchasing decision are determined by their behavior. Everyone makes decisions in a different way, which is why consumer behavior is so fascinating and significant. However, its uniqueness does not exclude study and prediction. It places the responsibility of conducting the necessary research and comprehending their target markets and individual consumers' habits on the product and service suppliers. Being aware of how customers behave is essential to operating a profitable company. Businesses must understand what customers want, when they want it, and where to get it. (Mukhopadhyay et al., 2023) Businesses may more accurately forecast market trends by having a greater understanding of consumer behavior, which enables them to make more educated decisions about how to run their operations. Businesses that are aware of their customers' demands will be better equipped to offer goods and services that better suit their needs. Understanding how customers engage with a business and how to enhance it can help businesses use consumer behavior to stay relevant. If you own a restaurant and you've noticed that guests who are situated on the left side of the restaurant tend to purchase more food, you can consider moving all of your tables to that location. Alternatively, you might increase the size of the plates or even switch up the kind of desserts you provide if you find that customers purchase more dessert when it's delivered on a smaller plate! These kinds of changes enable firms to continue producing money and keeping customers satisfied, which helps them stay relevant.

(Sawang, S. et al., 2023) The probability that a buying decision won't result in meaningful material advantages is referred to as financial risk. It's increasingly common for customers to base their selections on factors like cost-effectiveness, quality, durability, and quantity needed for a certain product. Consumer trust when making repeat purchases from a specific shop or brand might be impacted by temporal risk. Customers are reluctant to acquire from a supplier if it takes an excessive amount of time and effort. Performance risk is the term for goods or services that fall short of what customers had hoped to achieve since there is now a much higher chance of making poor purchasing selections. Because of the erratic state of the economy, consumers are making more efforts to make sure the goods or services they buy live up to their expectations. Customers have developed different purchase patterns both online and offline before and after the pandemic, and understanding consumer perceptions of service safety may be important for reducing perceived risk and increasing approach behaviors. It is shown in Figure 5.

Figure 5. Customer Purchase Behaviour

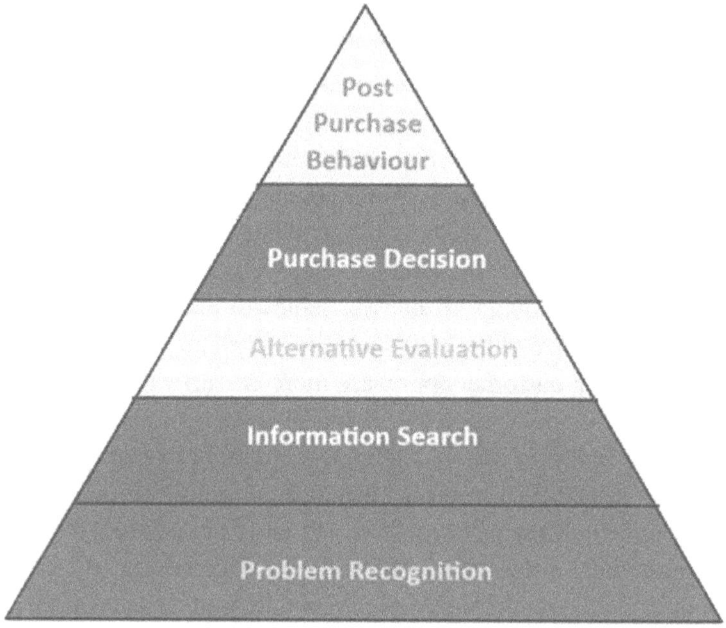

Real-Time Applications

The worldwide e-commerce sector stands out in the ever-expanding online retail space, with projections of an astonishing 8.1 trillion dollars by 2026. The U.S. Department of Commerce reports that in 2022, e-commerce sales made up 14.6% of retail sales, a 7.7% increase from the year. Kim et al., (2024) informs that the ease of purchasing online is largely responsible for the e-commerce market's rapid growth. In the past, many marketers concentrated on customer segmentation strategies by calculating the lifetime value of clients using data from prior purchases and characteristics encountered with the online store. Finding and influencing the factors that influence consumers' attitudes and behavior is one of the most difficult marketing tasks. Leaders have historically depended on a blend of quantitative information from surveys such as those measuring customer happiness and brand image and qualitative observations from focus groups and interviews. Two observations led to the creation of real-time experience tracking. First off, unlike human observers, customers' cell phones allow market researchers to simply track them around the clock without interfering with their views of the experiences they have. Despite the fact that customers may engage with a company in thousands of ways, the second insight was that you really only needed to know four details about each interaction: the brand, the type of touchpoint a TV ad, for example, or a call to the service center, the participant's sentiment regarding the experience, and the persuasiveness of the interaction.

Retail and E-Commerce: Tailoring Strategies

We learn from management mythology that strategy must come first with structure. Because of this, businesses have spent decades developing their strategy before deciding on the organizational structure and operating model that would best support it. In tailoring strategies, One of the most important parts of managing an online business is e-commerce. E-commerce has grown to be a vital component of the retail sector, and with the rise in online retailers, it is now more crucial than ever to customize your e-commerce website to differentiate it from the competitors. A crucial element in tailoring e-commerce for virtual prosperity is guaranteeing that your website is optimized for mobile devices. Your website must be mobile-friendly in order to guarantee that your clients can simply access it when they are on the go, given the increasing popularity of mobile devices. This entails creating a mobile-friendly style for your website that adjusts to various screen widths and making sure your website loads quickly. Firstly, understanding the fundamentals of e-commerce platforms refers to buying and selling products as well as dealing with services over the Internet.

Souitaris et al. (2007) ensure that it is necessary to offer an easy-to-use and user-friendly interface. Accurate product descriptions, excellent photos, and user-friendly navigation are all part of this. Enhancing the customer experience can also be achieved by making tailored recommendations based on the client's preferences and past purchases. The primary worries of many e-commerce sites are whether or not it works better as a method for generating loyalty among goal-oriented shoppers as opposed to experiencing shoppers. This observation could be explained by the curiosity-driven experiential shoppers' information- and variety-seeking behaviors. The results could offer a theoretically informed solution for the long-standing query of why fulfilled Consumers turn against us. Search engine optimization (SEO), social media marketing, and email marketing are some of the digital marketing strategies that can help tailor e-commerce. While social media marketing can assist in reaching a bigger audience and interact with customers, SEO can help the website appear more prominently in search engine results. Targeted messages about new products or exclusive deals can be sent to clients via email marketing. Finally, the tailoring strategies serve as a compass guiding the e-commerce business. The friendly format for strategies is as follows: Establish an All-Channel Presence, with customer experience as a top priority, utilize data to make informed decisions, and last, establish remarkable narratives and content.

Brand Management: Leveraging Insights

The term "brand management" describes the strategies used to gradually raise a product line's or brand's perceived worth. Building strong ties with your target market is essential to keeping your company at the top of consumers' minds. Using strategic brand management can help you get an edge over rivals. Additionally, it supports the development of enduring customer relationships and a strong brand image. In a competitive sector, marketers must use customer comments and reviews to gain insights into what customers need in order to match products and services. Piriyakul et al. (2024) state that reviews and comments given by customers affect the company brand. As we see in our daily lives, if the comments and reviews are high for any product, then the customer goes to buy that product and recommends others to buy the products. Touchpoints can range from the moment a potential customer searches for you on Google to the moment they make use of one of your offerings. Any moment a potential or existing customer interacts with a brand—before, during, or following their acquisition of a product from the brand owner. Customer satisfaction can be understood by their emotional response to fulfill their needs, expectations, and wishes. Emotions play an important role in influencing customer behavior to buy the products. Traditionally, any company hires employees to sell their products; they visit customers' places and sell their products and upgrade the brand image.

Sykora et al. (2022) deal emotionally with the customer to sell their products. Sentiment orientation pertains to categorizing user-generated content (UGC) emotion intensity as either positive or negative. A brief text segment, such as a tweet, is assigned a single sentiment score or a few ratings for the degree of expression, both positive and negative. Brand leveraging is the tactic of using the strength of an established brand name to assist a business's entry into a new but related product category by providing the customer with useful product information. As an example, the coffee maker's manufacturer launched a coffee vending machine by leveraging the strength of its brand name. Even though coffee and coffee vending machines are in different product categories, there is a significant association between the two, meaning that consumers in both categories are greatly influenced by the brand name. The few points related to leveraging insights to improve the brand name and product production are as follows:

- Effective brand leveraging gives customers a feeling of familiarity.
- Instant recognition of the brand through advertisement and customer feedback.

- More items provide the company with more shelf space, which increases sales prospects.
- A product that leverages a brand's reputation might be introduced for less money than a stand-alone new product.
- More products mean more efficient raw materials and manufacturing facilities.

Crisis Management: Navigating Negative Feedback

Negative comments from customers, experts, or the public can present serious difficulties for people and companies. Rather than reacting to crises right away, compelling crisis organizations to look into unfavorable feedback in order to minimize harm and advance profitable assurance. (Brown et al., 2023) The procedures for addressing negative evaluations include Attention. With caution, listen to the criticism effectively and express empathy and comprehension of the issues brought up.

- Remain Calm and Objective: Remain calm and objective when receiving unfavorable feedback, avoiding agitated emotions that could exacerbate the situation.
- Seek Clarification: To ensure a thorough comprehension of the issues stated, ask for clarification on whether the input is unclear or imprecise.
- Acknowledge and Prefer: Acknowledge the feedback and, with tact and respect, support the thoughts and opinions of the person providing it.

Sorry in the unlikely event that Important: If the feedback reveals a mistake or misconduct on your part, provide a heartfelt expression of regret to show responsibility and effort to rectify the issues.

- Offer solutions or Clarification: Provide workable solutions or explanations that address the issues brought up in the feedback and indicate a dedication to resolving the matter. (Stone, D. et. al., 2023) To acquire knowledge from the input, constructive criticism should be considered as a chance for advancement by considering the underlying reasons and identifying areas that could be used better.
- Effective Communication: Maintain open lines of communication, teach accomplices the plan, and give updates on exercises completed in response to criticism. Compile and
- Examine: Keep track of the feedback and any actions you take in response, and examine plans or trends to spot opportunities for organizational change.
- Follow-up: Consider what was said and inquire about any further comments or suggestions.

Predictive Analytics: Anticipating Future Trends

Provost et al. (2013) ensure that the basis for predicting design possibilities for unique firms and segments might be predictive analytics. By utilizing real data, advanced quantifiable computations, and machine-learning techniques, organizations can evaluate possible outcomes and designs. Carefully gathering and preparing data is the first step in this process, which yields insightful information on consumer behavior, marketing dynamics, and financial markets. Then, by carefully examining past data and making precise predictions, predictive models are trained to recognize patterns, correlations, and trends. These models produce projections based on identified patterns and relationships by using statistical techniques such as regression analysis and time series forecasting. This approach relies heavily on real-time monitoring and adaptation, which allows businesses to update projections, adjust strategy in response to emerging patterns, and continuously develop models. Ultimately, predictive analytics gives businesses the ability to make informed decisions, mitigate risks, and seize emerging opportunities, giving them a competitive advantage in the rapidly evolving business landscape of today.

A vital instrument for companies navigating the complexity of the modern marketplace is predictive analytics (James et al., 2013). Efficient analysis of countless archived data sets allows predictive models to uncover hidden future patterns and actions. This proactive strategy enables businesses to anticipate changes in consumer preferences, showcase components, and competitive landscapes, enabling them to stay ahead of the curve. Businesses may improve resource allocation, fine-tune promotional efforts, and streamline operations to adapt to projected designs by utilizing predictive analytics. It is shown in Figure 6.

Figure 6. Predictive Analytics Process

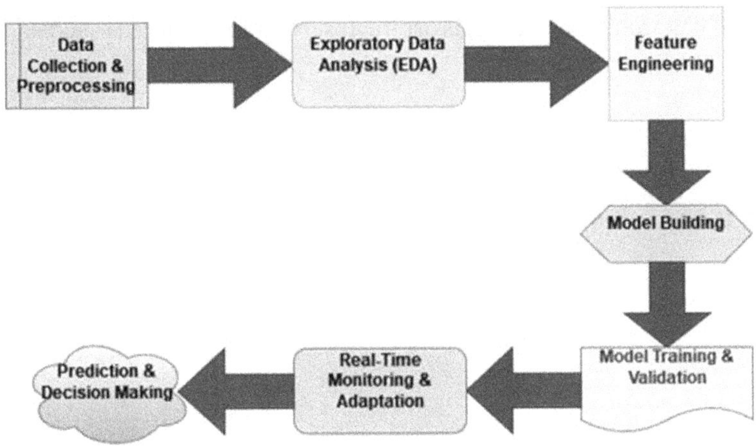

Furthermore, because predictive analytics is iterative, it allows for continuous improvement and modification, ensuring that models stay accurate and relevant in the face of shifting conditions. Organizations can eventually anticipate future designs, seize opportunities, and manage risks by taking on the management of proactive analytics. This will lead to financial development and success in a constantly changing commerce environment.

CONCLUSION AND FUTURE WORKS

This chapter summarizes the current scenario of customer behavior to buy the items online and offline. It promises to treat a customer as the highest priority to take the stake of the company at the highest peak level. Firstly, this chapter discussed the understanding of the customer in social media and how consumers manage purchasing items online as well as offline. Some reviews are based on consumer behavior in purchasing the items through the recent trends of the market. The ethics and privacy concerns of consumers are discussed in the chapter, which ensures the consumer buying behavior through their sentimental analysis. Finally, real-time applications in the retail market should be used to gain a better understanding of customer behavior, which can affect the accuracy of finding market revenue.

To mitigate the limitations of this chapter, Future studies should investigate possible differences in emotion and magnitude values across a wider range of languages and cultural contexts. Additionally, to create a more thorough customer journey map, researchers should look into methods for encouraging users to provide feedback at all touchpoints, even the ones that aren't used as much. Text mining techniques should be improved by investigating advancements in opinion mining, especially with regard to the use of electronic devices to assist the insertion of wildcards such as "emoji."

REFERENCES

Acquisti, A., & Gross, R. (2006, June). Imagined communities: Awareness, information sharing, and privacy on the Facebook. In *International workshop on privacy enhancing technologies* (pp. 36-58). Springer Berlin Heidelberg. DOI:10.1007/11957454_3

Ali, M., & Amir, H. (2024). Understanding consumer herding behavior in online purchases and its implications for online retailers and marketers. *Electronic Commerce Research and Applications*, 64, 101356. DOI:10.1016/j.elerap.2024.101356

Ausat, A. M. A., Permana, R. M., Angellia, F., Subagja, A. D., & Astutik, W. S. (2023). Utilisation of Social Media in Market Research and Business Decision Analysis. *Jurnal Minfo Polgan*, 12(1), 652–661. DOI:10.33395/jmp.v12i1.12485

Blömker, J., & Albrecht, C. M. (2024). Psychographic Segmentation of multichannel customers: Investigating the influence of individual differences on channel choice and switching behavior. *Journal of Retailing and Consumer Services*, 79, 103806. DOI:10.1016/j.jretconser.2024.103806

Boyd, D. (2014). *It's complicated: The social lives of networked teens.* Yale University Press.

Brown, B. (2022). *The gifts of imperfection: Let go of who you think you're supposed to be and embrace who you are.* Simon and Schuster.

Goh, K. Y., Heng, C. S., & Lin, Z. (2013). Social media brand community and consumer behavior: Quantifying the relative impact of user-and marketer-generated content. *Information Systems Research*, 24(1), 88–107. DOI:10.1287/isre.1120.0469

Goldberg, Y. (2016). A primer on neural network models for natural language processing. *Journal of Artificial Intelligence Research*, 57, 345–420. DOI:10.1613/jair.4992

Handler, I., & Kawaminami, J. (2023). Why do Japanese people visit hot springs during a pandemic? A psychographic segmentation analysis. *Journal of Outdoor Recreation and Tourism*, 41, 100530. DOI:10.1016/j.jort.2022.100530

Hickman, L., Thapa, S., Tay, L., Cao, M., & Srinivasan, P. (2022). Text preprocessing for text mining in organizational research: Review and recommendations. *Organizational Research Methods*, 25(1), 114–146. DOI:10.1177/1094428120971683

James, G., Witten, D., Hastie, T., & Tibshirani, R. (2013). *An introduction to statistical learning* (Vol. 112, p. 18). New York: springer.

Ju, W., Chen, Y., Chen, L., & Li, B. (2024). Locating Influence Sources in Social Network by Senders and Receivers Spaces Mapping. *Expert Systems with Applications*, 248, 123327. DOI:10.1016/j.eswa.2024.123327

Kietzmann, J. H., Hermkens, K., McCarthy, I. P., & Silvestre, B. S. (2011). Social media? Get serious! Understanding the functional building blocks of social media. *Business Horizons*, 54(3), 241–251. DOI:10.1016/j.bushor.2011.01.005

Kim, S., Shin, W., & Kim, H. W. (2024). Predicting online customer purchase: The integration of customer characteristics and browsing patterns. *Decision Support Systems*, 177, 114105. DOI:10.1016/j.dss.2023.114105

Kumar Singh, P., Setta, S., Singh, A. K., & Singh, A. P. (2024). Machine Learning and Sentiments Analysis: Analyzing Customer Reviews. In *Human-Centered Approaches in Industry 5.0: Human-Machine Interaction, Virtual Reality Training, and Customer Sentiment Analysis* (pp. 248-264). IGI Global.

Mao, Y., Liu, Q., & Zhang, Y. (2024). Sentiment analysis methods, applications, and challenges: A systematic literature review. *Journal of King Saud University. Computer and Information Sciences*, 36(4), 102048. DOI:10.1016/j.jksuci.2024.102048

Markham, A., Buchanan, E., & AOIR Ethics Working Committee. (2012). Ethical decision-making and internet research: Recommendations from the AoIR Ethics Working Committee (Version 2.0).

Mittelstadt, B. D., Allo, P., Taddeo, M., Wachter, S., & Floridi, L. (2016). The ethics of algorithms: Mapping the debate. *Big Data & Society*, 3(2), 2053951716679679. DOI:10.1177/2053951716679679

Mukhopadhyay, S., Vijayalakshmi, A., & Jain, S. P. (2023). Understanding consumers in-store behavior: The dual role of episode-specific motive adjustment and motive selection. *Journal of Retailing*, 99(3), 460–479. DOI:10.1016/j.jretai.2023.08.004

Piriyakul, I., Kunathikornkit, S., & Piriyakul, R. (2024). Evaluating brand equity in the hospitality industry: Insights from customer journeys and text mining. *International Journal of Information Management Data Insights*, 4(2), 100245. DOI:10.1016/j.jjimei.2024.100245

Provost, F., & Fawcett, T. (2013). *Data Science for Business: What you need to know about data mining and data-analytic thinking*. O'Reilly Media, Inc.

Sawang, S., Lee, C. C., Chou, C. Y., Vighnesh, N. V., & Chandrashekar, D. (2023). Understanding post-pandemic market segmentation through perceived risk, behavioural intention, and emotional wellbeing of consumers. *Journal of Retailing and Consumer Services*, 75, 103482. DOI:10.1016/j.jretconser.2023.103482

Shao, Y., Chen, L., Chen, Y., Liu, W., & Dai, C. (2023). Identifying multiple influence sources in social networks based on latent space mapping. *Information Sciences*, 635, 375–397. DOI:10.1016/j.ins.2023.01.127

Souitaris, V., & Balabanis, G. (2007). Tailoring online retail strategies to increase customer satisfaction and loyalty. *Long Range Planning*, 40(2), 244–261. DOI:10.1016/j.lrp.2006.11.006

Stone, D., Patton, B., & Heen, S. (2023). *Difficult conversations: How to discuss what matters most.* Penguin.

Sykora, M., Elayan, S., Hodgkinson, I. R., Jackson, T. W., & West, A. (2022). The power of emotions: Leveraging user generated content for customer experience management. *Journal of Business Research*, 144, 997–1006. DOI:10.1016/j.jbusres.2022.02.048

Varghese, S., & Agrawal, M. (2021). Impact of social media on Consumer Buying Behavior. *Saudi Journal of Business and Management Studies*, 6(3), 51–55. DOI:10.36348/sjbms.2021.v06i03.001

Wankhade, M., Rao, A. C. S., & Kulkarni, C. (2022). A survey on sentiment analysis methods, applications, and challenges. *Artificial Intelligence Review*, 55(7), 5731–5780. DOI:10.1007/s10462-022-10144-1

Yang, L., Zhang, Z., Song, Y., Hong, S., Xu, R., Zhao, Y., Zhang, W., Cui, B., & Yang, M. H. (2023). Diffusion models: A comprehensive survey of methods and applications. *ACM Computing Surveys*, 56(4), 1–39. DOI:10.1145/3626235

Zimmer, M. (2020). "But the data is already public": on the ethics of research in Facebook. In *The ethics of information technologies* (pp. 229–241). Routledge. DOI:10.4324/9781003075011-17

Chapter 5
Consumer Sentiment Analysis on Social Media for Business Intelligence

Sachin Mishra
VIT-AP University, India

A. Manimaran
https://orcid.org/0000-0002-5671-9466
VIT-AP University, India

Mohammad Gouse Galety
https://orcid.org/0000-0003-1666-2001
Samarkand International University of Technology, Uzbekistan

ABSTRACT

Previously, when looking towards the development of surveys to help us improve aftershave products, we relied on the traditional and often costly means of conducting comprehensive and expensive marketing surveys. Social media plays the role of a powerful tool for realizing consumer behavior and obtaining opinion of consumers of products that is referred to as consumer sentiment analysis. It observes and checks consumer preferences by investigating the content they are sharing, where they can be any images, videos, or simple text. Catchers get a new insight into target market needs and service adequacy. By tailoring to the consumers' needs as these demands keep changing, companies can make their operations better. Social media handling companies can gain rich insights from social media applications by using latest technologies like machine learning and natural language processing, especially through lexicon-based methods. Then, they can randomly select a few customers and conduct sentiment analysis on their feedback of the product which

DOI: 10.4018/979-8-3693-9246-1.ch005

is known as opinion mining.

INTRODUCTION

Purpose and Significance of Consumer Sentiment in Social Media Analytics

In social media analytics, customer sentiment has several applications, which are very important to decision-making based on consumer behavior and market dynamics (Neri et al., 2012). Individuals can be grouped into various categories according to their tastes, which, through product review research, is an essential step for obtaining information regarding target groups and customer preferences. As a result, the trends in seasonal products can be detected, so the marketing strategy would have to be set up in line with the trend (Neri et al., 2012). It boosts the business intelligence.

Customer feedback and reviews function like a compass for prospective buyers, enabling them to focus on the criteria required for buying the product. This changes the environment to become more productive for both consumers and businesses, providing the necessary information on customer needs and room for improvement. Social media can act as a marketing tool for companies to leverage. For example, companies can identify the shortcomings of their services or products, provide solutions to customers in a very short period, and increase customers' satisfaction.

Therefore, business competition became harder because of customer opinions given on social media. The role of social media analytics in determining the ability of businesses to monitor consumers' moods is certainly becoming very significant to their development and success as an institution today. These statistics perform very well as a measure of consumer satisfaction and product acceptance, enabling corporations to optimize their marketing activities. It is widely accepted that social media marketing is instrumental in businesses that aim to outperform their counterparts and create a meaningful relationship with their target audiences.

Overview of the Chapter

This chapter will discuss the importance of a business in seeing the sentiments about their products or services being offered on their social media. We will begin by defining consumer sentiment and why it's crucial in today's digital world. After that, we will give you instructions on breaking down the sentiment on social media and using the breakdown for your business. Using actual cases in our presentation will help demonstrate how other companies can achieve this effectively. We will discuss the possible future of sentiment analysis that might affect business. Besides, we will illustrate the impact of tracking online sentiment on business decisions and business evolving responsibly.

Definition of Consumer Sentiment

Consumer opinion designates the general attitude of consumers, their perspective toward the state of the economy and their financial condition, and their potential behavior in spending money on goods and services. Consumer attitude signifies the general feelings of individuals about the economy and the future of the financial situation. It represents one of the most prominent factors that drive people into certain production or consumption of material goods (Throop et al., 1992). The clear impact of public attitude on gross national product is beyond all doubt. The mechanism to measure consumer sentiment is composed of administering questionnaires that are designed to investigate the financial conditions of individuals, their near future outlooks, and judgments regarding whether now is the time to purchase important items. These investigations gather data and compute indices that are made up of numerical values that indicate the direction (Throop et al., 1992). In general, consumer confidence usually shoulders the higher consumer expenditure, which propels up economic growth; however, bad consumer confidence may mean a decline in economic activities.

Multiple aspects of consumer thinking depend on things like the condition of the job market, inflation, or news from the financial sector. In the case of the economy either getting better or stronger, consumers have been generally more confident about their spending habits (Throop et al., 1992). The Index of Consumer Sentiment (ICS), Katona's invention in 1951, is one of the most important measures of the changing opinions among consumers about whether the business conditions or employment conditions are good or bad. Consumer attitude, which has been the subject of research, proves to be a significant indicator of the economic macro data as well as an indicator of the expectations and views of consumers regarding their financial situations as well as the general economic environment (Lahiri et al., 2016).

Importance of Consumer Sentiment on Social Media

Sentiment analysis of social media data becomes a useful tool for marketing intelligence and the simultaneous investigation of customers' attitudes. In the digital era, media sites rest with the power to form public opinion and consumer behavior faster than it has done before in the shortest time possible. Vying the rightful place among digital marketing platforms, social media has become the number one favored source of knowledge for users checking about the products and services offered to them (Lahiri et al., 2016). Instead of relying entirely on traditional advertising mechanisms, consumers now turn to social media to assess word of mouth and the pulse of the peer group to make informed decisions.

Harnessing the large array of information on social media enables companies to find out consumer opinions, which translate to brand loyalty and overall product preferences (Lahiri et al., 2016), among other things. Social networks constitute a dynamic channel of feedback collection and opinion citing among customers, providing a range of data for sentiment analysis. To discover the hidden sentiments of consumers, sentiment analysis can now be used by businesses to decode the underlying emotion. Consequently, they understand the requirements and preferences of consumers deeply.

This chapter is about how sentiment analysis with social media data can be used for marketing and business growth development. By using case studies and real-life examples, we provide a practical understanding of how businesses can exploit social media data to better inform their strategic decisions and to gain a competitive edge. For example, extracting sentiments and actionable insights from social media to optimize marketing strategies, enhance engagement, and build brand loyalty are the places these methods can be used (Neri et al., 2012). Businesses that accept the terrible analysis of sentiments on social media data will be able to involve new directions of growth and prosperity in the digital field.

LITERATURE STUDY

Extracting Sentiment Insights From Social Media Posts

Customers' feelings about business intelligence matter a lot; that is why social media platforms are rich data resources for the best deals. Different approaches are used not only to explore sentiment but also to gain a better understanding of business through business intelligence. Such an approach uses natural language processing (NPL) and machine learning techniques that are powerful at deriving insights from textual data (Afful et al., 2014).

Sentiment analysis or opinion mining encompasses two main tasks: address issues such as subjectivity and polarity synthesis. Subjectivity detection seeks to distinguish objectivity from subjectivity, and it identifies sentences that carry opinions instead of real facts. On the contrary, sentiment detection is about getting the semantic meaning of the words and sentences of the messages and determining whether the emotions are positive, negative, or neutral. For example, polarity detection is used to measure the quantities of positive and negative sentiments expressed in the comments of individuals on a particular car brand. The approach is divided into three categories: Fundamentally, the approach is divided into three categories:

1. Machine learning approach
2. Deep learning approach
3. Lexicon based approach

MACHINE LEARNING APPROACH

In machine learning, labeled data serves as the cornerstone for predicting outcomes on unseen data. To effectively utilize machine learning approaches, preprocessing steps are necessary to prepare the data for model training. Within sentiment analysis, various classification algorithms are employed to categorize text into different sentiment classes (Colicev et al., 2017). These algorithms play a crucial role in deciphering the sentiment conveyed by textual data.

Naive Bayes (NB)

Naive Bayes is a popular classifier based on Bayes' theorem. It's widely used in text categorization, which judges documents based on word frequencies. It's simple and only needs a small amount of training data. Despite its strong assumptions, it works well in many domains.

Support Vector Machine (SVM)

Support Vector Machines (SVM) are recognized for their high performance in analyzing data and identifying patterns. They are utilized for both classification and regression tasks. The purpose of Support Vector Machines (SVM) is to identify the best position of the separating hyperplane by maximizing the distance between document vectors that are grouped into different classes (Chamlertwat et al., 2012).

Figure 1. Machine Learning Process

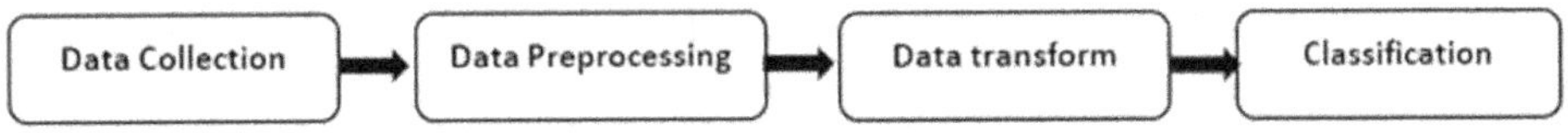

DEEP LEARNING APPROACH

Deep learning involves training neural networks to make predictions on unseen data, much like traditional machine learning approaches. Similarly, it begins with labeled data, and preprocessing steps are crucial to ensure the data is formatted appropriately for the model. Techniques such as one-hot encoding, word embedding, zero padding, and truncation are employed to convert textual data into a numeric format that neural networks can understand. In deep learning, the network structure is defined, often incorporating word embedding layers, which are particularly effective for natural language processing tasks. Deep learning approaches often demonstrate superior performance on large datasets, making them well-suited for tasks like sentiment analysis (Dang et al., 2020). Figure 2 illustrates the deep learning process.

Figure 2. Deep Learning Process

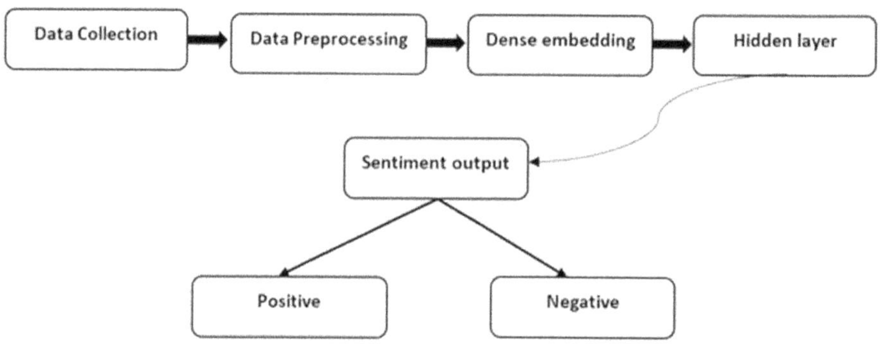

LEXICON-BASED APPROACH

It is a self-learning approach which is not supervised. In the lexicon-based approach, the target of the AI system is to check the meaning of the whole text and then tag the person. It is known as "rule-based classification with dictionary tagger" (Afful et al., 2014). The study of mood or emotion reflected in words is lexicon-based sentiment analysis and is one of the methods being used in natural language processing to understand the feelings expressed. The lexicon-based approach of sentiment analysis employs a word sense disambiguation technique, whose basic framework is inspecting the sentiment of expressions by identifying their semantic orientation. This method uses polarized dictionaries where words are labeled with their semantic orientation to determine the overall sentiment of a document.

There are several dictionaries frequently used, such as SentiWordNet and WordNet, but one can create one by certain requirements by adding more words and choosing them from the sentence-used words-based word lists. As with our research, we use a lexicon-based approach relying on the SentiWordNet, but it is possible to create it customized using the mentioned methods (Afful et al., 2014).

$$sentiment\ score = \frac{positive\ words - negative\ words}{total\ words}$$

SentiWordNet is indeed a lexical resource that assigns sentiment scores to synsets in WordNet, a large lexical database of English. These scores represent the positivity, negativity, and objectivity of each synset. They're useful for sentiment analysis and related natural language processing tasks (Chamlertwat et al., 2012).

PageRank WordNet, on the other hand, applies the concept of PageRank, which was originally developed by Google to rank web pages, to determine the importance or relevance of WordNet synsets based on their connectivity within the WordNet graph. This approach involves treating WordNet as a network where synsets are nodes and semantic relations like hypernyms, hyponyms, and meronyms are edges. By simulating random walks on this graph, PageRanking can determine the importance of synsets in terms of possessing certain semantic properties or representing specific concepts.

Both SentiWordNet and PageRanking WordNet are valuable tools for understanding and analyzing the semantic structure and sentiment of textual data. There are several popular lexicon models used for sentiment analysis. Figure 3 depicts the process involved in the lexicon-based approach.

Figure 3. Lexicon-Based Approach Process

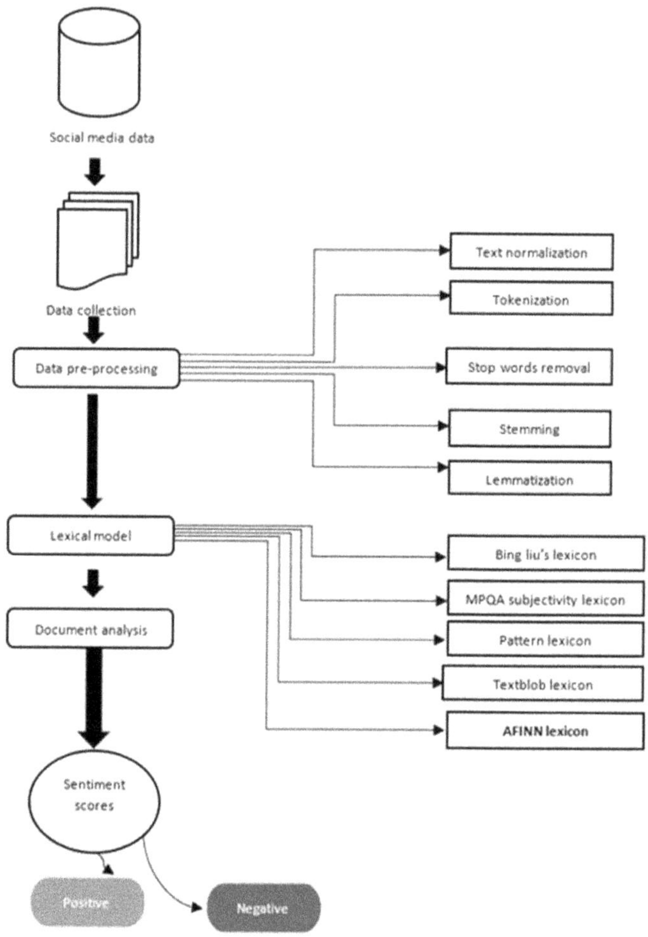

Steps for Sentiment Analysis

Problem Statement

At this point, businesses are making use of social media to get information about the comments they have received regarding their products. In the same manner, a business called "Apple". Using the information gathered from social media, a well-known brand is interested in learning what customers think about their business.

How people are thinking about their product and what people's perspectives are regarding their brand are being taken into consideration.

METHODOLOGY

Dataset Description

This study uses a dataset from Kaggle called "apple_twitter_sentiment_texts". This consists of 1630 instances and has a pair of columns for text (people's tweets) and sentiment (positive, negative, or neutral). The description of the Apple_twitter_sentiment_text Dataset is shown in Table 1.

Table 1. Apple_twitter_sentiment_text Dataset

INDEX	TEXT	SENTIMENT
0	@appl this is not expecting. #battery drainage. Not liked the product	-1
1	What is going with apple product?? #appleindia	0
2	Thanku apple now I can easily transfer photos with my collegues #easysharing	1
3	apple in now started manufacturing products in India #makeinindia	0
4	@appl can I change the color of my neW product to black color????	0

Within the context of this dataset, the sentiment column contains three distinct values, namely -1, 0 and 1. In this context, the value -1 indicates that individuals are receiving **negative** feedback, the value 0 indicates that individuals are receiving **neutral** feedback, and the value 1 indicates that individuals are receiving **positive** input.

Data Visualization

Visual perception is a cognitive process whereby objects are perceived through the senses, and internal connections produce images in the mind. Visualization serves his role in seeing the objects and building the cause-and-effects. It is through the graphical representation that information or data is conveyed from multiple perspectives, giving the stakeholders or the decision-makers a tool to rapidly and readily examine the data or information (Dang et al., 2020). It goes, indeed, from information visualization to data visualization, and it has done so mostly around the beginning of the century, and it has been quick-paced in the last decade. InfoVis technology is an expanding list of methods for data analysis that reflects the rapid pace of development of big data analytics. This area of machine learning has spawned many such scientific studies and is worked on by researchers from different disciplines of science. Data visualization recently developed into substantial research and decision-making approaches (Koto et al., 2015).

Bar charts are generated with the help of a Twitter date set. This allows for the progressive revelation of the trends and a deeper insight into the public's opinions, which is positive, negative or neutral towards the brand. Figure 4 illustrates the sentiment analysis of consumer reviews.

Figure 4. Sentiment Analysis of Consumer's Review

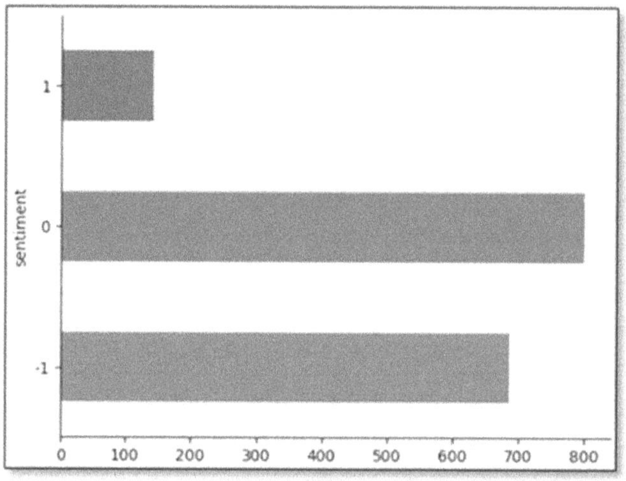

As can be seen from the bar chart in Figure 4:

- There are almost 700 bad reviews of the company.
- The neutral sentiment is the highest of all.
- And the positive sentiment is relatively low.

Data Preprocessing

The practice of removing extraneous words or data from a table is implemented by a technique known as data preprocessing. It is essential to exclude noisy words from the data to facilitate the process of text analysis. This will allow us to simply interpret the text into the machine and implement various algorithms to make predictions in the future (Chamlertwat et al., 2012).

If we want to know any context, we may analyze the document with the aid of keywords, and there is no need to utilize punctuation marks or grammar. This is similar to what a human being can comprehend. To a similar extent, to facilitate the comprehension of machines, we are focusing just on the keywords for further analysis. During the process of executing the data preprocessing, we are required to adhere to several significant procedures to achieve our objective in the preprocessing for sentiment analysis. The following is a list of the steps:

1. Text normalisation
2. Tokenisation
3. Stopword removal
4. Stemming
5. Lemmatisation

Figure 5. Data Preprocessing Major Steps

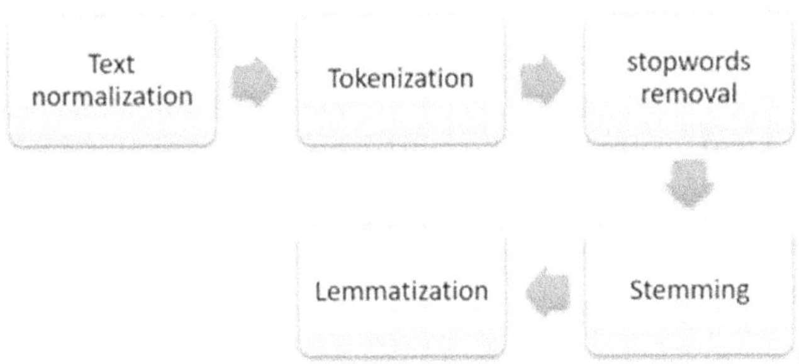

Figure 5 outlines the major steps involved in data preprocessing.

Text Normalization

To carry out the text analysis, it is necessary to first complete the text normalization (Chamlertwat et al., 2012). Bringing everything to a canonical (standard) form is what is meant by the term "normalization" in this context. Normalization of the text is the process of bringing the entire text to a canonical form of expression. The removal of punctuation, digits, and other elements is included in this (Chamlertwat et al., 2012). It makes our model less complicated, and if the complexity is reduced, then there is a greater likelihood that our model will learn more effectively.

Text is typically composed of abbreviations and letters written in capital letters. It is a frequent practice to overlook this stage, even though it is among the most straightforward and efficient steps in the process of text preprocessing. This is especially true in situations where the dataset is relatively sparse. It has been discovered that variations in capitalization (for example, "India" versus "india") have different outcomes. In other words, the computer will assume that two different words and two different word vectors are produced in the latter stages of word embeddings (Chamlertwat et al., 2012). This is because the computer will assume that the same words are two different words if they have one capital letter and one lowercase letter. As a result, the most effective method for text preparation has been to convert all of the words to lowercase.

For example:

- I am a hero -> "I am a hero"/ "I AM A HERO"

Tokenization

Natural language processing (NLP) preprocessing is the most fundamental one. In the process of tokenization, a phrase, sentence, or full text is portioned up into smaller units, which are then separated by commas. Through this process, we can obtain the individual word phrases (Chamlertwat et al., 2012). Every single one of these little units or individual words is referred to as a token. To maintain a wide vocabulary and intricate word constructions, it is utilized.

for example

- NLP is a field of artificial intelligence. => 'NLP', 'is', 'a', 'field', 'of', 'artificial', 'intelligence'

Stopwords Removal

In the process of removing stopwords, we eliminated all the frequently used words in phrases that contribute significance to the sentences. In this procedure, we eliminated all the stopwords, including words like "is," "am," "the," "their," and so on. It functions as a bridge and ensures that the sentences are grammatically correct. We utilize this tool to eliminate words before applying the NLP approach. For the implementation, we are using the stopwords library to eliminate the stopwords.

Stemming

For further analysis, stemming is a technique that involves removing the suffix from the term and using the base word as the keyword alternative. Stemming is a method of word shortening that is not just aggressive but also attempts to trace a term back to its foundational root. These are the suffixes: to bring it to the base root, it was sliced off when the semantic, There is no change in meaning for any of the several variants. To be sure, as a result, this may not always produce favorable outcomes because the word loses its meaning (Chamlertwat et al., 2012).

Through stemming, the input dimension is reduced. Methods of machine learning are more effective in making it work. While simultaneously reducing the size of the dictionary, it increases the density of the training data. In the end, it is beneficial to normalize the term in the paper.

For example, the details of the stemming are given in Table 2.

Table 2. Stemming Process

From	suffx	Stem
Cats	Removing 's'	Cat
Birds	Removing 's'	Bird

Lemmatization

A basic task in natural language processing (NLP) is called lemmatization, and it entails string-to-string transduction from an inflected word form to its citation(base) form, which is referred to as the lemma (Pavaloaia et al., 2019). This linguistic normalization is essential for several downstream applications of natural language processing, particularly for languages with a significant degree of inflection. In the past, it has been demonstrated that lemmatization can improve recall for information retrieval (Pavaloaia et al., 2019).

We can say that lemmatization works with the morphological meaning of the words and that it puts the word back to its base word by utilizing the grammar and vocabulary. This is a simple explanation of how lemmatization works. Table 3 details the lemmatization process.

For example:

Table 3. The Lemmatization Process

form	Meaning information	Lemma
Helps	Third person singular, present tense "help"	Help
Helping	"ing" form of the verb	Help

Difference Between Stemming and Lemmatization

Table 4 compares the differences between stemming and lemmatization.

Table 4. Stemming and Lemmatization

S no.	Stemming	Lemmatization
1.	Stemming cut off the final words to attain the purpose.	Lemmatization uses grammar and vocabulary to return words to their roots.
2.	Stemming operates on a single word without regard for context.	In lemmatization, it can use context to determine the right lemma of a word.
3.	Implementation is easy.	Implementation is difficult.
4.	Simple and faster.	More accurate but slower.
5.	Example: flies => flie flying => fly	Example: flies => fly flying => fly

Model Implementation

Lexicon modeling approaches and Natural Language Processing (NLP) modeling techniques, such as a bag of words, TF-IDF, and word embeddings, may be used for the modeling process. In this chapter, we will cover the lexicon-based approach in detail. Also, we will briefly discuss the bag of words, TF-IDF, and word embeddings.

lexicon Modeling

When it comes to unsupervised learning, we are employing the lexical model for sentiment extraction. It is usual practice to record sentiment associations in sentiment lexicons, which are collections of the word–sentiment pairings that are related to one another (and may also include a score that indicates the degree of association)

(Cambria et al., 2014). The procedure is based on the rules specific to the domain, which serve to assist with the extraction of features.

As an illustration, when the lexicon sentence is available, the features extraction process can place a greater emphasis on matching words and phrases inside a lexicon or on counting the number of times matches occur within the corpus, respectively. Simply put, a sentiment lexicon is a collection of sentiment words and phrases.

Table 5 presents several popular lexical models, including details on the total quantity of positive and negative words contained within each lexicon. They are listed as follows:

1. Bing Liu's lexicon
2. MPQA subjectivity lexicon
3. Pattern lexicon
4. Textblob pattern
5. AFINN lexicon

Table 5. Total Quantity of Positive and Negative Words in Each Lexicon

S NO.	lexicon	Positive words	Negative words
1.	Bing Liu's lexicon	2006	4784
2.	MPQA subjectivity lexicon	4000	2800
3.	Pattern lexicon	1800	2400
4.	Textblob pattern	4000	2000
5.	AFINN lexicon	1651	826

Bing Liu's Lexicon

Text analysis is a technique that involves examining the significance of a written document by utilizing predetermined word lists. The predetermined list consists of 2006 words with favorable connotations and 4784 terms with negative connotations (Cambria et al., 2014). The list has already been categorized according to its polarity. This method employs our existing lexicon, consisting of two lists with polarity, to compare it with the upcoming files or text, finally establishing the score (Cambria et al., 2014).

The primary goal is to tally the quantity of positive, negative, and neutral words to calculate the sentiment score of the document. The emotion score can be computed using a variety of approaches.

$$\textit{sentiment score} = \frac{\textit{positive words} - \textit{negative words}}{\textit{total words}}$$

MPQA Subjectivity Lexicon

Paul Hoffmann, Janyce Weibe, and Theresa Wilcon are the ones developing it. The subjectivity of a sentence plays a critical role in Opinion Mining and Sentiment Classification (Cambria et al., 2014). Hence, it is crucial to distinguish between subjective and objective sentences, giving subjective sentences greater significance than objective sentences. The act of determining subjective sentences is referred to as subjectivity categorization. Sentences that are subjective exhibit a certain degree of subjectivity, such as the inclusion of modifiers (adjectives), intensifiers, and diminishers, or the expression of a personal state, as stated in (Chamlertwat et al., 2012). Identifying subjective sentences from opinionated papers can be challenging due to the various ways subjectivity can be expressed.

It is extensively utilized in analysis tasks. MPQA (multi-perspective question answering) is a method that determines whether a document expresses feelings or not. In addition, the subjective terms are additionally marked with their polarity. It is an invaluable resource for conducting sentiment analysis, especially when faced with limitations in data availability and unique criteria related to a certain topic (Cambria et al., 2014).

Pattern Lexicon

This is referred to as lexicon-based sentiment analysis, in which previously defined dictionaries contain words and their sentimental values. The purpose of the dictionaries was to utilize machine learning with the labeled data. A polarity value is given to every word in the lexicon to express sentiment. The pattern lexicon method is based on the calculation of sentiment scores for words contained in texts being analyzed, and then the sentiment of the document or text is determined by the combination of the sentiment scores of each word.

The pattern lexicon approach is more straightforward and computationally efficient than other lexicon approaches, but it has some limitations: Thesauri fails to account for all the occurrences of words missing from a given document, which causes inaccuracies. It is true, especially when we are on the verge of handling sarcasm and irony either in reading or writing. Hence, it is imperative to blend the pattern lexicon method, which includes other key techniques, such as machine learning algorithms, into the solution to upscale the precision of sentiment analysis.

Textblob Pattern

Sentiment analysis is done using the text service offered by this tool. It is a textblob lexicon that includes a predefined lexicon or dictionary for polarity words and subjectivity. It refers not only to the support of polarization and opinion but also to the use of a text blob pattern. In this respect, press -1 stands for negative views, and press 1 implies positive feelings. TextBlob contains semantic tables so that systematic research becomes much simpler and more detailed (Dang et al., 2020). Subjective value refers to how much personal viewpoints and literary messages are in a piece of writing. The essay has a subjective tone, so the author states his/her opinion rather than dry facts.

TextBlob has one more setting: Unity can be achieved through resilience, as individuals come together and face challenges together, becoming stronger and able to tackle future difficulties effectively. TextBlob relies on the characteristic "intensity" to analyze the sentiment or subjectiveness of a text (Dang et al., 2020). On the other hand, when we analyzed the following statement, "I do not like this example; it is quite boring. " Here, polarity is -1, and subjectivity is 1, which is correct (Shayaa et al., 2018). However, for the sentence, "That was helpful, but I'd rather have another one"- degree 0. 0 for both polarity and objectivity - which is not very convincing when we have expected an option better than that one.

AFINN Lexicon

The AFINN lexicon is also predefined. The lexicon contains a collection of English terms for sentiment polarity with a score between -5 (negative) and +5 (positive). Finn Arup Nielson developed it between 2009 and 2011 (Cero et al., 2024). The AFINN lexicon contains a total of 3300 terms. AFINN is simple to use and does not require a machine-learning model to extract sentiment insights. We may readily understand the lexical score. It is quick to analyze the sentiment (Pavaloaia et al., 2019).

With the help of the AFINN lexicon, we can readily recognize neutral terms, too. It is used for generic sentiment analysis rather than domain-specific analysis. Researchers and scientists apply the AFINN lexicon to sentiment analysis in a variety of fields, including social media posts, consumer reviews, and other text data (Cero et al., 2024).

Sentiment Lexicon Generation Approaches

There are three approaches to developing the sentiment lexicon, which are given below:

1. Manual
2. Automated

Manual: We are manually developing/generating the lexicon with the assistance of scientists and researchers. It can be time-consuming and is caused by a large throng of individuals.

Automated: To save time and work efficiently, we are using the automated approach. There are two automated approaches:

1. Dictionary based
2. Corpus-based

Dictionary-based: In the dictionary-based approach, we use a well-known dictionary like WordNet. But before we use the dictionary, the method starts with some seed words (Pava). For example, the seed word for "sentiment" would be "senti," which could mean "good" or "bad." Iteratively add to the vocabulary by using dictionaries to find their opposites and similar words. It can be used for many things.

Corpus-based: It is bridging the possible shortcomings of sentiment generation based on dictionaries. Application specific to a domain works well with it. Example: microblogs and reviews. Both sentiment and aspect extracts benefit from it. In corpus-based, it can take on two forms:

- **Lexicon expansion:** Discover more sentiment words for the domain corpus, given a seed list of words that are used for general-purpose expressing sentiment. Additionally, we can integrate the dictionaries by utilizing association rules.
- **Lexicon adaptation:** The utilization of a domain corpus can modify a general-purpose sentiment lexicon within a new one. Words that convey sentiment, sentiment values, and other semantic information can be added or changed.

NLP Modelling

Natural language processing (NLP) is a technique developed to derive concepts and significance from text that is created by humans and lacks a specific structure. Executing a large number of documents rapidly and reliably is meant to be accomplished using computer algorithms (Juluru et al., 2021). In the field of artificial intelligence, natural language is used to try to understand and make useful sentences in human language. Over the years, natural language processing has been used for many things, such as speech recognition, language translation, information search, text summarization, and more. To understand the details better, let us first look at the steps of NLP, or the order in which a sentence is processed before it is turned into a parse tree (Ranjan et al., 2016).

There are Some NLP modeling techniques given below:

1. Bag of words
2. TF-IDF
3. Word embeddings

Bag of words: A bag of words Prepare the text or documents. It turns the text that has already been handled into a bag of words. It also keeps track of how many times the most common words are used. Most of the time, these ways are used to turn tokens into sets of features. As soon as we have a master collection of features, we can put together features from different works to make a feature vector. In the same way, we rated each fruit by asking, “Is this fruit red?” We look at every document in our collection and ask, “Does this document have the word 'a cut'?” instead of “Is this fruit round?” If that is the case, we put a value in the right cell of the feature vector.

We put a “0” or a “1” in which a term (feature) is missing or present in the simplest method. There are other ways to figure out how important words are and tell ML systems to give some words a greater significance than others (Ranjan et al., 2016).

Figure 6. Bag of Words from Document 1 and Document 2

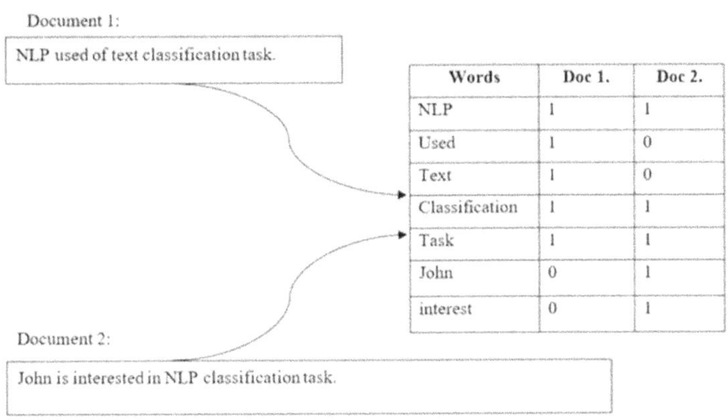

Words	Doc 1.	Doc 2.
NLP	1	1
Used	1	0
Text	1	0
Classification	1	1
Task	1	1
John	0	1
interest	0	1

Figure 6 illustrates the Bag of Words representation for Document 1 and Document 2.

TF-IDF

The abbreviation TF-IDF refers to the frequencies of terms and the inverse document frequency. The score can be measured with the assistance of tf-idf, which allows us to obtain information retrieval or summarization during the process. It also reflects the degree to which a term is significant in a particular publication. TF-IDF is a tool that assists in determining the significance of a particular word within the context of a repository of documents. The number of times the word appears in the document is taken into consideration, and the number of documents that are included in the corpus is used to ensure that the index is accurate. A term's frequency is calculated by dividing its frequency by the total number of terms in the document.

$$TF(t,d) = \frac{number\ of\ times\ t\ appears\ in\ d}{total\ number\ of\ terms\ in\ d}$$

IDF is obtained by first dividing the total number of documents by the number of documents that include the phrase and then taking the logarithm of the quotient that is generated from this division.

$$IDF(t) = log\frac{N}{1+df}$$

After that, TF-IDF is the result of multiplying two values, namely TF and IDF combined. Multiplication of two matrices is the process that is used to calculate the TF-IDF. The number of times a particular term appears in a document (TF). The inverse document frequency of the word across a collection of documents or documents (IDF).

$$TF - IDF(t,d) = TF(t,d)*IDF(t)$$

Word Embeddings

A word embedding is a way to express a word or a phrase as a vector, which is a vector (array) containing real values. It is used in natural language processing (NLP) and machine learning. There are word embedding models available that are capable of capturing the concept–semantic associations between words and phrases by converting them into a set of vectors with continuous vector space. In this area, words that bear resemblance (unveiled by vectors) are positioned close to each other.

However, in the past three years, deep learning techniques have been increasingly used in the domain of natural language processing (NLP).

Deep neural networks (DNN) are deep artificial neural networks (ANN) that contain multiple hidden layers that are stacked between the input and output layers. RNN can detect the sequences of some recognizable patterns, and CNN can comprehend the patterns that occur in the spatial dimensions. The latest progress in deep learning has tremendously developed the paring of some natural language processing (NLP) tasks related to text semantics analysis, such as text classification, sentiment analysis, etc. These NLP tasks involve the use of NER (Named Entity Recognition), recommendation systems, biomedical text mining, topic modeling, etc. As an example, pre-trained embeddings are raw vector representations of individual words where the word length is fixed, and its meaning is general phrase semantics and language patterns in natural language (Asudani et al., 2023). The different items on the list are put forward by scientists to seek such representations.

Fundamentally, word embedding can be incredibly helpful for understanding complex natural language processing (NLP) concepts. Word embedding approaches can be classified into three categories: conventional, distributional, and contextual substitutes for the word embedding model. Conventional word embedding, also known as count-based or frequency-based models, can be divided into three categories: Conventional word embedding, also known as count-based or frequency-based models, can be divided into three categories: BoW, n-gram, and TF-IDF being the crucial components for text analysis.

Document Analysis

When all of the relevant stages, such as data collecting and data preprocessing, have been finished. As part of our data processing, we implemented one of the lexical models that we developed. When working with this information, we utilized the AFINN lexicon, which is an effective tool for extracting social media postings and reviews from customers.

Following the implementation of the AFINN lexicon, we can quickly obtain the sentiment score of the respective tweets in terms of the polarity of the sentiment. The actual reviews, the actual sentiment that was expressed, and finally, the anticipated sentiment polarity that was produced by our lexical model are all displayed in Figure 8, which can be found below.

Figure 7. Screenshot of a Model Predicting Sentiment Polarity

```
REVIEW: cant use compon tcp instal disc due stupid os updat pay buck redownload
Actual Sentiment: negative
Predicted Sentiment polarity: -3.0
------------------------------------------------------------
REVIEW: media report say hire pro fashion retail industri fashion
Actual Sentiment: neutral
Predicted Sentiment polarity: 0.0
------------------------------------------------------------
REVIEW: whi group messag peopl
Actual Sentiment: neutral
Predicted Sentiment polarity: 0.0
```

Figure 7 shows a screenshot of the model predicting sentiment polarity.

Sentiment Score

During the last stage of the process, once we have obtained the polarity scores of the individual texts, we will now assign those scores to the column that is designated for expected feelings. Based on the polarity score, if it is less than zero, the predicted sentiment is 'negative'; if it is greater than 0 but less than one, the predicted sentiment is 'neutral'; and if it is equal to one and greater than one, the predicted sentiment is 'positive.' We are currently checking the accuracy of our model once we have assigned all of the polarity for the model. The classification report and confusion matrix were utilized in this process.

Figure 8 displays the classification report, which includes the precision, recall, f1-score, and accuracy of the model, as well as the confusion matrix. In every single one of these evaluations, the actual sentiment and the expected sentiment are taken into consideration.

Figure 8. Classification Report and Confusion Matrix

	precision	recall	f1-score	support
negative	0.45	0.83	0.59	507
neutral	0.61	0.07	0.13	615
positive	0.29	0.59	0.39	108
accuracy			0.43	1230
macro avg	0.45	0.50	0.37	1230
weighted avg	0.52	0.43	0.34	1230

	negative	positive	neutral
negative	421	21	65
positive	475	45	95
neutral	36	8	64

Here, in this confusion matrix, we are getting less accuracy because we have skewed data for the analysis. We can overcome this problem by adding more data to our model.

Expected Future Trends in Consumer Sentiment Analysis

The company can determine the efficacy of its different marketing efforts by conducting sentiment analysis. It helps them in making better decisions and improving the growth of their business. The business can demonstrate how it catches up with the popular mood of consumers and predicts their future preferences; therefore, the company will certainly grow. The participation of new technologies such as artificial intelligence and continuous new studies will permit us to use predictive tools to estimate future consumer sentiment. There are several potential trends to monitor:

1. The development of progressive sentiment analysis approaches is pivotal.
2. Multimodal sentiment analysis
3. In parallel with the data sources, capture the sentiments.
4. Personalized sentiment analysis
5. Domain-specific sentiment analysis
6. It is what helps make the decision-making system more practical, effortless, and efficient.
7. Wielding decision-making mechanism.

Generally, the next generation of consumer sentiment research is based on the amalgamation of technological transformations and its industry-specific custom-made applications, which depend on the changing patterns of consumers' interests and modes of interaction.

CONCLUSION

Ultimately, we have come to terms with the relationship between business development and sentiment analysis. What factors drive business intelligence from the impact perspective? With the help of modern data analysis facilities in sentimental analysis, we can go far enough to make the intellectual leader of our business operations. Sentiment analysis serves as the basis for transparent communication between consumers and businesses, which are becoming more dynamic. Over the entire of this chapter, we have a journey through the complexity of getting sentiment and defined ways of winning the consumer's emotions. This has helped us a lot to be experts in different ways to be precise in appraising sentiment. Our investigation has been implemented using the lexical technique. As for business intelligence, numerous elements can make a difference. They can determine whether or not the course of business intelligence initiatives will become successful. Through knowledge and judgment of the effects of these factors, companies can formulate clear strategies and learn appropriate approaches to advance their business.

An important variable that goes hand in hand with data quality is data quality. Accuracy, integrity, and data reliability set the foundation for a successful perception of business intelligence initiatives. When the data used as a basis for analysis is low-quality and not reliable, the results of the analysis made on the data might be misleading and/or wrong. That is why organizations have to monitor public sentiment in real-time, which can be regulated using the latest technology, for example, sentiment analysis. For this, data scientists exploit sentiment analysis to keep consumers up to date with transparent transactions. It is one of the strongest tools in a marketer's armory that allows a business to acquire valuable customer sentiment and opinion knowledge. Through consumers' language and emotional translation of their product usage, businesses can develop their understanding of consumers' needs and choices.

Consequently, companies benefit from this by making careful decisions; hence, they are in a position to improve their products and services. Knowing that one of the typical advantages of sentiment analysis is that it can reveal concealed patterns and trends, which are contained in customer feedback, is among the common characteristics of sentiment analysis. Analyzing a great deal of customer data implies discovering general topics and feelings that customers may have. This data can be subsequently used to recognize key improvement areas and to pinpoint and resolve any concerns that the customers might face. In addition, sentiment analysis can facilitate our mission to measure customers' sentiment and the essential moves to govern this situation.

Sentiment analysis was something that we were quite versed in. We got a clear picture of the various approaches and techniques that could be used in textual analysis. We have a method called "Lexical Swedish," which we used in our investigation. In the future, sentiment analysis will be extremely important for business intelligence as this method can provide rich information on consumer feelings and desires. Financial modeling in this form will serve as a decision-making tool and will make it possible for a business to remain competitive. By examining the sentiments in customer feedback and internet conversations, the companies can investigate the target audience and use statistical information to make reasonable conclusions as to how best to meet the client's needs. Sentimental analysis will be the key to better understanding consumers' feelings and will let companies improve products, services, and customer satisfaction. Analyzing the vast volume of data rapidly and precisely, sentiment analysis will play a decisive role in transforming the way businesses conduct and perform their tasks as far as customer communication is concerned. Utilizing the effectiveness of sentiment analysis as a business tool gives enterprises the leading edge and positions them high above competitors in the fast-paced and worldwide arena.

REFERENCES

Afful-Dadzie, E., Nabareseh, S., Oplatková, Z. K., & Klímek, P. (2014, August). Enterprise competitive analysis and consumer sentiments on social media-insights from telecommunication companies. In *International Conference on Data Management Technologies and Applications* (Vol. 2, pp. 22-32). SciTePress. DOI: 10.5220/0004991300220032

Asudani, D. S., Nagwani, N. K., & Singh, P. (2023). Impact of word embedding models on text analytics in deep learning environment: A review. *Artificial Intelligence Review*, 56(9), 10345–10425. DOI: 10.1007/s10462-023-10419-1 PMID: 36844886

Cambria, E., & White, B. (2014). Jumping NLP curves: A review of natural language processing research. *IEEE Computational Intelligence Magazine*, 9(2), 48–57. DOI: 10.1109/MCI.2014.2307227

Cero, I., Luo, J., & Falligant, J. M. (2024). Lexicon-Based Sentiment Analysis in Behavioral Research. *Perspectives on Behavior Science*, 1–28. PMID: 38660506

Chamlertwat, W., Bhattarakosol, P., Rungkasiri, T., & Haruechaiyasak, C. (2012). Discovering consumer insight from Twitter via sentiment analysis. *Journal of Universal Computer Science*, 18(8), 973–992.

Colicev, A., Kübler, R., & Pauwels, K. (2017). Social Media's Impact on Consumer Mindset: When to Use Which Sentiment Extraction Tool? *MSI Working Paper Series*.

Dang, N. C., Moreno-García, M. N., & De la Prieta, F. (2020). Sentiment analysis based on deep learning: A comparative study. *Electronics (Basel)*, 9(3), 483. DOI: 10.3390/electronics9030483

Dey, L., Chakraborty, S., Biswas, A., Bose, B., & Tiwari, S. (2016). Sentiment analysis of review datasets using naive bayes and k-nn classifier. *arXiv preprint arXiv:1610.09982*.

Juluru, K., Shih, H. H., Keshava Murthy, K. N., & Elnajjar, P. (2021). Bag-of-words technique in natural language processing: A primer for radiologists. *Radiographics*, 41(5), 1420–1426. DOI: 10.1148/rg.2021210025 PMID: 34388050

Karami, M., Mosallanezhad, A., Sheth, P., & Liu, H. (2022). Estimating topic exposure for under-represented users on social media. *arXiv preprint arXiv:2208.03796*.

Khoo, C. S., & Johnkhan, S. B. (2018). Lexicon-based sentiment analysis: Comparative evaluation of six sentiment lexicons. *Journal of Information Science*, 44(4), 491–511. DOI: 10.1177/0165551517703514

Kiritchenko, S., & Mohammad, S. M. (2017). The effect of negators, modals, and degree adverbs on sentiment composition. *arXiv preprint arXiv:1712.01794*.

Koto, F., & Adriani, M. (2015). A comparative study on twitter sentiment analysis: Which features are good? *Natural Language Processing and Information Systems: 20th International Conference on Applications of Natural Language to Information Systems, NLDB 2015, Passau, Germany, June 17-19, 2015Proceedings*, 20, 453–457.

Lahiri, K., & Zhao, Y. (2016). Determinants of consumer sentiment over business cycles: Evidence from the US surveys of consumers. *Journal of Business Cycle Research*, 12(2), 187–215. DOI: 10.1007/s41549-016-0010-5

Malaviya, C., Wu, S., & Cotterell, R. (2019). A simple joint model for improved contextual neural lemmatization. *arXiv preprint arXiv:1904.02306*. DOI: 10.18653/v1/N19-1155

Neri, F., Aliprandi, C., Capeci, F., & Cuadros, M. (2012, August). *Sentiment analysis on social media. In 2012 IEEE/ACM international conference on advances in social networks analysis and mining*. IEEE.

Orimaye, S. O. (2013). Learning to classify subjective sentences from multiple domains using extended subjectivity lexicon and subjective predicates. *Information Retrieval Technology: 9th Asia Information Retrieval Societies Conference, AIRS 2013, Singapore, December 9-11, 2013 Proceedings*, 9, 191–202.

Pavaloaia, V. D., Teodor, E. M., Fotache, D., & Danileţ, M. (2019). Opinion mining on social media data: Sentiment analysis of user preferences. *Sustainability (Basel)*, 11(16), 4459. DOI: 10.3390/su11164459

Ranjan, N., Mundada, K., Phaltane, K., & Ahmad, S. (2016). A Survey on Techniques in NLP. *International Journal of Computer Applications*, 134(8), 6–9. DOI: 10.5120/ijca2016907355

Shah, P. (2020). Sentiment analysis using textblob. *Diambil kembali dari Medium:*https://towardsdatascience.com/my-absolute-go-to-for-sentimentanalysis-textblob-3ac3a11d524

Shayaa, S., Ainin, S., Jaafar, N. I., Zakaria, S. B., Phoong, S. W., Yeong, W. C., Al-Garadi, M. A., Muhammad, A., & Zahid Piprani, A. (2018). Linking consumer confidence index and social media sentiment analysis. *Cogent Business & Management*, 5(1), 1509424. DOI: 10.1080/23311975.2018.1509424

Tabassum, A., & Patil, R. R. (2020). A survey on text preprocessing & feature extraction techniques in natural language processing. *International Research Journal of Engineering and Technology*, 7(06), 4864–4867.

Throop, A. W. (1992). Consumer sentiment: Its causes and effects. *Federal Reserve Bank of San Francisco Economic Review*, 1, 35–59.

Chapter 6
Predicting Consumer Vision: An AI-NLP Bot on Various Communication Platforms

J. Jayapriya
https://orcid.org/0000-0001-6672-6865
Christ University, India

M. Vinay
https://orcid.org/0000-0003-0297-3597
Christ University, India

Blessy Louis
Christ University, India

C. Balakrishnan
Christ University, India

R. Sindhu
Christ University, India

ABSTRACT

Combining NLP technology into communication platforms offers noteworthy opportunities for businesses to gain actionable insights into consumer behaviour and tailor their marketing strategies accordingly. By leveraging the predictive capabilities of NLP bots, organizations can enhance customer engagement, improve product offerings, and eventually drive business growth in a progressively competitive market landscape. With this context, this chapter focuses on developing an AI-NLP bot to understand consumer or customer behaviour patterns using data from various

DOI: 10.4018/979-8-3693-9246-1.ch006

communication platforms. AI-NLP bot is used to predict and analyze the customer's thoughts about the product through their opinion sharing in various sources. Results indicate that the NLP bot accurately predicts consumer sentiment, identifies emerging trends, and anticipates market preferences across different communication platforms. Moreover, the system's malleability and scalability make it appropriate for positioning in various business contexts, extending from e-commerce and retail to hospitality and healthcare.

INTRODUCTION

In the current era of swiftly embryonic digital landscape, businesses are increasingly contingent on effective communication to employ with consumers. With the proliferation of communication platforms such as social media, messaging apps, and email, understanding user sentiment and preferences has become dominant for fruitful publicizing stratagems. Natural Language Processing (NLP) techniques offer an auspicious possibility for analyzing vast amounts of textual data to extract insights and envisage consumer behavior. This chapter presents a different methodology for prophesying consumer vision using an NLP bot deployed across various communication platforms. The anticipated system connects the influence of machine learning algorithms to analyze text data from diverse sources, including customer reviews, social media posts, and email correspondence. The bot gains deep insights into consumer preferences, sentiments, and trends by leveraging cutting-edge NLP schemes such as sentiment analysis, topic modeling, and named entity recognition. Key modules of the NLP bot include data pre-processing, feature extraction, model training, and deployment across multiple communication platforms.

Distributional semantics focuses on representing the meaning of words and phrases based on their distributional patterns in a large corpus of text. Methods like word embeddings (e.g., Word2Vec, GloVe, FastText) and contextual embeddings (e.g., ELMo, BERT, GPT) are used to capture semantic relationships and similarities between words and phrases. With the dawn of deep learning, neural networks have become prevalent in NLP. Architectures like Recurrent Neural Networks (RNNs), Long Short-Term Memory (LSTM) networks, Convolutional Neural Networks (CNNs), and Transformer models have shown remarkable performance in tasks such as machine translation, text generation, sentiment analysis, and named entity recognition.

The following Figure 1 represents the general outline of natural language processing.

Figure 1. The Outline of the NLP

The bot utilizes state-of-the-art machine learning models, such as Long Short-Term Memory (LSTM) and transformer-based architectures like BERT, to accurately predict consumer behavior. Long Short-Term Memory (LSTM) networks, a type of recurrent neural network (RNN) architecture, offer several advantages, particularly in tasks involving sequential data processing, such as natural language processing, speech recognition, and time series forecasting. One noteworthy benefit of LSTM networks is their skill in handling long-term dependencies in sequential data. Traditional RNNs often scrap with capturing dependencies over long sequences due to the vanishing gradient problem, where gradients reduce exponentially or go off during training. LSTMs address this issue by introducing specialized memory cells and gating mechanisms that enable the network to retain or discard information selectively over multiple time steps.

This capability allows LSTMs to apprehend and employ relative data from distant past states, making them well-suited for tasks requiring an understanding of temporal relationships and complex patterns. Additionally, LSTMs exhibit robustness to noise and irrelevant inputs. The gating mechanisms in LSTM cells enable the network to filter out irrelevant information while retaining essential context adaptively. This feature is advantageous in real-world applications where data may contain noise, missing values, or irrelevant features. For instance, LSTMs effectively process text with grammatical errors, typos, or varying lengths in natural language processing tasks, leading to more robust and accurate models. Likewise, LSTMs offer flexibility and adaptability, allowing them to be integrated into diverse neural network architectures and combined with other types of layers, such as convolutional layers, to form hybrid models tailored to specific tasks. This flexibility enables researchers and practitioners to design sophisticated architectures that leverage the strengths of different neural network components. Furthermore, LSTMs excel at modeling complex temporal dynamics present in sequential data. By learning meaningful representations of sequential patterns, LSTMs capture subtle nuances and dependencies within the data, improving performance in tasks such as language modeling, sentiment analysis, and speech recognition. The advantages of LSTM networks, including their ability to handle long-term dependencies, robustness to noise, flexibility, and capability to model complex temporal dynamics, make them a powerful tool for a wide range of sequential data processing tasks, contributing to advancements in various fields of artificial intelligence and machine learning. The structure of LSTM is shown in Figure 2.

Figure 2. Basic Structure of LSTM

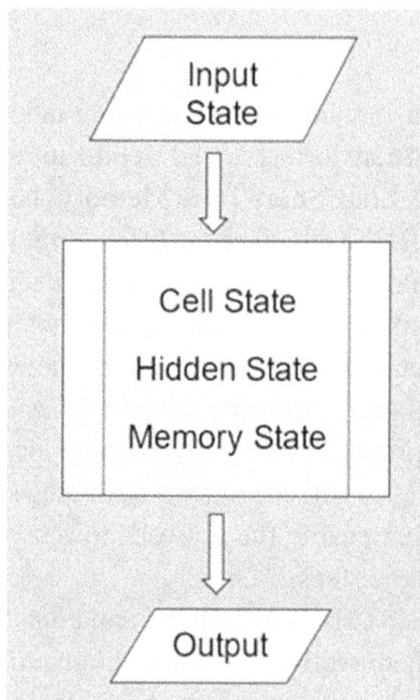

Bidirectional Encoder Representations from Transformers (BERT) is a groundbreaking language representation model. BERT exhibits a deep understanding of context in natural language. Unlike traditional models like word embeddings or recurrent neural networks (RNNs) that process text sequentially or in fixed-size windows, BERT employs a transformer architecture to capture bidirectional context. By considering both left and right context during training, BERT generate contextualized word embeddings that better capture the nuances of language semantics and syntax. BERT excels at capturing long-range dependencies in text. The self-attention mechanism in transformer architectures enables BERT to attend to relevant words across the entire input sequence, irrespective of distance. This capability allows BERT to model complex linguistic structures and dependencies, making it particularly effective for tasks requiring understanding of contextually rich language, such as question answering, natural language inference, and sentiment analysis.

BERT is pre-trained on large-scale corpora of text data using unsupervised learning techniques, such as masked language modeling and next-sentence prediction. This pre-training enables BERT to learn general-purpose language representations that capture rich semantic information. Fine-tuning BERT on downstream tasks with task-specific labeled data further enhances its performance and adaptability across various NLP tasks, ranging from text classification and named entity recognition to machine translation and summarization. BERT is highly versatile and transferable across different languages and domains. While pre-trained BERT models are initially trained on English text, they are adapted to other languages using cross-lingual transfer learning techniques.

Moreover, BERT-based models can be fine-tuned on domain-specific datasets to improve their performance in specialized domains, making them applicable to various real-world applications. The advantages of BERT, including its ability to capture contextual information, model long-range dependencies, leverage pre-training for general-purpose representations, and adaptability to diverse languages and domains, have propelled it to the forefront of NLP research and applications, enabling significant advancements in various fields such as information retrieval, text understanding, and conversational AI. Additionally, the system employs techniques for handling unstructured data and managing real-time interactions with consumers.

The framework of the BERT is shown in the following Figure 3.

A series of experiments conducted on real-world datasets from diverse industries demonstrate the effectiveness of the proposed approach.

The Objectives are as follows:

1. Preprocessing techniques
2. Prediction model for customer feedback pattern with additional information

Figure 3. Framework of BERT

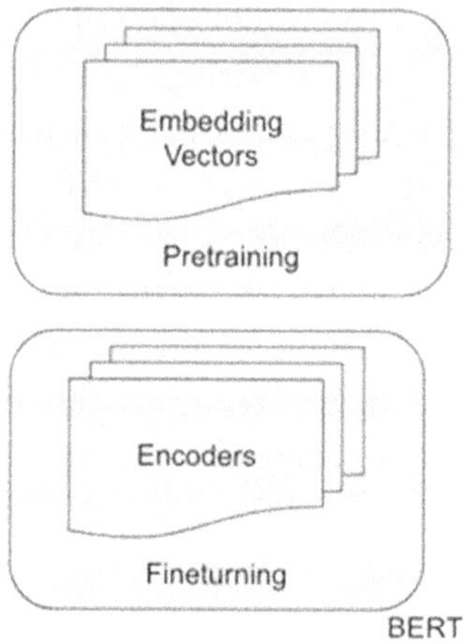

LITERATURE REVIEW

This section examines the related works of customer prediction using artificial intelligence to find the gaps. Various researchers identify customer behavior patterns by employing different algorithms and models. Each researcher has contributed to each model depending on the objective framed or according to the target.

Figure 4 depicts the representation of the literature of related papers on customer behavior depending on the products. The papers included range from 2015 to 2024. In this, nodes represent the paper, and edges depict the relationship between the papers. The node's color representing the paper indicates the year it was published; lighter shades are considered old paper, and darker shades are recent papers. The node size shows the number of citations; when more, it is big, and less, it is small.

Figure 4. Literature Review Representation through Connected Papers

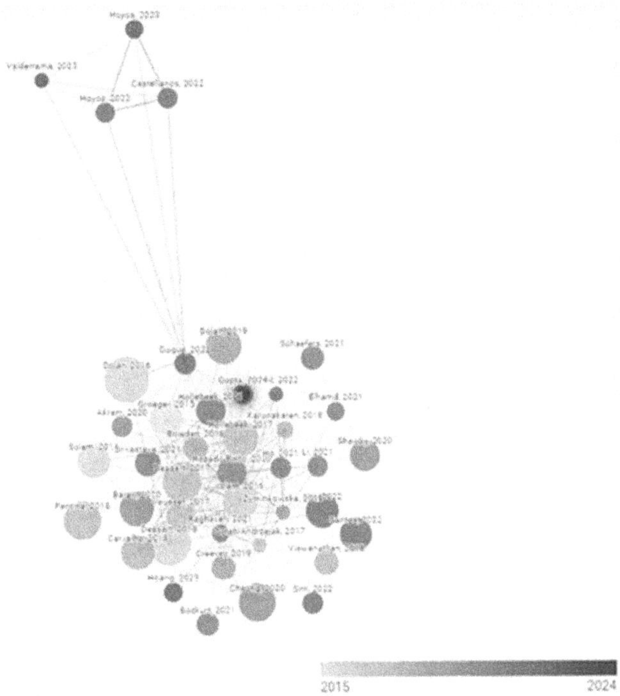

AI is increasingly being realized to have potentially transformative effects across many domains, notably in the personalization of marketing strategies and customer interactions. By applying complex algorithms to mountains of data on customers, AI can, in a very accurate way, predict preferences and provide personalized experiences that make a big difference in customer satisfaction and loyalty (Babatunde et al., 2024). For example, AI-driven personalization in marketing offers recommendations and offers based on not just the individual consumer but also their future needs and trends, driving engagement and retention. However, while its power in this regard is immense, there is much to debate about the ethical considerations regarding data privacy and fairness. There needs to be transparency and fairness in dealing with data to ensure consumer confidence and compliance with regulatory requirements.

In e-commerce, AI's impact on consumer behavior and market dynamics is particularly notable Shaik Vadla et al. (2024). AI-powered personalization tools have revolutionized how businesses understand and interact with customers, leading to more efficient and effective marketing strategies Pasupulati et al. (2024). Studies have shown that AI significantly enhances product design and development by accurately detecting customer emotions and preferences, as demonstrated by using pre-trained BERT and T5 models, achieving high accuracy rates in emotion detection for eco-friendly products (Marti et al., 2024). Moreover, AI's role in retail predictive analytics is vital in forecasting consumer behavior, optimizing inventory management, and enhancing overall customer engagement (Chiu et al., 2024). The successful implementation of AI in retail has led to increased operational efficiency, sales, and customer retention, highlighting its indispensable role in modern commerce (Ajiga et al., 2024).

Furthermore, AI's integration into online customer communities and chatbots has the potential to revolutionize customer service and interaction. By providing a conceptual framework for AI integration, businesses better understand and harness its capabilities to foster loyalty and improve service delivery (Ogundipe et al., 2024). Cross-country evaluations of AI-infused chatbots, such as those conducted in the USA and the UK, reveal that while chatbots significantly enhance user satisfaction, they must be optimized to effectively address cultural nuances and diverse customer expectations (Ekechi et al., 2024). Additionally, the advent of smart product-service systems (SPSS) that combine multiple AI models for personalized services exemplifies how AI captures and utilizes customer characteristics and preferences to deliver superior user experiences (Zlatanova-Pazheva et al., 2024).

The transformative power of AI extends to various business processes, including product management, where AI enhances creativity, market research, and quality assurance (Kumar et al., 2024). By automating content creation and adjusting marketing strategies based on real-time data analysis, AI enables product managers to drive growth and maintain competitive advantage (Rosário et al., 2024). The integration of AI in product development has also been shown to increase efficiency in data analysis, management, and supply processes. AI's ability to optimize business workflows and create long-term value underscores its pivotal role in shaping the future of commerce and consumer behavior (Ozen et al., 2024). As businesses continue to adopt and refine AI technologies, the focus on ethical considerations, such as data privacy and equitable treatment of consumers, will be crucial in sustaining the positive impact of AI on the market (Shankar et al., 2024) (Haldorai et al., 2024).

This comprehensive study introduces an integrated AI-based model to address customer churn in the telecom sector, using K-nearest neighbors, Decision Trees, Random Forest, and Support Vector Machine methods for churn prediction and regression models for optimal personalized service fees. The Random Forest method excels, achieving a predictive accuracy of 94% and an AUC of 98%, preventing churn, increasing profitability by 36%, and reducing data collection by 20%. The model's robustness and generalizability apply to various sectors (Ortakci et al., 2024). The study also develops a scale to measure customer experience (CX) in AI-enabled products based on Churchill's (1979) framework. It identifies five dimensions: data capture, classification, delegation, social, and anthropomorphic experiences, providing marketers with a framework to enhance CX in evolving consumer environments

Wang et al. (2024). Additionally, the paper proposes a smart retail analytics system using machine learning to improve retail efficiency and customer engagement. This system integrates predictive models and a hybrid architecture for customer tracking, enhancing store operations, and understanding consumer behavior (Hossam et al., 2024). Moreover, the research predicts the next market basket for loyalty program users through AI, improving purchasing strategies and customer satisfaction by analyzing transaction data (Meftah et al., 2024)

Lastly, the study explores AI's impact on CX for Maxim transportation in Surabaya, showing that AI enhances personalization and user experience, opening new business opportunities and highlighting the importance of addressing ethical and data protection issues (Terenggana et al., 2024). From the above literature review, it is clear that many algorithms and models are employed to predict customer patterns using various sources of information. After this study, it is understood that we need to predict the customer's thoughts with artificial intelligence, which helps to improve the product and its sales.

PROPOSED WORK

The main idea of this proposed work is to understand customer sentiment and predict future feedback patterns using artificial intelligence (AI) and natural language processing (NLP). The process is given below:

- **Data Collection:** Companies gather customer opinions from various sources, such as social media posts, surveys, and support chats.
- **Data Preprocessing:** The raw data might be messy and unusable for analysis. This step involves cleaning the text data, removing irrelevant information, and formatting it for the NLP models.
    NLP Models: Here, powerful AI models trained on vast amounts of text data come into play. These models analyze the preprocessed text and identify sentiment (positive, negative, neutral) and even extract specific topics or themes from the feedback.
- **Predicting Feedback Patterns:** The model predicts future feedback patterns by analyzing historical customer feedback and identifying trends. This allows businesses to anticipate customer concerns and proactively address them.

Figure 5. AI-NL-CV Bot

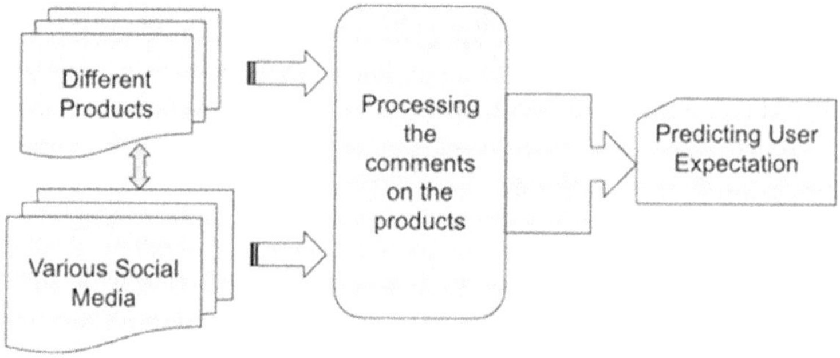

Figure 5 shows the outline of the proposed AI-NLP-CV bot, which is based on artificial intelligence with natural language processing for consumer vision. Regarding customer comments and feedback for the products, data is collected from various sources, such as social media and communication media. The collected data is preprocessed with various natural language processing methods.

Figure 6. Dataset Collection

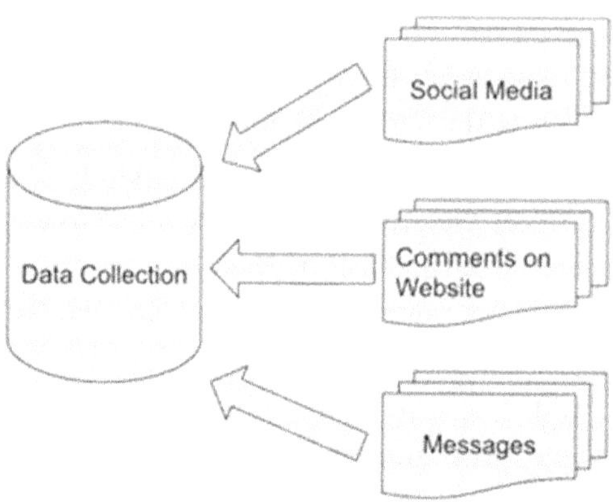

Dataset Description

Figure 6 represents the data collection from various sources, and the description is as follows. Social media and communication channels have become a goldmine for understanding customer sentiment. By analyzing comments, reviews, and mentions across these platforms, companies gain valuable insights into what customers love, hate, and everything about their products. The comments are in the form of positive feedback, negative feedback, and neural feedback. Foremost feedback is given as Customers often praise features that solve their problems or make their lives easier. For example, a fitness tracker user might tweet, "Loving the new sleep tracking feature on my [Brand X] watch! Helps me understand my sleep patterns." Positive comments about a product's lasting quality or strong performance are valuable. A customer review on an e-commerce site might say, "[Brand Y] vacuum cleaner is a beast! Still going strong after 5 years.", positive interactions with customer support build brand loyalty.

A Facebook post might state, "Huge shoutout to [Brand Z]'s customer service! They were super helpful and resolved my issue quickly." Furthermore, with negative feedback, customers can quickly voice their frustration with bugs or glitches in a product. A tweet directed at a software company might say, "Ugh, so annoyed with this latest update on [Software X]. It keeps crashing! "customers may express disappointment about missing features they believe would enhance the product. An Instagram comment on a phone review might say, "[Brand Y] phone is great, but wish it had a built-in wide-angle lens.", negative experiences with customer service damage a brand's reputation. A customer service email complaint might read, "Been waiting for a response from [Brand Z] for days! Their customer service is terrible." Eventually, some customer gives neural comments, for example, "it's okay," or "somewhat it is okay."

Apart from these comments social media also offers insights into customer emotions and brand perception. Customers who become brand advocates organically promote the product through positive mentions and recommendations. A glowing product recommendation on a forum could look like, "If you're looking for a reliable laptop, check out [Brand X]. They're the best!". Humorous comments and memes reveal unexpected customer perspectives on a product. A funny meme about a product's quirk might spark discussions and highlight areas for improvement. Social media allows you to track conversations around your product and identify emerging trends and competitor analysis. Focus resources on fixing bugs, developing missing features, and improving areas with consistent customer complaints. Identify areas for improvement in customer service interactions and tailor training programs accordingly. To build trust and credibility, leverage positive customer feedback and brand advocacy in marketing campaigns. Social listening is a continuous process. By actively monitoring conversations and incorporating customer feedback, companies create products that resonate with their audience and build stronger customer relationships.

Algorithm 1: Data Preprocessing

Input: Data Collected in different forms
Output: Information

1. Data cleaning- text data, emoji data, code word data
2. Normalization - transforming all information into feedback data
3. Tokenization - break the data into basic words
4. Lemmatization / Stemming - context-based words, stemming the words

Figure 7. Comments Processing Steps

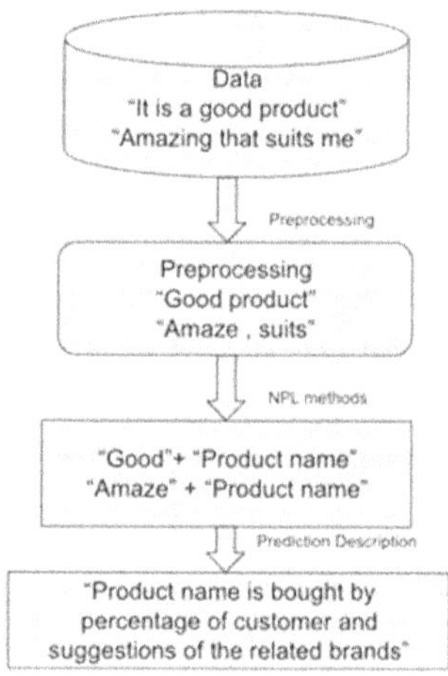

The world of data is a messy place, especially when it comes to text, emojis, and codewords. These data types offer rich insights, but their unstructured nature makes them challenging to analyze. Data cleaning is the essential first step in transforming your raw information into usable feedback data, as demonstrated in Figure 7. Here, the data is a room filled with books in various languages, some upside down and others with missing pages. Normalization is the process of creating an order of words from messy junks of words i.e., organizing books in the library. It aims to bring consistency to data, making it easier to analyze and compare. The collected data is in different formats like text data, emoji data, codeword data, etc. Text data are in different forms, like case sensitive, that are converted into lowercase to remove the case-sensitivity barrier, allowing for easier comparisons. Removing unnecessary punctuation helps us focus on the core meaning of the text. Multiple spaces between words create unnecessary clutter. Normalization ensures consistent spacing, making the data more streamlined. Typos and grammatical errors throw a wrench into the analysis. Tools like spell checkers or pre-defined dictionaries help to iron out these inconsistencies. Next, emoji data is converted into text data. Emojis, those little digital expressions, are tricky and are converted into descriptive words that capture their sentiment.

For example, a smiley face becomes "happy" while a thumbs-down becomes "disappointed." This allows the analysis to understand the emotional context conveyed by the emoji. Removing unwanted data for the analysis. Codewords are like secret codes used within a group. Normalization involves creating a reference list that translates these codewords into their corresponding meanings. For all this data, tokenization and lemmatization are employed. In tokenization, text data is segmented into individual words, emoji becomes a separate token, and codeword becomes a single token. Here, numerical data is rated into text data. After this data transformation, raw, messy data becomes a structured format, inconsistent text becomes standardized and comparable, Unclear emojis are translated or removed, and codewords reveal their true meaning. With information, the analysis is done, and the complete summary feedback of the product is given. In our proposed idea, apart from predicting the feedback, it also gives suggestions by comparing it with other products, considering the customer comparison feedback comments.

Figure 8. Process of Prediction

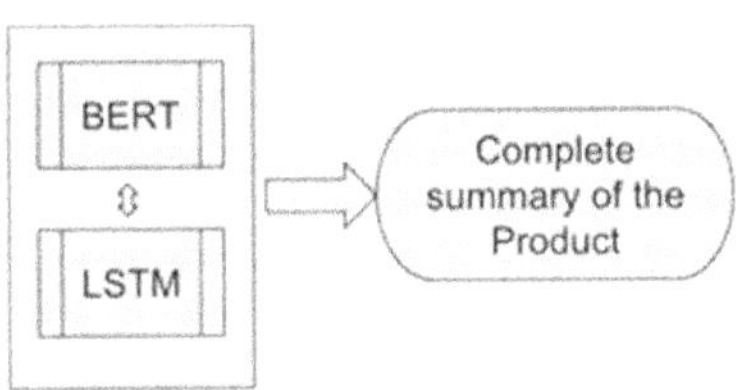

Figure 8 shows the final step process of the AI bot, which gives a summary prediction of the product. This combines Figures 2 and 3, shown in Section 1. The following section discusses the proposed work.

DISCUSSIONS

In today's digital age, understanding customer sentiment is no longer a luxury; it's a necessity. Fueled by social media, online reviews, and ever-expanding communication channels, customer feedback is a constant stream of valuable data. The powerful duo of artificial intelligence (AI) and natural language processing (NLP) allows companies to understand current customer sentiment and predict future feedback patterns, empowering them to stay ahead of the curve and deliver exceptional customer experiences.

The proposed model helps generate a feedback summary of a particular product. Highlights of the proposed model are

1. Preprocessing the collected data for necessary feature selection
2. Predicting the customer suggestions using the feedbacks
3. In addition, giving more insights about the product irrespective of the customer from the customer history data.

Figure 9. Data from Various Sources

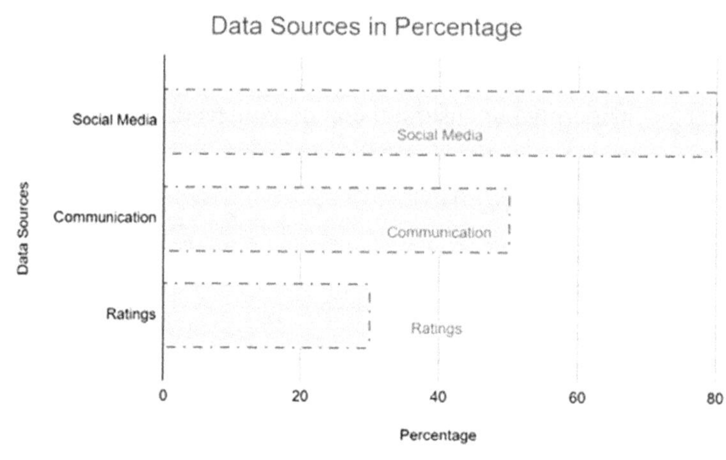

Data from various sources in percentage is depicted in Figure 9. It is understood that most of the data taken from comments is shared through social media. But still the communication messages and ratings feedback data are also considered.

Figure 10. Different Data Formats

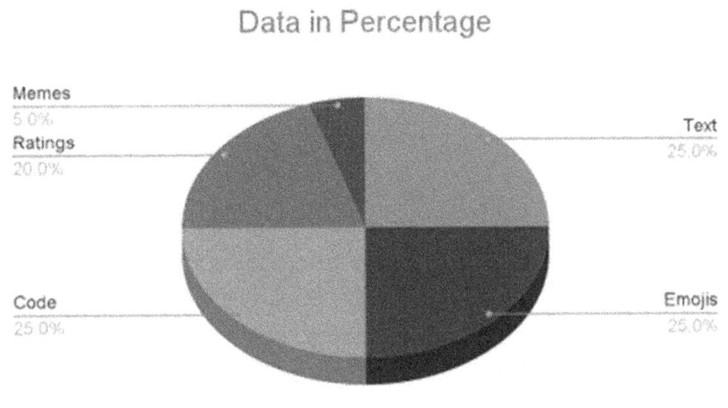

Figure 10 above shows the percentage of different types of data taken from various sources for this proposed work.

Figure 11. Comparison of ML, DL and AI-NLP-CV Bot

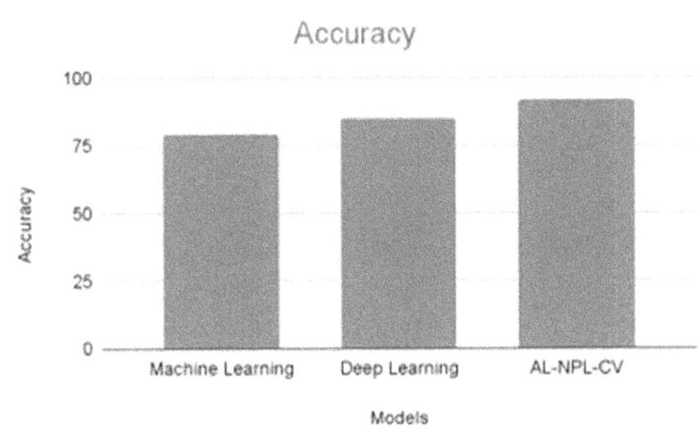

Figure 11 shows the comparison of average accuracy of the machine learning algorithms, deep learning models and the proposed AI-NLP-CV bot. It is clear that the proposed bot gives more accurate predictions and summary feedback on the product.

The benefits of the results obtained by this model are used in capturing the complex customer behavior and preferences that improve the product recommendation. Sales are increased by offering relevant products that lead to customer satisfaction. The results predicted are also used for personalized marketing campaigns of the products.

CONCLUSION

In conclusion, NLP technology integrated within communication platforms empowers businesses to unlock a treasure trove of actionable customer insights. By harnessing the predictive power of NLP bots, organizations elevate customer engagement, refine product offerings, and propel business growth in an increasingly competitive marketplace. This chapter has delved into the development of such an AI-NLP Bot, specifically designed to analyze customer behavior patterns gleaned from diverse communication platforms. The bot serves as a powerful tool to predict and understand customer sentiment based on their product opinions shared across various channels. The results demonstrate the bot's effectiveness in accurately gauging consumer sentiment, identifying emerging trends, and anticipating market preferences.

REFERENCES

Ajiga, D. I., Ndubuisi, N. L., Asuzu, O. F., Owolabi, O. R., Tubokirifuruar, T. S., & Adeleye, R. A.. (2024). AI-driven predictive analytics in retail: A review of emerging trends and customer engagement strategies. *International Journal of Management & Entrepreneurship Research*, 6(2), 307–321. DOI:10.51594/ijmer.v6i2.772

Babatunde, S. O., Odejide, O. A., Edunjobi, T. E., & Ogundipe, D. O. (2024). The role of AI in marketing personalization: A theoretical exploration of consumer engagement strategies. *International Journal of Management & Entrepreneurship Research*, 6(3), 936–949. DOI:10.51594/ijmer.v6i3.964

Chiu, M. C., Tsai, C. Z., & Huang, Y. C. (2024). Integrating object detection and natural language processing models to build a personalized attraction recommendation agent in a smart product service system. *Advanced Engineering Informatics*, 61, 102484. DOI:10.1016/j.aei.2024.102484

Ekechi, C. C., Chukwurah, E. G., Oyeniyi, L. D., & Okeke, C. D. (2024). AI-infused chatbots for customer support: A cross-country evaluation of user satisfaction in the USA and the UK. *International Journal of Management & Entrepreneurship Research*, 6(4), 1259–1272. DOI:10.51594/ijmer.v6i4.1057

Haldorai, A., Murugan, S., & Balakrishnan, M. (2024). Harnessing Intelligent AI to Elevate Business Modeling: A Perspective. In *Artificial Intelligence for Sustainable Development* (pp. 429–440). Springer Nature Switzerland. DOI:10.1007/978-3-031-53972-5_22

Hossam, A., Ramadan, A., Magdy, M., Abdelwahab, R., Ashraf, S., & Mohamed, Z. (2024). Revolutionizing Retail Analytics: Advancing Inventory and Customer Insight with AI. arXiv preprint arXiv:2405.00023. DOI:10.1109/ICMISI61517.2024.10580424

Kumar, V., Ashraf, A. R., & Nadeem, W. (2024). AI-powered marketing: What, where, and how? *International Journal of Information Management*, 77, 102783. DOI:10.1016/j.ijinfomgt.2024.102783

Marti, C. L., Liu, H., Kour, G., Bilgihan, A., & Xu, Y. (2024). Leveraging artificial intelligence in firm-generated online customer communities: A framework and future research agenda. *Journal of Service Management*, 35(3), 438–458. DOI:10.1108/JOSM-10-2023-0443

Meftah, M., Ounacer, S., & Azzouazi, M. (2024). Enhancing Customer Engagement in Loyalty Programs Through AI-Powered Market Basket Prediction Using Machine Learning Algorithms. In *Engineering Applications of Artificial Intelligence* (pp. 319–338). Springer Nature Switzerland. DOI:10.1007/978-3-031-50300-9_18

Ogundipe, D. O., Babatunde, S. O., & Abaku, E. A. (2024). AI and product management: A theoretical overview from idea to market. *International Journal of Management & Entrepreneurship Research*, 6(3), 950–969. DOI:10.51594/ijmer.v6i3.965

Ortakci, Y., & Seker, H. (2024). Optimising customer retention: An AI-driven personalised pricing approach. *Computers & Industrial Engineering*, 188, 109920. DOI:10.1016/j.cie.2024.109920

Ozen, C., & Tas, Z. M. (2024). The Role of Artificial Intelligence (AI) in Improving Product Development Efficiency. In Digital Technologies in Modeling and Management: Insights in Education and Industry (pp. 313-335). IGI Global.

Pasupulati, R., Mahendran, J., & Majumdar, A. (2024). Predictive Analysis of Digital Consumer Behaviour: Key to Understanding Success and Failure of a Brand. In Enhancing and Predicting Digital Consumer Behavior with AI (pp. 238-267). IGI Global.

Raji, M. A., Olodo, H. B., Oke, T. T., Addy, W. A., Ofodile, O. C., & Oyewole, A. T. (2024). E-commerce and consumer behavior: A review of AI-powered personalization and market trends. GSC Advanced Research and Reviews, 18(3), 66-77.

Rosário, A. T. (2024). Artificial Intelligence in the Consumer Behavior Process in Business. In Enhancing and Predicting Digital Consumer Behavior with AI (pp. 39-73). IGI Global. DOI:10.4018/979-8-3693-4453-8.ch004

Shaik Vadla, M. K., Suresh, M. A., & Viswanathan, V. K. (2024). Enhancing Product Design through AI-Driven Sentiment Analysis of Amazon Reviews Using BERT. *Algorithms*, 17(2), 59. DOI:10.3390/a17020059

Shankar, M. S. S. N., Kishore, G. D. K., Anudeep, N. L. N. C., Bhavya, R., & Sankar, R. S. E-commerce Chatbot For Price Negotiation Using NLP.

Terenggana, C. A. (2024). The Influence of Artificial Intelligence on Customer Experience (Study of Maxim Users in Surabaya, East Java). *Economics Studies and Banking Journal*, 1(1), 37–45. DOI:10.62207/jhctec97

Wang, P., Li, K., Du, Q., & Wang, J. (2024). Customer experience in AI-enabled products: Scale development and validation. *Journal of Retailing and Consumer Services*, 76, 103578. DOI:10.1016/j.jretconser.2023.103578

Zlatanova-Pazheva, E. (2024). Model of consumer behavior when applying ai at every point of consumer contact. *International Journal of Engineering Technologies and Management Research*, 11(3), 35–46. DOI:10.29121/ijetmr.v11.i3.2024.1433

Chapter 7
Investigating the Relationship Between Twitter Sentiment and Bitcoin Price Movements

Sapternab Chatterjee
Christ University, India

Rupali Sunil Wagh
Christ University, India

ABSTRACT

Bitcoin, launched in 2009 as the pioneer cryptocurrency, has caught the eye of investors worldwide due to its nature and remarkable growth. Recent studies indicate that elements, like announcements, news sentiments, government regulations, and overall market sentiments significantly influence the trajectory of bitcoin price. This work presents a comprehensive analysis to provide deeper insights into the relationship between user sentiments expressed as tweets and bitcoin prices (daily price changes). As a significant contribution, the authors propose multiple approaches to performing the non-trivial task of integrating tweet sentiment with bitcoin price data. The chapter also presents the incorporation of inherent lag in the expression of sentiments and their impact on price change by demonstrating lagged sentiment analysis. The work employs state-of-the-art machine learning and deep learning models for analyzing the data both as classification and regression tasks to uncover hidden patterns.

DOI: 10.4018/979-8-3693-9246-1.ch007

INTRODUCTION

Due to their decentralized nature and ability to replace traditional fiat currencies, cryptocurrencies—such as the groundbreaking Bitcoin—have entirely changed the global financial landscape. Introduced in 2009 by the enigmatic Satoshi Nakamoto, Bitcoin revolutionized digital finance by introducing blockchain technology and spurring the development of other rival cryptocurrencies. Due to its fixed amount, decentralized structure, and cryptographic security, Bitcoin has become a disruptive force in finance, upending conventional ideas about money and investing.

The cryptocurrency market is infamous for its extreme volatility, with values subject to swings based on various factors, including regulation changes, technological advancements, macroeconomic patterns, and investor sentiment. Particularly concerning Bitcoin, there have been significant price fluctuations ranging from sharp increases to sharp decreases, offering investors both a profitable opportunity and a dangerous risk. Because of the market's inherent volatility, there is intense interest in comprehending the fundamental reasons behind price fluctuations and creating prediction models to assist investors in managing its complexity.

The attempt to anticipate bitcoin values has sparked a flurry of research, with experts investigating various approaches and data sources to get insights into market dynamics. This research journey begins with essential publications such as those by Lamon et al., (2017), who pioneered the development of prediction models based on news and social media sentiment. Their findings proved the possibility of using supervised learning algorithms for text-based sentiment classification, setting the framework for future research. With social media becoming ubiquitous in the last decade, this research has paved the way for a multitude of studies presenting approaches to integrate public sentiments with traditional analysis methods for a comprehensive understanding of the bitcoin market.

By examining the price movement prediction of more cryptocurrencies, such as Bitcoin, Ethereum, Ripple, and Litecoin, Valencia et al., (2019) broadened the scope. Their study demonstrated the potential of sentiment analysis and machine learning in predicting market trends, with neural networks emerging as a viable technique for predictive modeling.

Building on this foundation, Raju and Tarif (2020) looked into the use of machine learning algorithms and sentiment analysis of social media data to predict the price of Bitcoin in real time. Their findings emphasized the need to incorporate sentiment research into predictive models, particularly given Bitcoin's high volatility and the rapid transmission of information via social media platforms. Research like that of Kurniasari and Setyanto (2020), who examined sentiment analysis utilizing deep learning and neural networks, has allowed the investigation of sentiment analysis approaches to continue to advance. Their study, which concentrated on the Indonesian language, demonstrated the ability of word2vec algorithms and recurrent neural networks (RNNs) to obtain high sentiment classification accuracy rates.

Tan et al., (2022) proposed a hybrid model that combines Transformer and Recurrent Neural Network (RNN) architectures to effectively capture the nuances of sentiment analysis. Their solution performed better than conventional techniques in addressing lexical variety and imbalanced datasets. In natural language processing, Mathew and Bindu (2020) explored applying pre-trained models in sentiment analysis, highlighting the effectiveness of models such as BERT and BiGRU in extracting polarity from textual data. Their comparative investigation demonstrated how pre-trained models can improve sentiment analysis's precision and effectiveness across a range of applications.

Stenqvist and Lönnö (2017) examined the potential of predicting Bitcoin price fluctuations using Twitter sentiment analysis. They found that sentiment changes in millions of tweets about Bitcoin could be a sign of future price swings. This research provides important new information about the function of social media sentiment in cryptocurrency markets. Critien et al., (2022) used data volume analysis and sentiment analysis on Twitter to study Bitcoin price fluctuations and trends. With neural network models based on recurrent and convolutional architectures, their research went beyond predicting the direction of price changes to forecasting the size of these changes.

Using Twitter sentiment analysis, Haritha and Sahana (2023) presented an end-to-end approach for predicting bitcoin prices. Using cutting-edge machine learning approaches like Bidirectional Encoder Representations from Transformers (BERT) and Gated Recurrent Units (GRU) for sentiment prediction and price forecasting, their study combined historical price data with sentiment analysis of tweets.

This study intends to provide new insights into the ever-changing field of cryptocurrency price prediction research by utilizing cutting-edge machine learning techniques and incorporating data from the literature. Examining the effect of sentiment analysis with time lags on patterns in Bitcoin prices is one of the main goals. By analyzing sentiment data at varying temporal offsets relative to Bitcoin price movements, the research seeks to uncover potential patterns and causal relationships that may take time. Additionally, using sentiment analysis as a basis, this study assesses how well different machine learning algorithms anticipate changes in bitcoin prices. By comparing the effectiveness of regression and classification algorithms, including cutting-edge deep learning architectures like recurrent neural networks (RNNs) and long short-term memory (LSTM) networks, as well as traditional methods like logistic regression and linear regression, the study seeks to identify the most suitable models for cryptocurrency price prediction.

This research intends to further our understanding of the intricate interaction between sentiment analysis and bitcoin price movements and improve prediction algorithms. The research attempts to elucidate the intricate relationship between price fluctuations, social media dynamics, and market sentiment by combining knowledge from other domains, such as finance, machine learning, and natural language processing. The goal of the research is to uncover the basic principles underlying cryptocurrency markets and offer investors useful information for navigating this unpredictable environment through in-depth empirical analysis and theoretical investigation. This research attempts to further the conversation about cryptocurrency price prediction by creating more reliable prediction models based on sentiment analysis. The research intends to provide investors with actionable information for navigating the unpredictable cryptocurrency markets by clarifying the temporal dynamics of sentiment-price interactions and assessing the effectiveness of different machine learning methods.

DATA DESCRIPTION

Our work investigates the relationship between sentiment analysis and Bitcoin prices by considering the multiple facets of the following two datasets –

1) Bitcoin price data
2) User Sentiments

Due to the temporal nature of the problem, both datasets are obtained from the period of February 5, 2021, to January 9, 2023.

Table 1. Bitcoin Historical Price Data

Date	Price	Open	High	Low	Vol.	Change %
03-02-2021	37,646.80	35,486.90	37,646.80	35,401.50	131.03K	6.09%
04-02-2021	36,982.10	37,648.90	38,707.90	36,380.80	151.44K	-1.77%
05-02-2021	38,297.60	36,976.40	38,303.60	36,640.00	100.60K	3.56%
06-02-2021	39,256.60	38,300.60	40,939.70	38,275.30	156.02K	2.50%
07-02-2021	38,852.90	39,228.90	39,691.70	37,381.60	125.37K	-1.03%

Bitcoin Historical Price Data which has the following columns -

- Date: The date of the Bitcoin price.
- Price: The Bitcoin price in USD.

- Open: The opening price of Bitcoin on the day.
- High: The day's highest price for Bitcoin.
- Low: The day's lowest price for Bitcoin.
- Volume: The total amount of Bitcoin traded during the day.
- Change%: The proportion of the price of Bitcoin that has changed from the day before.

A subset of the data used for our study shown in Table 1, which spans the periods from February 5, 2021, to January 9, 2023, offers a historical record of the price of Bitcoin. It was sourced from the Investing.com website. Bitcoin Tweets Data is given in Table 2.

Table 2. Bitcoin Tweets Data

date	text
10-02-2021	$BTC A big chance in a billion! Price: \4872644.0 (2021/02/11 08:51) #Bitcoin #FX #BTC #crypto
10-02-2021	This network is secured by 9 508 nodes as of today. Soon, the biggest bears will recognise: #BTC in too big to failâ€¦ https://t.co/1XovDA8rKw

Note: Only the columns utilized for the study are included in the table

The Bitcoin Tweets Data additionally consists of the following columns

- user_name: The name of the user, as they've defined it.
- user_location: The user-defined location for this account's profile.
- user_description: The user-defined UTF-8 string describing their account.
- user_created: Time and date, when the account was created.
- user_followers: The current count of followers for an account.
- user_friends: The total count of friends that an account presently possesses.
- user_favourites: The quantity of favorites now associated with an account
- user_verified: denotes a verified account; date: the time and date the tweet was created in UTC; text: the tweet's actual UTF-8 text.
- hashtags: all other hashtags included in the tweet, along with #Bitcoin & #btc.
- source: the utility used to post the tweet; tweets from the Twitter website have a source value of web.
- is_retweet: Indicates if the authenticating user has retweeted this tweet.

This data provides a record of the sentiment of Bitcoin tweets which are analyzed for their Sentiments to get the polarity scores and sentiment labels. This data is obtained from Kaggle. It has the same date range as the Historical Price Data and contains 46,89,354 tweets.

METHODOLOGY

The impact of sentiment analysis on various business processes across domains has been an extensively researched field in the recent past. Sentiment analysis for share market prediction which is very similar to bitcoin prediction is a very popular field of study as emphasized in the review of literature. While the relationship between user sentiments and price fluctuations in the bitcoin markets has been explored in previous studies, studies focusing on the integration of historical prices of bitcoins with user sentiments to get a comprehensive understanding are relatively sparse. As depicted in Figure 1 our work employs the following 4 distinguished approaches.

Figure 1. Approaches for Analysis

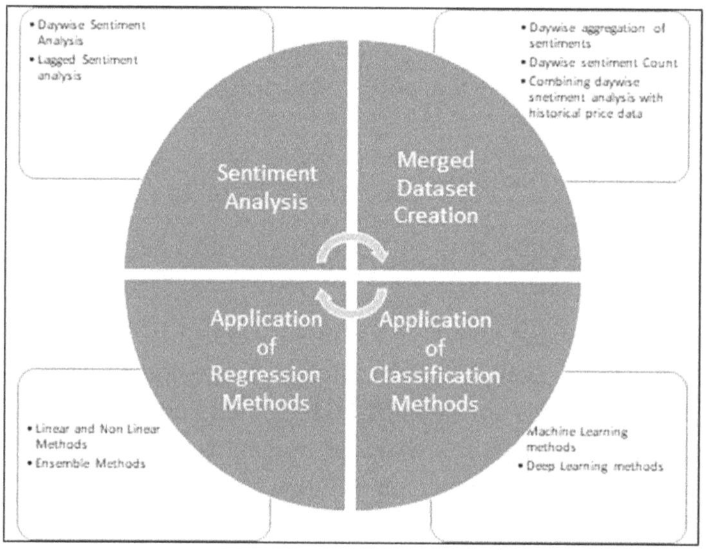

Sentiment Analysis

Sentiment analysis is important to our purpose for conducting research. Sentiment analysis is a fast-expanding area that uses natural language processing and machine learning techniques to extract and quantify sentiment, emotions, and views expressed in textual data, notably on social media platforms. Drus and Khalid (2019) conducted a thorough literature review on the many methods used for sentiment analysis in social media, including lexicon-based approaches like SentiWordNet and TF-IDF, as well as machine learning techniques like Naive Bayes and Support Vector Machine.

The authors emphasize Twitter as the most popular social media site for sentiment research due to its rich content and accessibility. Meanwhile, Wankhade et al., (2022) present a complete overview of sentiment analysis methods, categorizing them as machine learning, lexicon-based, and hybrid approaches.

The paper by Wankhade et al., (2022) discusses the applications of sentiment analysis across diverse domains, such as marketing, finance, and healthcare, while also addressing the challenges involved, including domain dependence, sarcasm detection, and multilingual sentiment analysis. Together, these two studies provide a comprehensive overview of the present level of sentiment analysis research and its practical applications in social media data. The rise of Web 2.0 has substantially altered the social media environment, with people actively exchanging information, ideas, and personal experiences across multiple platforms. The number of social media users worldwide is expected to exceed 2.77 billion by 2019, resulting in massive amounts of raw data in the form of text, videos, images, and audio. This abundance of user-generated content poses both obstacles and opportunities for corporations, researchers, and politicians seeking meaningful insights through sentiment analysis. By studying the sentiment encoded in social media posts, researchers and practitioners can acquire a better knowledge of public opinions, customer preferences, and the influence of events or campaigns.

While combining sentiment analysis with Bitcoin price data on a specific day can help in understanding the immediate impact of user sentiments, there might also be a lag between the expression of sentiment and its subsequent impact on price movements. To investigate it further, we experimented with lagged sentiment analysis. This involved collocating sentiment data with lagged price changes over intervals of 1, 3, and 7 days. For each lag interval, we meticulously conducted regression and classification tasks, allowing us to understand the relationship between sentiment and subsequent price changes. By accounting for these lagged effects, we aimed to uncover deeper insights into the temporal dynamics governing sentiment-driven price movements in the Bitcoin market.

Merged Dataset Creation

Integrating heterogeneous datasets, particularly with the significant disparity in volume between daily historical price data and tweets, posed a formidable challenge. When analyzing the association between user sentiments and bitcoin prices, it is critical to apply the same time factor to both datasets, which necessitates translating many tweets to a single sentiment value. To tackle this challenge comprehensively, we employed the following two distinct approaches to dataset merging, each offering unique insights and advantages.

1. Day Wise Tweet Aggregation- Initially, we opted for the aggregation approach, wherein all tweets corresponding to a specific day were concatenated into a single entity. Subsequently, sentiment analysis was conducted on this aggregated dataset. This method allowed us to capture the collective sentiment of Twitter users towards Bitcoin daily, providing a broad overview of market sentiment trends. An extract of Merged Dataset of Aggregate Sentiment for tweets in each day is given in Table 3.

Table 3. Extract of Merged Dataset of Aggregate Sentiment for Tweets in Each Day

textblob_ sentiment_ label	Date	Price	Open	High	Low	Vol.	Change %	Shift
Negative	05-02-2021	38297.6	36976.4	38303.6	36640	100600	3.56	1
Positive	06-02-2021	39256.6	38300.6	40939.7	38275.3	156020	2.5	1
Positive	07-02-2021	38852.9	39228.9	39691.7	37381.6	125370	-1.03	0
Positive	08-02-2021	46395.7	38853.6	46596.4	38057	248470	19.41	1

2. Daywise Sentiment Count- However, recognizing the potential loss of granularity inherent in the aggregation approach, we also explored a granular approach to dataset merging. Here, instead of aggregating all tweets for a day, we created separate columns to house the count of positive(npos), negative(nneg), and neutral(nneut) tweets for each day as shown in Table 4. This granular approach preserved the nuanced variations in sentiment over time, enabling us to capture subtle shifts and fluctuations in market sentiment with greater precision.

Table 4. Extract of Merged Dataset of Count Sentiment Label for Each Day

nneg	nneut	npos	Date	Price	Open	High	Low	Vol.	Change %	Shift
169	895	630	05-02-2021	38297.6	36976.4	38303.6	36640	100600	3.56	1
262	1735	1281	06-02-2021	39256.6	38300.6	40939.7	38275.3	156020	2.5	1
237	1400	1393	07-02-2021	38852.9	39228.9	39691.7	37381.6	125370	-1.03	0
504	2885	2258	08-02-2021	46395.7	38853.6	46596.4	38057	248470	19.41	1

Application of Classification Machine Learning Methods

The classification task was formulated to predict the directional shifts in Bitcoin price (positive or negative). We used various classification machine learning techniques, including Logistic Regression, Naïve Bayes, Support Vector Classifier, Decision Tree Classifier, and Random Forest Classifier, to identify patterns that could predict future price movements. By combining sentiment analysis characteristics with historical price data, we hoped to create robust classification models capable of accurately forecasting the direction of price movements based on market sentiment.

Application of Regression Machine Learning Methods

Complementary to the classification task, the regression task is designed to estimate the precise percentage change in Bitcoin's price. We attempted to capture the complex interplay between sentiment and price dynamics using a variety of regression machine learning algorithms such as Linear Regression, Support Vector Regressor, Decision Tree Regressor, and Random Forest Regressor. By incorporating sentiment analysis features alongside historical price data, we sought to develop robust regression models capable of accurately predicting price changes based on prevailing market sentiment.

Application of Deep Learning Methods

In addition to typical machine learning algorithms, we investigated the use of deep learning approaches to identify complicated temporal correlations in our data. Specifically, we used cutting-edge deep learning architectures such as LSTM (Long Short-Term Memory) and RNN. By processing sequential data, where each time step represents a distinct point in time, LSTM and RNN architectures allowed us to capture subtle patterns and trends that could inform future price movements.

EXPERIMENTAL SETUP

Figure 2. Flowchart of the Research Study

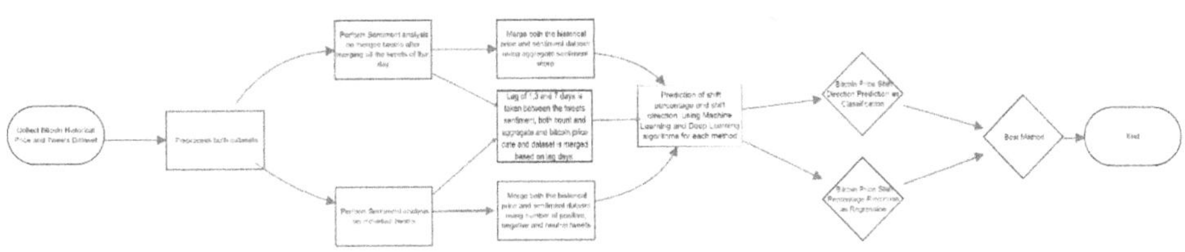

Utilizing the datasets as created above we experimented with state-of-the-art algorithms for performing the regression and classification tasks. Figure 2 depicts the steps followed during the analysis.

DATA PREPROCESSING

The data preprocessing phase is crucial for ensuring the cleanliness and integrity of our datasets. The preprocessing of the Bitcoin Price Dataset was necessary because even the numerical values were in object format and wrongly represented which the algorithms wouldn't understand. For our work, the following pre-processing tasks were carried out

- Handling missing values by removing them as they were consisting of approximately 0.65% of the total Bitcoin Tweets dataset.
- Converting data types to their appropriate formats.
- Scaling the price data as everything was given in K or M we had to remove that and multiply it by 1000 and 100000 respectively.
- Removing the commas between the prices as they were represented as 41,567 and they had to be taken as 41567 so that the algorithm would work.
- Removing the percentage from all the numbers in the Change% column.

The following subsections describe various algorithms used for the analysis

Linear Regression

The paper by Rathan et al., (2019) explored the use of linear regression for Bitcoin price prediction. The authors stated that linear regression is a simple and widely used regression algorithm that models the relationship between input features and the output variable (in this case, Bitcoin price) as a linear function in the form y = mx + b, where y is the output, x is the input feature, m is the slope, and b is the intercept. This linear model is simple to read and execute, and it has a closed-form solution, making it easy to compute the coefficients that best suit the data. While linear regression presupposes a linear relationship between input and output properties, the authors demonstrated its effectiveness in predicting Bitcoin prices, which can be extremely unpredictable and complex.

Logistic Regression

Logistic regression is a popular classification approach that is especially well-suited for binary classification problems like forecasting whether a price change will be up or down. Despite its name, logistic regression is utilized for classification rather than regression. It uses a logistic function to model the probability that a given input belongs to a specific class, mapping input attributes to values ranging from 0 to 1. The algorithm may forecast the class depending on whether the probability is more than or less than a certain threshold, which is commonly set at 0.5.

Lamon et al., (2017) recommend utilizing logistic regression for Bitcoin price prediction since it is simple and easy to grasp, allowing for quick error analysis and determination of the next steps. Logistic regression is a classification algorithm that uses the sigmoid function to output the probability of an example being '1'. It is chosen because it can handle binary classification problems effectively and provides the best results on the development set for Bitcoin price predictions.

Naive Bayes

Naive Bayes is a probabilistic classification technique that relies on Bayes' theorem and the "naive" assumption of feature independence. Despite its simplicity, Naive Bayes is well-known for its performance in text classification and other applications. It computes the likelihood of each class based on the input features and uses the class with the highest probability as the prediction. Naive Bayes can be beneficial in predicting the direction of price changes because of its speed and efficiency, especially when dealing with high-dimensional data. It works particularly effectively with categorical features and can tolerate missing data. Naive Bayes is also resistant to irrelevant characteristics, making it an excellent candidate for tasks in which feature selection is difficult. However, the assumption of feature independence may not always be valid in practice, affecting prediction accuracy.

Support Vector Machine

Support vector regression is a sort of support vector machine that may be applied to regression tasks. It operates by transferring the input features to a higher-dimensional space and determining which hyperplane best separates the data points. SVR is effective for regression applications because it can describe non-linear correlations between input characteristics and output variables using a kernel function to translate the data to a higher-dimensional space. The Support Vector Classifier (SVC) is a sophisticated classification technique that works by determining which hyperplane best splits data points into different classes. SVC seeks to increase the margin between the hyperplane and the nearest data points, making it resistant to outliers and noise in the data. It can do both linear and nonlinear classification tasks by mapping the data to a higher-dimensional space using various kernel functions.

SVC can be beneficial in forecasting the direction of price changes because of its capacity to deal with complex decision boundaries and non-linear correlations between input features and target variables. It is effective in high-dimensional domains and performs well with small to medium-sized datasets. SVC is also noted for its versatility, as it can be tailored with multiple kernel functions to handle different sorts of data. However, SVC can be computationally demanding, particularly with large datasets, and tweaking the hyperparameters can be difficult.

According to Auti et al. (2022), Support Vector Machine (SVM) is a better alternative for Bitcoin price prediction because of its success in high-dimensional spaces, which are common in financial datasets such as Bitcoin. SVM is adept at determining the best hyperplane to separate classes, making it appropriate for classification issues. Furthermore, it performs well when there are more dimensions than samples, which is common with Bitcoin data. The paper indicates that SVM surpasses other algorithms in terms of accuracy, including Linear Regression, LSTM, and Random Forest, making it an important tool for investors attempting to forecast market values and improve their investment methods. SVM's capacity to handle complicated, non-linear correlations in data makes it an effective model for forecasting the turbulent and unpredictable Bitcoin market.

Decision Tree

Decision tree regression is a sort of decision tree that may be applied to regression tasks. It functions by recursively separating the data into smaller and smaller groups, based on the characteristic that offers the best split. The tree stops growing when each group is homogeneous, meaning that the average value of the output variable is the same for all data points in the group. Decision tree regression is beneficial for regression applications because it can detect nonlinear correlations between input features and output variables. It is also interpretable, which means you can easily understand how the algorithm makes predictions. However, it is susceptible to overfitting, particularly if the tree is too deep or there are too many decision trees in the random forest.

In the context of predicting price change direction, a decision tree classifier can be useful for its simplicity and interpretability. It can process both numerical and categorical data and is resistant to outliers and missing values. Decision trees can capture nonlinear relationships between input features and target variables, making them appropriate for jobs with complex decision limits. However, decision trees are prone to overfitting, particularly deep trees, which can result in poor generalization of new data. Rathan et al., (2019) address the use of multiple machine-learning techniques to predict Bitcoin prices, with a Decision Tree model being one of the possibilities. Decision trees are preferred in this application because they can handle complicated datasets by dividing them down into smaller groups while simultaneously generating an associated decision tree. The tree's nodes reflect the dataset's features, while the branches indicate the features' outcomes, resulting in a clear and interpretable model structure. This strategy is especially effective in the volatile Bitcoin market, which exhibits non-linear price movements. The Decision Tree model can capture non-linear correlations between multiple variables and the Bitcoin price, making it an appropriate choice for investors and analysts seeking a more sophisticated understanding of cryptocurrency price fluctuations. Furthermore, the interpretability of Decision Trees enables simple comprehension and explanation of prediction outcomes, which is critical for making sound investment decisions.

Random Forest

Random forest regression is an ensemble of decision trees that are trained on diverse subsets of data and averaged to generate a final prediction. It is beneficial for regression tasks because it can capture non-linear correlations between input characteristics and output variables, and it is less likely to overfit than a single decision tree. Random forest regression can be used for regression tasks including non-linear relationships between input characteristics and output variables, as well as data with a high level of noise. It is also beneficial for feature selection because it can determine which features are most essential in predicting the output variable.

A random forest classifier is an ensemble learning system that combines numerous decision trees trained on distinct subsets of data. The forecasts of individual trees are merged to get the final prediction. The random forest classifier overcomes the limitations of a single decision tree by avoiding overfitting while enhancing the model's accuracy and robustness. A random forest classifier, due to its capacity to handle complex relationships and prevent overfitting, can be beneficial in forecasting the direction of price changes. It is resistant to noise and outliers in the data and can handle high-dimensional data efficiently. Random forest classifiers are also less sensitive to hyperparameters than single decision trees, making them easier to modify and optimize. However, a random forest classifier can be computationally expensive, especially with a large number of trees in the ensemble, and it may not be as interpretable as a single decision tree.

The paper by Chen (2023) presents a compelling case for using Random Forest in Bitcoin price prediction. Random Forest Regression stands out for its great prediction accuracy and explainability. Unlike black-box models such as neural networks, Random Forest provides unambiguous insights into the significance of each explanatory variable in the prediction process2. The paper reveals that Random Forest's prediction errors, RMSE and MAPE, outperform those of LSTM, another prominent deep learning model. Furthermore, Random Forest effectively captures the changes in variables influencing Bitcoin prices over different periods, making it a robust choice for predicting Bitcoin's highly volatile prices.

Recurrent Neural Network

Recurrent Neural Networks (RNNs) are a type of neural network that processes sequential data by retaining a recollection of previous inputs. In regression tasks such as forecasting price change percentages, RNNs can capture temporal dependencies in the data, making them ideal for time series analysis. By progressively processing past pricing data, RNNs can learn patterns and trends that influence price fluctuations over time. RNNs' capacity to handle sequential data makes them useful for capturing the dynamic character of financial markets and forecasting future price changes based on historical trends.

Pant et al. (2018) discuss the application of Recurrent Neural Networks (RNN) for Bitcoin price prediction, citing their capacity to analyze sequential input and grasp temporal connections. RNNs are well-suited for time series prediction because they can recall prior information and utilize it to impact present output, which is critical for dealing with the volatility and time-dependent character of Bitcoin pricing. By studying Twitter sentiment and past price data, RNNs may efficiently forecast future prices by learning from data trends.

Long Short-Term Memory

Long Short-Term Memory (LSTM) networks are a sort of RNN architecture that aims to solve the vanishing gradient problem and capture long-term dependencies in sequential data. In regression tasks such as forecasting price change percentages, LSTMs excel at detecting complicated patterns and correlations in time series data. Because of their ability to retain long-term dependencies, LSTMs can simulate complex links between past price movements and future changes, making them highly effective at forecasting price swings.

In classification tasks such as predicting price change direction, LSTMs can leverage their memory cells to retain information about past price movements and use this context to make predictions about future price directions. By learning from sequential data, LSTMs can detect minor patterns and trends that indicate whether the price will rise or fall. This makes LSTMs valuable for capturing the nuanced relationships between historical price data and predicting the direction of future price movements.

Hua's (2020) study examines the use of Long Short-Term Memory (LSTM) networks for Bitcoin price prediction and explains why LSTM is appropriate for the purpose. To begin, LSTMs are a more advanced version of Recurrent Neural Networks (RNNs) designed to address the vanishing gradient problem, allowing them to efficiently capture long-term dependencies in time-series datasets. This is crucial for financial data like Bitcoin prices, which are influenced by past events over varying time intervals. Secondly, LSTMs incorporate input, output, and forget gates within their cell units, enabling them to selectively remember and discard information through the training process. This selective memory is beneficial for managing the chaotic and non-linear nature of Bitcoin prices. Lastly, the paper's results indicate that LSTMs can outperform traditional time-series models like ARIMA in terms of efficiency and precision, especially when trained with appropriate data and for a sufficient number of epochs. Therefore, LSTMs offer a more robust and accurate approach to predicting the highly volatile and unpredictable Bitcoin market.

RESULTS

Following the previously described complete methodology, we carried out several regression and classification tasks using various machine learning algorithms and deep learning architectures. We meticulously explored the impact of sentiment analysis on Bitcoin price fluctuations over a variety of durations, from aggregate to lagged sentiment data. The subsequent analysis and model evaluations shed light on the effectiveness of each approach in forecasting cryptocurrency market trends. Let's delve into the results to get insights into the performance of each model.

Table 5. Evaluation of Regression Task

Method Name	Model Name							
	Linear Regression		Support Vector Regressor		Decision Tree		Random Forest	
	Train Acc.	Test Acc.	Train Acc.	Test Acc.	Train Acc.	Test Acc.	Train Acc.	Test Acc.
Aggregate Reg.	**0.961**	**0.693**	0.011	-0.018	1.000	-0.460	0.889	-0.047
Count Reg.	**0.961**	**0.691**	0.011	-0.018	1.000	-0.521	0.879	0.094
Lag 1 Aggregate Reg.	**0.963**	**0.686**	0.003	-0.021	1.000	-0.705	0.879	-0.051
Lag 1 Count Reg.	**0.962**	**0.684**	0.003	-0.021	1.000	-1.053	0.867	-0.052
Lag 3 Aggregate Reg.	**0.961**	**0.699**	-0.003	-0.018	1.000	-0.369	0.872	-0.297
Lag 3 Count Reg.	**0.960**	**0.700**	-0.003	-0.018	1.000	-3.314	0.860	-0.054
Lag 7 Aggregate Reg.	**0.959**	**0.721**	0.022	-0.004	1.000	-3.971	0.889	-0.027
Lag 7 Count Reg.	**0.959**	**0.713**	0.022	-0.004	1.000	-4.075	0.896	-0.138
Historical Data Reg.	**0.962**	**0.703**	0.011	-0.018	1.000	0.006	0.940	0.340

Table 5 shows that Linear Regression consistently achieves the highest test accuracy when compared to other regression models such as Support Vector Regressor, Decision Tree Regressor, and Random Forest Regressor across all of the methods we tried - Aggregate Regression, Count Regression, Lag 1/3/7 Aggregate/Count

Regression, and Historical Data Regression. The test accuracies for the other models are frequently negative, indicating extremely low performance. Linear Regression's greater performance can be attributed to its suitability for modeling linear connections between input features (sentiment data) and goal variables (Bitcoin price movement). Linear Regression makes simple, interpretable assumptions about the data, which allows it to capture the underlying patterns effectively.

Other more complicated regression models, such as Decision Trees and Random Forests, may overfit the training data, resulting in poor generalization to the test set. Support Vector Regression also struggles to handle the nonlinear and noisy nature of the Bitcoin price-sentiment relationship. Additionally, the search results suggest that incorporating lagged sentiment features can help improve the predictive power of the models, with the Lag 7 Aggregate Regression method achieving the highest individual test accuracy. This indicates that accounting for the delayed impact of sentiment on future prices is an important consideration. Overall, the search results show that a well-designed Linear Regression model, possibly with lagged sentiment features, is the most successful method for forecasting Bitcoin price fluctuations using sentiment analysis on tweets. Linear Regression is a better choice than the other regression approaches you tried because of its simplicity and resilience.

Table 6. Evaluation of Classification Task

Method Name	Model Name									
	Logistic Regression		Naïve Bayes		Support Vector Classification		Decision Tree		Random Forest	
	Train Acc.	Test Acc.	Train Acc.	Test Acc.	Train Acc.	Test Acc.	Train Acc.	Test Acc.	Train Acc.	Test Acc.
Aggregate Clas.	0.510	0.508	0.510	0.493	**0.658**	**0.537**	1.000	0.552	1.000	0.522
Count Clas.	0.510	0.508	0.510	0.493	**0.658**	**0.493**	1.000	0.522	1.000	0.508
Lag 1 Aggregate Clas.	0.548	0.522	0.548	0.522	**0.619**	**0.522**	1.000	0.478	1.000	0.567
Lag 1 Count Clas.	0.548	0.522	0.542	0.522	**0.568**	**0.537**	1.000	0.567	1.000	0.552
Lag 3 Aggregate Clas.	0.497	0.522	0.529	0.478	**0.587**	**0.448**	1.000	0.433	1.000	0.493
Lag 3 Count Clas.	0.497	0.522	0.529	0.478	**0.684**	**0.448**	1.000	0.522	1.000	0.478
Lag 7 Aggregate Clas.	0.477	0.478	0.523	0.522	**0.645**	**0.478**	1.000	0.508	1.000	0.493
Lag 7 Count Clas.	0.477	0.478	0.523	0.522	**0.626**	**0.537**	1.000	0.508	1.000	0.493
Historical Data Clas.	0.510	0.507	0.510	0.493	**1.000**	**1.000**	**1.000**	**1.000**	**1.000**	**1.000**

As we can see from Table 6, the Support Vector Classifier (SVC) emerges as the best-performing model across most of the methods. In the Aggregate Classification, Count Classification, Lag 1 Aggregate Classification, Lag 3 Count Classification, Lag 7 Aggregate Classification, and Lag 7 Count Classification methods, SVC outperforms other classifiers such as Logistic Regression, Naive Bayes, Decision Tree, and Random Forest.

For Historical Data Classification method, SVC, Decision Tree, and Random Forest all achieve perfect 1.000 test accuracies, indicating they are tied for the best model. SVC's better performance can be attributed to its ability to successfully manage the nonlinear and complex interactions between sentiment data and the direction of Bitcoin price movement. SVC creates an ideal hyperplane in the high-dimensional feature space to optimize the margin between positive and negative classes, resulting in robust classification. In contrast, simpler models like Logistic Regression and Naive Bayes may struggle to capture the nuances in the data, resulting in lower accuracies.

Decision Trees and Random Forests, while powerful, can be prone to overfitting if not tuned properly. The search results also suggest that incorporating lagged sentiment features can improve the predictive power of the models, with the Lag 1 and Lag 7 methods achieving some of the highest individual test accuracies. This indicates that accounting for the delayed impact of sentiment on future price direction is an important consideration. Overall, the search results show that a well-designed SVC model, possibly with lagged sentiment features, is the most successful method for forecasting the direction of Bitcoin price movements based on sentiment analysis of tweets. SVC's ability to handle nonlinearity and its robustness to noise make it the superior choice compared to the other classification techniques you explored.

Table 7. Loss Metrics for Deep Learning Algorithms

Method Name	Model Name			
	RNN		LSTM	
	Val. Loss	Test Loss	Val. Loss	Test Loss
Aggregate Regression	**0.014**	**0.020**	0.007	0.011
Count Regression	**0.004**	**0.010**	0.007	0.011
Aggregate Classification	0.658	0.559	**0.693**	**0.471**
Count Classification	0.969	0.529	**0.705**	**0.471**

Table 7 shows that the LSTM (Long Short-Term Memory) network outperforms the majority of the other approaches. In the Aggregate Regression, Aggregate Classification, and Count Classification algorithms, LSTM has the lowest validation and test losses when compared to RNN. Similarly, in the Aggregate Classification and Count Classification tasks, LSTM outperforms RNN by a large margin, with test losses of 0.471 versus 0.559 and 0.529, respectively.

LSTM's high performance can be due to its capacity to properly manage long-term dependencies in sequential data. LSTM networks contain a distinct memory cell layout that includes input, forget, and output gates, allowing them to selectively recall or forget information from an input sequence. This gating technique allows LSTM networks to bypass the vanishing gradient problem that can arise in ordinary RNNs, making them more successful for tasks with long-term dependencies. However, it's worth noting that in the Count Regression method, RNN achieves a lower test loss of 0.010 compared to LSTM's 0.011. This suggests that for certain specific tasks, RNN may still be a competitive choice. Overall, the search results show that LSTM networks are the best deep learning models for tasks that need sequential data, such as regression and classification issues. The capacity of LSTM to capture long-term dependencies, as well as its robustness to vanishing gradients, making it the preferred choice over traditional RNN architectures.

CONCLUSION

Based on the findings of the research on cryptocurrency price prediction using sentiment analysis, it is clear that a variety of approaches and algorithms have been used to estimate Bitcoin values. However, determining the most successful approach and its subsequent algorithm is a difficult process. The literature review identified numerous approaches, ranging from traditional machine learning techniques to advanced deep learning models like recurrent neural networks (RNNs) and transformers. This research has shown how sentiment analysis might improve forecast models for cryptocurrency price movements.

For instance, Lamon et al. (2017) pioneered the development of prediction models based on news and social media sentiment, laying the groundwork for future research in this area. Raju and Tarif (2020) further emphasized the importance of incorporating sentiment analysis into predictive models, particularly given Bitcoin's high volatility and the rapid dissemination of information through social media platforms. However, despite efforts to integrate sentiment analysis with historical data, the results show no substantial gain in predicting accuracy. This finding emphasizes the difficulties of combining sentiment analysis and historical data for Bitcoin price prediction.

One of the difficulties in merging sentiment with historical data lies in the dynamic and volatile nature of cryptocurrency markets, especially during periods of uncertainty and instability, such as the aftermath of the COVID-19 pandemic. The selected period for analysis, from February 5, 2021, to January 9, 2023, coincides with one of the most volatile periods in the markets, further complicating the prediction process. The complexity of capturing and interpreting sentiment signals from social media and news sources adds another layer of challenge. While sentiment analysis techniques have shown promise in extracting sentiment polarity from textual data, incorporating these signals into predictive models alongside historical data poses significant computational and methodological challenges.

Additionally, relying solely on Twitter sentiment alongside historical data poses limitations in capturing the full spectrum of factors influencing Bitcoin fluctuations. While social media sites such as Twitter can give useful information about market mood and investor sentiment, they also have inherent biases and limitations.

Twitter, as a source of information, lacks authenticity and reliability compared to traditional financial data sources. The rapid dissemination of information on Twitter, coupled with the prevalence of misinformation and rumors, can lead to distorted perceptions of market sentiment. Individuals may propagate unsubstantiated claims or speculative narratives without credible evidence, leading to market distortions and irrational price movements. Furthermore, the herd mentality prevalent on social media platforms like Twitter can amplify market volatility, as investors may react impulsively to trending topics or sensationalized news stories without conducting thorough research or analysis. This phenomenon underscores the importance of cross-validating sentiment analysis from multiple sources and incorporating robust risk management strategies to mitigate the impact of market uncertainty.

In conclusion, while sentiment analysis techniques, particularly those leveraging Twitter data, offer valuable insights into market sentiment, they represent only one piece of the puzzle in predicting Bitcoin price fluctuations. The volatile nature of Bitcoin prices, coupled with the inherent limitations of social media data, underscores the complexity of forecasting cryptocurrency markets. Future research efforts should adopt a multi-faceted approach, incorporating diverse data sources, advanced modeling techniques, and comprehensive risk management strategies to enhance predictive accuracy and resilience in volatile market conditions

FUTURE SCOPE

The literature review in the subject of cryptocurrency price prediction using sentiment analysis revealed a diverse set of approaches and algorithms, paving the path for future research efforts. We recognize that addressing the challenges related to computational capacity, data collection limitations, and algorithmic advancements is crucial for unlocking new insights into market dynamics and improving predictive accuracy. One of the primary limitations faced by current research, as highlighted in our study, is computational capacity. The utilization of primitive sentiment analysis algorithms like Textblob underscores the need for more sophisticated techniques capable of capturing nuanced sentiment signals. Authors such as Mathew and

Bindu (2020) have explored the effectiveness of pre-trained models like BERT and BiGRU in sentiment analysis, demonstrating their potential to enhance accuracy and efficiency. Future research could delve deeper into leveraging state-of-the-art deep learning architectures and pre-trained models to improve sentiment analysis performance.

Moreover, the restriction on data collection from platforms like Twitter and Reddit due to scraping limitations and API restrictions presents a significant challenge for researchers. As we mentioned, accessing new data without incurring hefty subscription fees is increasingly difficult. Therefore, exploring alternative data sources or devising innovative data collection strategies is crucial for overcoming this obstacle. Authors such as Haritha and Sahana (2023) have recommended combining historical pricing data with sentiment analysis of tweets, demonstrating the value of social media data despite the obstacles associated with data collecting. Future studies could explore novel approaches for data acquisition, such as leveraging alternative social media platforms, news sources, or proprietary datasets, to enrich the predictive modeling process.

Furthermore, the literature review reveals a trend toward the development of newer sentiment analysis models and hybrid approaches combining multiple techniques. Tan et al. (2022) suggested hybrid models that combine Transformer and Recurrent Neural Network (RNN) architectures to overcome the constraints of traditional approaches. Kurniasari and Setyanto (2020) investigated sentiment analysis with deep learning and neural networks, demonstrating the power of sophisticated approaches in sentiment categorization. Future research could improve on these efforts by looking into novel hybrid models that combine the strengths of various approaches, such as deep learning and classic machine learning algorithms or ensemble methods.

Additionally, there is a need for research that goes beyond traditional sentiment analysis by considering contextual factors and domain-specific knowledge. Authors such as Stenqvist and Lönnö (2017) investigated the feasibility of predicting Bitcoin price variations using Twitter sentiment analysis, emphasizing the significance of knowing the context in which sentiment signals emerge. Future studies could explore domain-specific sentiment lexicons, sentiment drift detection techniques, or sentiment analysis frameworks tailored specifically for cryptocurrency markets to improve predictive accuracy.

The future scope of research in cryptocurrency price prediction using sentiment analysis is expansive and multifaceted. We recognize that addressing the challenges related to computational capacity, data collection limitations, and algorithmic advancements is crucial for unlocking new insights into market dynamics and improving predictive accuracy. By leveraging state-of-the-art deep learning architectures, exploring alternative data sources, and developing innovative hybrid models, we can enhance our understanding of market sentiment and its impact on cryptocurrency prices.

Moreover, we acknowledge the importance of integrating contextual information and domain-specific knowledge into sentiment analysis frameworks. By taking into account market conditions, regulatory developments, and investor behavior, we can improve our models' ability to reflect the intricate interplay of factors that influence bitcoin price fluctuations. This comprehensive approach to sentiment analysis will allow us to create more powerful predictive models capable of dealing with the inherent volatility of cryptocurrency markets. Furthermore, we emphasize the need for continued interdisciplinary collaboration and innovation in the field. By fostering partnerships between researchers from diverse backgrounds, including computer science, finance, economics, and psychology, we can leverage a wide range of expertise to tackle the complex challenges of cryptocurrency price prediction. Through collaborative efforts, we can explore novel methodologies, validate findings across different datasets, and develop best practices for integrating sentiment analysis into predictive modeling frameworks.

Overall, we envision a future where advances in computational techniques, data collection methods, and algorithmic innovations converge to drive significant progress in cryptocurrency price prediction. By embracing these opportunities and overcoming existing limitations, we can unlock new frontiers in predictive analytics and empower stakeholders with actionable insights into cryptocurrency market dynamics.

REFERENCES

Auti, A., Patil, D., Zagade, O., Bhosale, P., & Ahire, P. (2022). Bitcoin Price Prediction Using Svm. *Int. J. Eng. Appl. Sci. Technol*, 6(11), 226–229. https://www.ijeast.com/papers/226-229,%20Tesma611,IJEAST.pdf

Chen, J. (2023). Analysis of Bitcoin price prediction using machine learning. *Journal of Risk and Financial Management*, 16(1), 51. DOI: 10.3390/jrfm16010051

Critien, J. V., Gatt, A., & Ellul, J. (2022). Bitcoin price change and trend prediction through Twitter sentiment and data volume. *Financial Innovation*, 8(1), 1–20. DOI: 10.1186/s40854-022-00352-7

Drus, Z., & Khalid, H. (2019). Sentiment analysis in social media and its application: Systematic literature review. *Procedia Computer Science*, 161, 707–714. DOI: 10.1016/j.procs.2019.11.174

Haritha, G., & Sahana, N. (2023, February). Cryptocurrency price prediction using Twitter sentiment analysis. In *CS & IT conference proceedings* (Vol. 13, No. 3). CS & IT Conference Proceedings. https://csitcp.com/paper/13/133csit02.pdf

Hua, Y. (2020). Bitcoin price prediction using ARIMA and LSTM. In *E3S web of conferences* (Vol. 218, p. 01050). EDP Sciences. DOI: 10.1051/e3sconf/202021801050

Kurniasari, L., & Setyanto, A. (2020, February). Sentiment analysis using recurrent neural network. In *Journal of Physics: Conference Series* (Vol. 1471, No. 1, p. 012018). IOP Publishing. https://iopscience.iop.org/article/10.1088/1742-6596/1471/1/012018/pdf

Lamon, C., Nielsen, E., & Redondo, E. (2017). Cryptocurrency price prediction using news and social media sentiment. *SMU Data Sci. Rev, 1*(3), 1-22. https://cs229.stanford.edu/proj2017/final-reports/5237280.pdf

Mathew, L., & Bindu, V. R. (2020, March). A review of natural language processing techniques for sentiment analysis using pre-trained models. In *2020 Fourth international conference on computing methodologies and communication (ICCMC)* (pp. 340-345). IEEE. DOI: 10.1109/ICCMC48092.2020.ICCMC-00064

Pant, D. R., Neupane, P., Poudel, A., Pokhrel, A. K., & Lama, B. K. (2018, October). Recurrent neural network based bitcoin price prediction by twitter sentiment analysis. In *2018 IEEE 3rd international conference on computing, communication and security (ICCCS)* (pp. 128-132). IEEE. DOI: 10.1109/CCCS.2018.8586824

Raju, S. M., & Tarif, A. M. (2020). Real-time prediction of BITCOIN price using machine learning techniques and public sentiment analysis. *arXiv preprint arXiv:2006.14473*. https://doi.org//arXiv.2006.14473DOI: 10.48550

Rathan, K., Sai, S. V., & Manikanta, T. S. (2019, April). Crypto-currency price prediction using decision tree and regression techniques. In 2019 3rd International Conference on Trends in Electronics and Informatics (ICOEI) (pp. 190-194). IEEE. DOI: 10.1109/ICOEI.2019.8862585

Stenqvist, E., & Lönnö, J. (2017). Predicting Bitcoin price fluctuation with Twitter sentiment analysis. https://www.diva-portal.org/smash/get/diva2:1110776/FULLTEXT01.pdf

Tan, K. L., Lee, C. P., Anbananthen, K. S. M., & Lim, K. M. (2022). RoBERTa-LSTM: A hybrid model for sentiment analysis with transformer and recurrent neural network. *IEEE Access : Practical Innovations, Open Solutions*, 10, 21517–21525. DOI: 10.1109/ACCESS.2022.3152828

Valencia, F., Gómez-Espinosa, A., & Valdés-Aguirre, B. (2019). Price movement prediction of cryptocurrencies using sentiment analysis and machine learning. *Entropy (Basel, Switzerland)*, 21(6), 589. DOI: 10.3390/e21060589 PMID: 33267303

Wankhade, M., Rao, A. C. S., & Kulkarni, C. (2022). A survey on sentiment analysis methods, applications, and challenges. *Artificial Intelligence Review*, 55(7), 5731–5780. DOI: 10.1007/s10462-022-10144-1

Chapter 8
Scoping Review on Financial Distress Prediction in India Using Modern Methods of Machine Learning:
Scope for Future Research

Reynal Lavita Fernandes
Christ University, India

Veerta Tantia
https://orcid.org/0000-0002-8487-5175
Christ University, India

ABSTRACT

Every business has its objective to attain, along with earning profits. A business needs finances and also throws up difficulties in meeting its financial requirements. If ignored, the challenges could affect the business. It is crucial for the business to keep a regular check, necessitating the right, analytical methods. A large number of works of literature have focussed on the traditional methods of predicting financial distress. Recently, a few studies evolved using modern methods. This study has a review of the literature regarding modern methods used in predicting financial distress. The present study has adopted the structure of a scoping review developed by Arksey and O'Malley and the main aim is to showcase the importance of predicting financial distress with modern methods through the machine learning approach. It also aims to highlight the drawbacks of statistical methods while predicting financial

DOI: 10.4018/979-8-3693-9246-1.ch008

distress and covering the reasons for them.

INTRODUCTION

A disaster in business is the premier sporadic yet ceaseless hitch. A business faces tremendous risk in its operations from investment through every stage of its development. Financial risk can lead to distress, and for many years, the muddle of financial distress has been an open-ended problem. Kashyap and Bansal (2019) state that financial distress is a colossal impediment to the economy. A company must check its performance regularly, as the performance yardstick is essential to understanding the scope of improvement. However, companies must follow the right methods to check their progress, as performance and methods work together to understand the company's stand.

Modern methods are increasing due to distinct advancements, and technology in every field plays a vital role in the functionality of business (Srivastava & Bagga, 2014). Business houses are expected to upgrade in all aspects to match the growing global pace. Through their large share of profits, big business houses can cope with change. However, medium and small-scale companies find it difficult to match the growing scenarios. Hence, businesses face challenges and tough competition from their counterparts in the same industry. In terms of technology, there is an upswing in sophisticated and high-end technologies at a faster rate throughout developed nations. This adversely affects many businesses in developing nations, as they lack advanced technology. Srivastava and Bagga (2014) opined through a study on the usage of information technology that the latest technologies have freed human resources. Thus, ample time is gained in developing strategies for improving a company.

Sustainability is a big challenge, as businesses work in a dynamic environment and are exposed to various ups and downs. Internally, the business has to face the challenges of the management, board of directors, employees, and shareholders. However, a higher risk is involved when companies face the external environment. The company has to survive tides from various competitors, face challenges, meet the customers' requirements, and follow the rules and regulations of the regulators, government, and other related authorities. The business should plan its short-term and long-term requirements. Hence, it needs finance, which is the lifeblood of the business and plays a major role in its survival.

Financial distress is overwhelming when a company is low on finance and fails to meet its obligations on time. The company's reputation and success are directly based on its meeting the requirements of its creditors and lenders. If the company meets these requirements, it will create goodwill and not face adversities when it needs credit in the future. On the other hand, if the company cannot meet its creditors' obligations on time, it can also face the dire consequence of sustaining the business for an extended period.

Financial distress prediction has been a long-term research project by several scholars and researchers. Stakeholders of the company do not favor financial distress as per the studies conducted by Alaka et al. (2015); Tinoco et al. (2013); Hafiz et al. (2015); Xu and Zhang (2009), as it takes on high costs (Bauweraerts, 2016; Waqas & Md-Rus, 2018; Senbet & Wang, 2012). Bankruptcies would be faced by business firms that follow the firm's operations in financial distress (Doumpos & Zopounidis, 1999). Also, there is no standard definition for financial distress in the literature. Many researchers have studied it, taking the meaning of bankruptcy into context (Altman, 1968; Nyitrai, 2019); (Sun et al., 2013); (Khoja et al., 2019); (Alifiah & Tahir, 2019). It is studied using many terminologies, such as default, corporate distress, bankruptcy prediction, etc. Financial distress can be defined based on the business's profitability, market movements of shares, the economic status of the country, and changes in dividend policies. Nevertheless, no standard model in the literature predicts financial distress. Hence, the researchers developed the prediction model based on the distinct phases of business activities, different theoretical frameworks, and various definitions related to corporate distress, which have challenged the comparisons to extant literature. Jones and Hensher (2004) have stated the significance of forecasting bankruptcy on different grounds.

In a developing country like India, timely financial distress prediction is needed. As per the recent updates from "The Ministry of Company Affairs," it is clear that many companies are facing financial distress. Its correct prediction requires the right models. So, it is always necessary to have the latest ones, as these would be more accurate in the changing business environment with advanced technology. The latest models have proved to be appropriate and less rigid. Xiao et al. (2012); Chen & Du (2009); Ding et al. (2008); Sun and Li (2008); Alifiah & Tahir (2019); Fawzi et al. (2015) in terms of their performance when compared to the traditional models Al-khatib & Al-Horani (2012); Gregova et al. (2020); Basak et al. (2018). There are many studies on the variables of financial distress prediction, but not many on modern methods and their comparison with traditional methods of adopting financial distress prediction in India. The present study adds novelty to the existing literature by adopting Arksey & O'Malley's (2005) scoping review. This showcases the importance of predicting through modern methods of the machine learning approach, highlighting the drawbacks found in statistical methods and the reasons for the predictions. This seems to be the first scoping review study that concludes that it is important for companies to analyze the importance of predictions and adopt modern methods that are accurate and flexible.

METHODOLOGY

This study has implemented a scoping review technique introduced by Arksey & O'Malley (2005) on a methodological structure to review the literature on modern methods of predicting financial distress in India. The main focus of this paper is to showcase the noteworthiness of prediction using modern methods compared to traditional methods and evaluate the main aim and reasons for predicting financial distress.

A scoping review is a structure for integrating articles that fall under a particular area or discipline, further supporting emerging, new ideas (Daudt et al., 2013). It is a path for finding the research gap in the literature (Mays et al., 2001). A scoping review scrutinizes extant literature and various information databases, consolidating findings from various studies (Davis et al., 2009).

As stated by Mays et al. (2001) and Daudt et al. (2013), the scoping review is applied to the advanced and most difficult area of research or to that area that has not been reviewed much. A scoping review paves the way for further systematic and meta-analysis, demanding appropriate reviews and low bias (Sucharew & Macaluso, 2019).

The scoping review involves five phases, as introduced by Arksey & O'Malley (2005)

(i) Phase 1: Identification of research question/s
(ii) Phase 2: Identification of appropriate articles
(iii) Phase 3: Selection of the study
(iv) Phase 4: Charting the data and
(v) Phase 5: Collation, encapsulation, and chronicle of research results.

Phase 1: Identification of Research Question/s

As per the structure laid by Arksey & O'Malley (2005), the first phase comprises identifying the research question/s. The research is possible only when there is a question. This questioning ability leads the way to finding the answer and makes the study valuable. In this current scoping review, an attempt is made to answer the research question/s: What is the reason for predicting financial distress using the latest ML techniques? How do statistical methods fail compared to modern methods in predicting financial distress? What are the various reasons for researchers to focus on financial distress predictions? The author believes the research question covers a substantial aspect to get a sizeable further scope.

Phase 2: Relevant Article Identification

Appropriate articles that suit the title of the study were pooled from various databases comprising the below sources:

- Google Scholar
- Google web browser
- Wikipedia search engine
- ResearchGate
- ScienceDirect
- Jstor
- Proquest
- Institutional databases

The above-mentioned databases have immeasurable articles related to the topic selection. To ensure repeatability, reliability, consistency, and quality, the articles are thoroughly examined after pooling from the above sources. In this way, source bias is guaranteed, and the majority of the studies across the globe are covered, eliminating geographical biases.

Due to the factors posed by COVID-19, the review has included only articles between 2010 and 2019. Due to the latest advancement in AI, the articles for the scoping review are post-2010. A few articles between 1966 and 2000 are also included to highlight the traditional methods used in predicting financial distress. There are two reasons for selecting the starting date in 2010: The first reason is that the study of AI was conducted only for a few years, and even if it had been done before, it was primarily not used in the financial domain. Inter-disciplinary research gained importance only two decades ago. The other reason is the prominence of financial distress in the post-2007 recession. Hence, bankruptcy and financial distress are still the creamy layers for many researchers.

In general, companies have to face external and internal forces. However, the business world will be much more deeply affected when a tough recession or pandemic attacks the economy. Even healthy companies will be affected and in distress. Despite these adversities, companies suffering financial distress can be turned to return to a healthy stage if the prediction is conducted at the right time. The models developed for predicting financial distress during this particular time would only fit that particular period and face high chances of failing during another period.

Given these conclusions, the researcher felt that it was inappropriate to incorporate the articles related to the 2007 recession. Post-recession, companies need some time to recover and reach a state of normal operations. Hence, the articles compiled between 2007 and 2010 were not included, as they would affect the study.

The articles from 2010 onwards are included, assuming that the companies would have recovered to a certain extent and resumed normal operations.

Along similar lines, the study ends in December 2018, factoring in the pandemic, which reduced many healthy companies to dire states. Predicting financial distress during this period is only temporarily suitable, so it has not been included. Research articles other than in English have been discarded.

The search query for the articles in the electronic databases is initiated from the research question. Random search queries are conducted to check the display of the articles. Improvement has been made to the pilot's search queries. Once the right form of search queries is fixed, the actual hunting of the research articles for review began, using phrases like financial distress prediction, financial distress prediction in India, methods adopted for distress prediction, bankruptcy prediction, methods of bankruptcy prediction, financial crises, and their prediction, prediction of financial difficulties for companies, prediction of financial distress in developed countries, modern methods v/s traditional methods, machine learning in the financial distress prediction, machine learning in bankruptcy prediction and so on. The search was first conducted on the Google Scholar database, and subsequent searchers were taken from other sources. In the beginning, a general search was initiated. However, subsequent searches went through filtering for the year, subject, and area.

The first search was initiated on June 3, 2022. On June 14, follow-up literature searches were conducted to upgrade and refine the scoping review. Five more pieces of literature were added to the review process. After scrutiny, seven articles were considered relevant to the study.

Phase 3: Study Selection

Due to the broad phrases used in the search engine, the search included several unrelated articles from the databases. The irrelevant articles that mismatched the title and did not support the research question were pulled from the scoping review of the topic and research question. The repeated duplicates were also removed through a physical and thorough check to ensure the scoping review's unbiasedness.

Eligibility Criteria

The eligibility criteria were based on the main part of the research question. The decision on eligibility criteria was taken very carefully to ensure quality articles. Only peer-reviewed articles were included, as these ensured the quality and the research was considered valid, which would eliminate the incorrect outcomes. The eligibility criteria support Arksey & O'Malley's (2005) reliability steps in decision-making. Studies have included articles on bankruptcy or financial distress prediction using traditional and modern methods. Though this study focuses on modern methods, studies on prediction using traditional methods are included for comparison with studies using modern methods. Articles in languages other than English are also excluded from the scoping review.

Title and Abstract Suitability Evaluation

The title and abstract of the article should be reviewed to save time while searching and to avoid the collection and wastage of unrelated articles for scoping review (Pham et al., 2014). A suitable design for the title and abstract, as required for the search, is designed prior to the application. Only the review of the title and abstract is insufficient. In such cases, the introduction, the methodology adopted, and the conclusion were also checked, and, in a few cases, when needed, even the complete article was read to decide the suitability of the article for a scoping review. The full-text articles are available for the literature included in the study.

Data Collection

All suitable citations post the examination process were obtained for complete article analysis. The author's institutions mostly collect the articles. A literature matrix was built to collect the information about the articles in one place. This enabled the filtration process. The literature matrix was built in Excel and included the headings, such as title, author, year of publication, objectives of the study, methodology, type of sample, sample size, data analysis tools, methods applied, volume, page numbers, doi, results, and conclusion (see Appendix B). The literature review matrix was pretested and then revised as per the requirement before its application. The eligibility and attributes of every complete paper are assessed and extracted. The entire article is scrutinized, and the decision to include or discard is made before entering the template. A few articles have been published on the research on financial distress prediction using modern methods, especially in developed nations; this scoping study has adopted restricted articles for the scoping review (Charalambakis & Garrett, 2019). The following articles only have been included in the scoping review: (2016) (Monti & Garcia, 2010); (Farooq et al., 2018); (Li & Du, 2011); (Khoja et al., 2019); (Ray, 2011), (AlAli, 2019), (Trujillo-Ponce et al., 2013), Nair et al. (2018).

Phase 4: Data Charting

In this stage, the data collected through several sources are charted through the flow chart. As described by Ritchie and Spencer (1994), charting is defined as organizing information after examining the core theme of the topic and presenting it in a form that is easy to interpret. This study follows the data charting in the form of Appendix B. The chart contains information about the author, the year, and the title. Every heading of the article forms the beginning of the scoping review. Figure 1 outlines the complete flow of the selection of literature studies from the identification to the conclusion, inclusion, and exclusion criteria.

Phase 5: Collate, Encapsulate, and Chronicle Research Results

At this stage, the review is collated, and the results are summarized. Sucharew and Macaluso (2019) state that the scoping review results are broad and disbursed widely. This section presents the review of the literature selected and charted in Phase 4. The findings of the review are summarized, and the information is outlined in the form of notes.

Figure 1. Article Selection Process Flow Chart

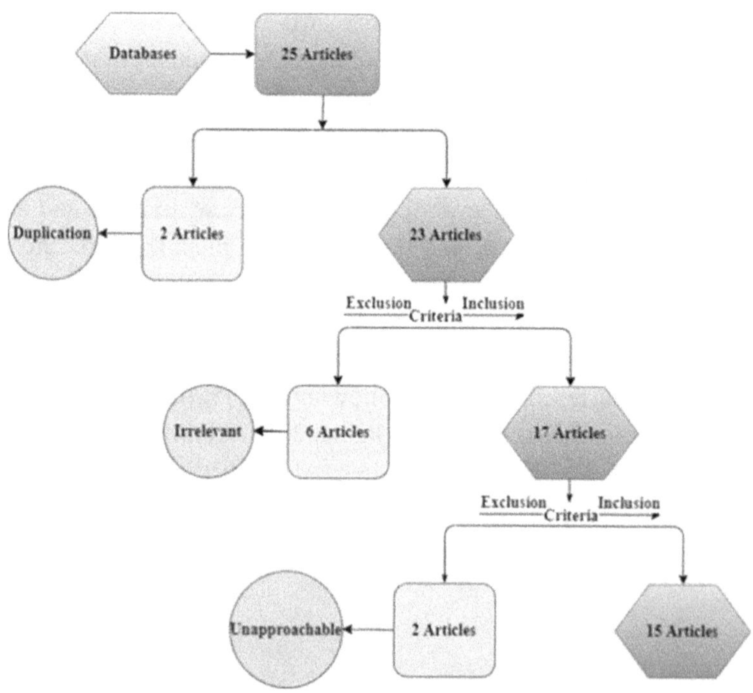

LITERATURE REVIEW

The models predicting financial distress largely contribute to the classification, categorizing companies as financially distressed or healthy and supported, with reduced failure (Ray, 2011). Predicting financial distress for companies has been a concept that has ruled for decades (Karas et al., 2015). In the recent past, modern models have replaced traditional models (Xiao et al., 2012). The prediction models will be of great use to the various interested stakeholders of the company, who can make timely decisions on their investments (Ray, 2011). Constructing accurate prediction models is vital to reduce business distress (Xiao et al., 2012). A model that can predict business failures in advance would enable the company to protect itself from falling into financial disaster (Deakin, 1972). The models and the variables for predicting financial distress at the right time among the companies are the most important subject of research (Xiao et al., 2012). The method used to predict financial distress among corporates concerns the companies' stakeholders (Baimwera et al., 2014).

For years, many bankruptcy models have been created by researchers using accounting information. The models classified the companies as either failed or non-failed. The concept of distress and its prediction was pioneered by Fisher (1936), who tried to understand the best combinations of linear ratios and developed the Linear Discriminant Algorithm. The light on the financial distress prediction was thrown after Beaver (1966) made a noticeable prediction, attaining 78% accuracy five years before the failure, using univariate analysis. Altman (1968) took it further and developed a model using five variables, adopting Multiple Discriminant Analysis, which could predict 95% accuracy one year after the failure. Altman's (1968) model has been the base for many other models and researchers. Many academists have validated the Altman model. Several studies state that the Multivariate Discriminant Analysis (MDA) method is more robust in prediction than the bivariate analysis (Kashyap & Bansal, 2019; Baimwera et al., 2014; Shilpa & Amulya, 2017).

Kashyap & Bansal (2019) have stated that many academicians, scholars, industrialists, and researchers have extensively used the Altman Model for studies, such as Shilpa & Amulya (2017); Ray (2011); Rao et al. (2013), for distress predictions and validation of the performance of the model. Studies by Kashyap and Bansal (2019), Baimwera et al. (2014), and Shilpa & Amulya (2017) have found that Multivariate Discriminant Analysis (MDA) is a good method to be applied in distress prediction rather than the dichotomous variable. However, the studies by Perry et al. (1984) using MDA yield good results only when used in a single industry. Hence, static models outperform when used during different periods, as these models work well for a particular period during which it is developed. Therefore, the static models are inappropriate and unsuitable for predicting bankruptcy, which occurs intermittently.

The firm's characteristics change year by year, and the data of healthy firms are eventually ignored when they go bankrupt (Shumway, 2001).

Conversely, a few researchers studied distress predictions and built the models on market information (Black & Scholes, 1973; Merton, 1974). These researchers claimed that accounting models do not take the volatile data of the business but estimate only the records, which would lead to the failure of distress predictions. The researchers (Bharath & Shumway, 2008; Mamatzakis & Koutsomanoli-Filippaki, 2009) tested the market-based models and option-based and Merton-type models to predict using the volatile information of the business. Black-Scholes-Merton's (BSM) bankruptcy model is a market value-based analysis and found good results in predicting financial distress (Charitou et al., 2013). Maheshwari and Agrawal (2016) analyzed financial distress prediction by combining MDA and logistic regression along with the Merton Distance-to-Default. The stock market-based models performed well when Merton (1974), the original model, paved the way for many other research models that used market data.

As research developed over the years, researchers started to use accounting and market information in developing models, which can have a higher predictive capacity (Fawzi et al., 2015 & Zmijewski, 1984). This gave rise to parametric tests like the Probit model, Logit model (Ohlson, 1980; Zmijewski, 1984; Martin, 1977), and Cox model (Henebry, 1996) (Shumway, 1999) with fewer statistical assumptions.

Through the years, many non-parametric tests are linked to Artificial Intelligence. Artificial Neural Network and Data Envelopment Analysis (DEA) methods are used by Al-Shayea & El-Refae (2012; Cielen et al., 2004; Back, Laitinen, & Sere (1996;Ding et al., 2008; Cút, 2014). Contemporary research uses advanced methods with the coding language of a computer network to analyze financial distress data for prediction. Strangely, these methods are free from sophisticated statistical assumptions.

A study using Neural Network, Support Vector Machine, and Decision Tree on the Shenzhen Stock Exchange and Shanghai Stock Exchange found that a neural network is more accurate than other classifiers. A majority of the study votes high on the prediction capacity of the Artificial Neural Network (ANN) model (Liang et al., 2015; Niemann et al., 2008; Psillaki et al., 2010; Tseng & Hu, 2010; Cheng et al., 2006). (Ohlson, 1980) The o-score model is also considered the best-known multi-discriminant model.

(Chen & Du, 2009) has tried to predict financial distress by adopting Artificial Intelligence (AI) and Data Mining (DM) techniques to predict financial distress. Factor analysis was used in the selection of adaptive variables. This research used the modern techniques of AI and DM due to their radical changes and the dynamic business environment, as the statistical methods have strict assumptions such as linearity, normality, and independence for the variables. This study claims that it is vital to develop models that include the evolutionary approach, which can cope with the dynamic variables and hence use the AI and DM approaches to construct the model. It validated its accuracy after being compared with the traditional factor analysis approach. The study made good contributions to the literature. More usage of factor analysis will result in less accuracy by the ANN and DM techniques and high errors in distressed companies compared to normal companies. The closer to the distress, the higher the accuracy of the models – almost 82.14%. ANN has better prediction than the DM approach compared to the two modern techniques. Overall, modern techniques are better and more robust in predicting financial distress, especially for dynamic variables in the evolutionary phase of business. In contrast, traditional approaches are taken with strict assumptions and unsuitable for dynamic variables. Oz et al. (2017) conducted studies to examine if a prediction can be made considering the accounting theory based on different countries, using two techniques - Panel Logit and Neural Network. The developed model enhanced the predictability of financial distress, contributing to generalisability throughout distinct countries and industry sectors.

Recent studies in many developed nations have proved that modern models have higher accuracy rates and are free from sophisticated, rigid assumptions, unlike statistical methods. This confirms the importance of adopting modern methods for financial distress predictions. Table 1 provides a review of selected studies.

Table 1. Review of Selected Studies

Author/s	Variables	Year	Result	Statistical Model	Country
Jones	Cash flow ratios	2016	The model could predict 81.34% of non-failed companies and 82.26% of failed companies.	Logistic Regression	Australia
Monti & Garcia	Financial and stock market variables	2010	Both tools performed well and could assess the financial distress of the firm within just six months	Statistical Inference of Risk and Return	Argentina
Farooq and Qamar	Financial and stock market variables	2019	Provided a base to develop a vast area of research on financial distress.	Three-stage Dynamic Model	Pakistan
Li, J., & Du, W.	Financial and non-financial variables	2011	Non-financial predictors improve the predictive capacity	Logistic Analysis	China
Khoja et al.	Macro-economic indicators	2019	Similarity in solvent and individual differences in insolvent companies	Multi-Dimensional Scaling and Cluster Analysis	UK, GCC, and USA

continued on following page

Table 1. Continued

Author/s	Variables	Year	Result	Statistical Model	Country
Karas & Režňáková	smof ratios	2015	The results prove that different branches have specific predictors	Bankruptcy Prediction Models	Czech
Ray	Altman Model ratios	2011	Can predict the financial distress of the company	Altman model	India
Al Ali	Share prices of two companies - Yiaco and Advanced Technology,	2019	investors should pay attention to the financial indicators, which have a significant influence on the share prices of companies	Altman Zeta Model and Regression Analysis	Kuwait
Trujillo-Ponce et al	Accounting and stock market information	2013	The comprehensive model is more robust than single-built models in predicting credit risks.	Regression Analysis	European countries
Wang	Financial Variables	2016	The predictive accuracy of both the model's logistic regression and neural network is similar in a short time horizon of one to three years.	Neural Networks and Logistic Regression	
Kashyap & Bansal	Financial ratios	2019	The model could predict 95% of cases accurately before the distress	Statistical Model	
Chen and Du	Financial ratios	2009	ANN's techniques give better prediction with higher accuracy than the data mining technique.	Data Mining Techniques and Artificial Neural Networks	
Khademolqorani et al.	Financial ratios	2015	It found that modern methods are robust and have a higher accuracy rate in predicting financial distress than traditional models.	Statistical and Neural Network and Decision Tree	Tehran
Devi et al.	Financial ratios	2018	Machine learning methods provide a higher prediction accuracy rate than traditional statistical methods with smaller data sets.	Statistical Techniques and Machine Learning	
Malakauskas, A., & Lakštutienė, A	Financial ratios	2021	An Artificial Neural Network is the best model for predicting financial distress.	Logistic Regression, Artificial Neural Networks, and Random Forest	
Kang et al.	Merton model's determinants	2022		Traditional Models	

continued on following page

Table 1. Continued

Author/s	Variables	Year	Result	Statistical Model	Country
Sun et al	Financial ratios	2021	Financial distress prediction models are sensitive when there is an imbalance in the training sample size.	SVM model integrated with decomposition and fusion methods	China
Zhang et al.	Financial ratios	2022	DGAMLPE has very good results in accuracy, F1-score, and AUC, which shows that we can predict the financial distress of Chinese-listed companies very well.	GA-MLP ensemble algorithm (DGAMLPE)	China
Li et al.	Lexicon Sentiments and Vocabulary	2021	A sentimental lexicon can independently give satisfactory results in financial distress predictions four years prior to actual distress.	SVM, DT, Xgboost, and DNN	China

RESULTS AND DISCUSSION

Financial distress prediction is exclusively needed, as it may bring a company to a closure. Financial distress happens when a company faces difficulties meeting its obligations or even its day-to-day expenses. It will not happen quickly, but it is a slow and long process that takes years. Most directors ignore the beginning or are unaware of the company's distress. This may lead to severe difficulties, and many companies will become bankrupt. Managers and the company's board of directors need to check the performance of the business on time. This requirement can be fulfilled through the models that predict business distress in advance. Hence, prediction models play a significant role in safeguarding and supporting the company to remain healthy.

Models developed using the latest technology would support companies in predicting and sustaining their business. Traditional models are period-specific. They also have drawbacks, such as low accuracy and less precision in prediction. In addition, the old traditional models have strict assumptions that must be followed in development. In contrast, modern models overcome the limitations of traditional models in terms of accuracy and precision in predicting power without rigid assumptions, and using the latest technology would disclose results faster.

The results of the present scoping review are showcased based on the outcome of the financial distress predictions. It is based on analyzing the main outcome, such as financial distress through machine learning, the drawbacks of traditional methods, and the major reasons for financial distress predictions. The obtainable literature study from various sources underpins the scoping review of this study. The initial part of this section focuses on predicting financial distress with modern methods, the second section deals with the drawbacks of a traditional model, and the final section deals with the necessity of financial distress predictions and the research gaps observed.

(a) Predicting financial distress using the modern models

According to the study in the current scoping review, it is not just advantageous but also a necessity for businesses to predict financial distress using modern models. The review clearly states that most developed countries are using different modern models of AI to arrive at accurate predictions. Recent studies in developed nations focus on different sets of analyses to determine the best variables in diverse environments and develop various models using computer network algorithms. As advancements occur at a rapid speed, companies must also upgrade their businesses to an advanced level. With the advancements in business, companies can face tough competition at the global level. It has to satisfy its stakeholders. The company should upgrade its business operations to achieve its business objectives and sustain itself at the corporate level. To cater to the operational needs and all other requirements to sustain and enjoy higher profits, the company should include advanced methods that give quick results and can accommodate large datasets. Hence, for all these, the company must adopt modern technology.

Moreover, the company has to check its performance at regular intervals. Due to large and complex business activities, the business must adopt the latest methods to test the performance of the business houses. This necessitates the adoption of the latest technology comprising modern methods of prediction. Most developed nations successfully bring business to greater heights by adopting modern methods. They play a greater role in accurately predicting financial distress, so business heads should adopt them in their companies. A few recent studies (Li et al., 2021; Zhang et al., 2022; Sun et al., 2021; Kang et al., 2022; Malakauskas et al., 2021) have indicated that modern models perform well in predicting financial distress. The models adopt the latest techniques, give better accuracy, and, with certain feature validation, improve the performance of the models. Hence, modern models offer optimum solutions for predicting financial distress.

(b) Drawbacks in Traditional Models

Traditional financial distress prediction models, mostly based on statistical analysis, have been extensively used by earlier researchers and have given the best results. Every statistical analysis has its assumptions. They are strict and rigid, not flexible, making the model very stringent and impossible to adopt in a dynamic environment.

On the other side, the business environment is very volatile and dynamic. The models adopted in such an environment should be flexible for changing situations. The statistical models are not flexible and may not work in the changing scenarios of the business. That is why most statistical methods fail to predict the financial distress of companies today. The prediction models should be innovative and flexible, which traditional models lack. Hence, developing the statistical model for predicting companies in today's fast-moving business activities is a challenge and a major drawback of traditional methods.

Traditional models can predict when sample data is limited. However, they cannot handle huge amounts of data. Today's businesses have enormous amounts of data and need the right models to handle vast datasets. So, traditional models cannot be used to predict big businesses that use advanced technology.

The traditional model uses statistical assumptions. The developed model, with many assumptions and adjustments, would not bring out accurate results. Hence, the predictions may not be accurate. If the industry uses inaccurate models or those that are not remarkable, then the company will end up with wrong decisions that will hamper the business. It is of paramount importance for a business to show accurate results, especially in ascertaining its performance, as the top level has to make further decisions. Therefore, the model should accurately predict the company's performance. The accuracy is high in modern models as they do not require any assumptions. Modern models are trained to use the company's real data; thus, their accuracy is higher than traditional models.

Earlier researchers developed the traditional models that were required for a short period. The models thus developed gave out the best results only during that period, but after a certain time, they became obsolete. Hence, the traditional model cannot be used when the business situation changes. Various forces influence the business environment and tend to change depending on the economy. In such cases, the traditional model becomes useless. On the other hand, modern models are developed through trained data, and since they are flexible, they can be modified according to the changing environment. Modern models can be generalized, whereas traditional models cannot be.

Due to these factors, a business must use modern models to predict its financial distress to understand its performance and take corrective actions wherever necessary.

(c) Reasons for Financial Distress Prediction

A business's primary motive is to earn profit and sustain itself in the market for a long time. Every business desires to expand and has objectives. Hence, the business has to keep a check on its performance regularly. Financial distress should be predicted regularly to avoid dire consequences for the business.

Investors need to check the performance of the businesses to decide on their investments. To infuse confidence into investors, the company should check the ongoing activities of the business on time and predict financial distress to help it avoid such discrepancies. Similarly, other business stakeholders are interested in knowing the business's operations. Hence, financial distress prediction is essential to earn confidence and satisfy stakeholders. Noronha and Mehta (2012) studied the importance of corporate governance in making investment decisions. The study was conducted through a structured questionnaire to find the connection between corporate governance and investors. The study found that investment increases when the company maintains integrity in its financial statements. Integrity can be achieved through the implementation of corporate governance. As per this study, the implementation of corporate governance takes first place. Ethical behavior took second place, and maximization of shareholders' wealth was given third place, followed by managing risk to protect minorities' interests. All these make investors gain confidence in the business and protect the company from financial distress. In their study, Rajpurohit and Rijwani (2020) revealed that practicing CG will enhance financial reporting so companies can improve their business. Neglecting this would impact the business negatively, leading to financial distress.

Financial distress can be caused when corporations are affected by fraud, as studied by P. and Chandrasekhar (2016), who found that fraudulent activities are one of the crises of sustainability and would significantly impact the business.

Attri and Pahwa (2012) state that brand building is essential for the company to avoid financial distress. This has been proved by a study conducted on brand building on petrol. Building long-lasting customer relationships can avoid financial distress, as the customers would be loyal to the company even during turbulent times. In their study on cross-training, Ninan et al. (2019) opined that it reduces the attrition rate of employees, thus reducing the cost of recruitment, selection, staffing, training, and development. The study also found that cross-training can secure employees' job and motivate them to manage their idle time efficiently. This will indirectly bring in more profits and earnings to the company, avoiding financial distress. Sam (2011) argued through a case study that corporate managers' failure to fulfill their duties toward the shareholders is a major mistake, and thus, a company faces a crisis. Greed and self-interest without any interest in the maximization of the long-term wealth of shareholders can cause severe financial distress for businesses.

CONCLUSION

Earlier, financial distress prediction was mostly developed through traditional models that performed well during its development period. But later, the same model outperformed. Hence, most developed nations are making modern models of machine learning. This study made a scoping review of the modern models for financial distress prediction and stated the drawbacks found in the traditional model. As half of the world's GDP is due to emerging markets and India is an emerging country (Shivdas & Chandrasekhar, 2016), the emerging economy is developing traditional models. Therefore, the focus on developing modern models for financial distress predictions is paramount. Hence, a huge research gap exists as financial distress prediction using modern models shows very minimal study in India. Also, financial distress is not the same in all situations, so in future studies, a model for predicting financial distress at different time intervals can be developed. The prediction of financial distress at various levels can also be studied, as the distress of every company is different. However, financial distress prediction is important for a company and its stakeholders, the economy, and the country. Hence, the methods applied to predict financial distress play a major role. In our advanced era, companies must upgrade themselves with modern models to safeguard businesses from tragic falls and liquidation.

REFERENCES

Alaka, H., Oyedele, L., Toriola-Coker, O., Owolabi, H., Akinade, O., & Bilal, M. (2015) Methodological approach of construction businesses failure prediction studies: a review. In: Raiden, AB, & Aboagye-Nimo, E (eds) *Proceedings 31st annual ARCOM conference*, 7–9 September 2015. Association of Researchers in Construction Management, Lincoln UK, pp 1291–1300

AlAli, M. S. (2019). Examining the Effect of Altman's Zeta Model Score on the Share Price of Healthcare Companies Listed at Kuwait Stock Exchange. In International Journal of Economics and Finance (Vol. 11, Issue 4, p. 25). DOI: 10.5539/ijef.v11n4p25

Alifiah, M. N., & Tahir, M. S. (2019). Macroeconomic Variables and the Prediction of Financial Distress Companies in the Manufacturing Sector in Malaysia. *International Journal of Recent Technology and Engineering*, 8(1S).

Al-Shayea, Q. K., & El-Refae, G. A. (2012). Evaluation of Banks Insolvency Using Artificial Neural Networks. *In Proceedings of the 11th WSEAS international conference on Artificial Intelligence, Knowledge Engineering and Data Bases (AIKED'12)*, (pp. 22-24). Cambridge.

Altman, E. I. (1968). Financial Ratios, Discriminant Analysis and the Prediction of Corporate Bankruptcy. *The Journal of Finance*, 23(4), 589–609. DOI: 10.1111/j.1540-6261.1968.tb00843.x

Arksey, H., & O'Malley, L. (2005). Scoping studies: towards a methodological framework. In International Journal of Social Research Methodology (Vol. 8, Issue 1, pp. 19–32). DOI: 10.1080/1364557032000119616

Attri, R., & Pahwa, M. S. (2012). Learning for public sector oil marketing companies in India to manage business in turbulent times. *Prabandhan & Taqniki. Management Research Journal*, 6, 29–36.

Back, B., Laitinen, T., & Sere, K. (1996). Neural Networks and Genetic Algorithms for Bankruptcy Predictions. *Expert Systems with Applications*, 11(4), 407–413. DOI: 10.1016/S0957-4174(96)00055-3

Baimwera, B., & Muriuki, A. M. (2014). Analysis of corporate financial distress determinants: A survey of non-financial firms listed in the NSE. *International Journal of Current Business and Social Sciences*, 1(2), 58–80.

Basak, S., Kar, S., Saha, S., & Khaidema, L. (2019). Predicting the direction of stock market prices using tree-based classifiers. *The North American Journal of Economics and Finance*, 47, 552–567. DOI: 10.1016/j.najef.2018.06.013

Bauweraerts, J. (2016). Predicting bankruptcy in private frms: Towards a stepwise regression procedure. *Int J Financ Res*, 7(2), 147–153. DOI: 10.5430/ijfr.v7n2p147

Beaver, W. H. (1966). Financial Ratios As Predictors of Failure. *Journal of Accounting Research*, 4, 71–111. DOI: 10.2307/2490171

Bharath, S. T., & Shumway, T. (2008). Forecasting Default with the Merton Distance to Default Model. *Review of Financial Studies*, 21(3), 1339–1369. DOI: 10.1093/rfs/hhn044

Black, F., & Scholes, M. (1973). The Pricing of Options and Corporate Liabilities. *Journal of Political Economy*, 81(3), 637–654. DOI: 10.1086/260062

Charalambakis, E. C., & Garrett, I. (2019). On corporate fnancial distress prediction: What can we learn from private frms in a developing economy? Evidence from Greece. *Review of Quantitative Finance and Accounting*, 52(2), 467–497. DOI: 10.1007/s11156-018-0716-7

Charitou, A., Dionysiou, D., Lambertides, N., & Trigeorgis, L. (2013). Alternative bankruptcy prediction models using option-pricing theory. *Journal of Banking & Finance*, 37(7), 2329–2341. DOI: 10.1016/j.jbankfin.2013.01.020

Chen, W.-S., & Du, Y.-K. (2009). Using neural networks and data mining techniques for the financial distress prediction model. In *Expert Systems with Applications* (Vol. 36, Issue 2, pp. 4075–4086). DOI: 10.1016/j.eswa.2008.03.020

Chen, W.-S., & Du, Y.-K. (2009). Using neural networks and data mining techniques for the financial distress prediction model. *Expert Systems with Applications*, 36(2), 4075–4086. DOI: 10.1016/j.eswa.2008.03.020

Cheng, C.-B., Chen, C.-L., & Fu, C.-J. (2006, February). Financial Distress Prediction by a Radial Basis Function Network with Logit Analysis Learning. *Computers & Mathematics with Applications (Oxford, England)*, 51(3-4), 579–588. DOI: 10.1016/j.camwa.2005.07.016

Cielen, A., Peeters, L., & Vanhoof, K. (2004). Bankruptcy prediction using a data envelopment analysis. *European Journal of Operational Research*, 154(2), 526–532. DOI: 10.1016/S0377-2217(03)00186-3

Cút, S. (2014). Prediction of Company Financial Distress Using Neural Network Based on the Radial Basis Function. *Lecture Notes in Management Science - 2014 2nd International Conference in Humanities, Social Sciences and Global Business Management (ISSGBM 2014).*

Daudt, H. M., van Mossel, C., & Scott, S. J. (2013). Enhancing the scoping study methodology: A large, interprofessional team's experience with Arksey and O'Malley's Framework. *BMC Medical Research Methodology*, 13(1), 48. DOI: 10.1186/1471-2288-13-48 PMID: 23522333

Davis, K., Drey, N., & Gould, D. (2009). What are scoping studies? A review of the nursing literature. *International Journal of Nursing Studies*, 46(10), 1386–1400. DOI: 10.1016/j.ijnurstu.2009.02.010 PMID: 19328488

Deakin, E. B. (1972). A Discriminant Analysis of Predictors of Business Failure. *Journal of Accounting Research*, 10(1), 167–179. DOI: 10.2307/2490225

Devi, S. S., & Radhika, Y. (2018). A survey on machine learning and statistical techniques in bankruptcy prediction. *International Journal of Machine Learning and Computing*, 8(2), 133–139. DOI: 10.18178/ijmlc.2018.8.2.676

Ding, Y., Song, X., & Zen, Y. (2008). Forecasting financial condition of Chinese listed companies based on support vector machine. *Expert Systems with Applications*, 34(4), 3081–3089. DOI: 10.1016/j.eswa.2007.06.037

Ding, Y., Song, X., & Zen, Y. (2008). Forecasting financial condition of Chinese listed companies based on support vector machine. *Expert Systems with Applications*, 34(4), 3081–3089. DOI: 10.1016/j.eswa.2007.06.037

Doumpos, M., & Zopounidis, C. (1999). A Multicriteria Discrimination Method for the Prediction of Financial Distress: The Case of Greece. *Multinational Finance Journal*, 3(2), 71–101. DOI: 10.17578/3-2-1

Farooq, U., & Qamar, M. A. J. (2019). Predicting multistage financial distress: Reflections on sampling, feature and model selection criteria. In *Journal of Forecasting* (Vol. 38, Issue 7, pp. 632–648). DOI: 10.1002/for.2588

Farooq, U., Jibran Qamar, M. A., & Haque, A. (2018). A three-stage dynamic model of financial distress. *Managerial Finance*, 44(9), 1101–1116. DOI: 10.1108/MF-07-2017-0244

Fawzi, N. S., Kamaluddin, A., & Sanusi, Z. M. (2015). *Monitoring Distressed Companies through Cash Flow Analysis. 7th International Conference on Financial Criminology 2015. 28.* ScienceDirect., DOI: 10.1016/S2212-5671(15)01092-8

Fisher, R. (1936). The use of multiple measurements in taxonomic problems. *Annals of Eugenics*, 7(2), 179–188. DOI: 10.1111/j.1469-1809.1936.tb02137.x

Gregova, E., Valaskova, K., Adamko, P., Tumpach, M., & Jaros, J. (2020). Predicting financial distress of slovak enterprises: Comparison of selected traditional and learning algorithms methods. *Sustainability (Basel)*, 12(10), 3954. DOI: 10.3390/su12103954

Hafiz, A., Lukumon, O., Muhammad, B., Olugbenga, A., Hakeem, O., & Sa-heed, A. (2015) Bankruptcy prediction of construction businesses: towards a big data analytics approach. In: Big data computing service and applications (Big-DataService), 2015 IEEE first international conference, Redwood City, CA, USA, 347–352. DOI: 10.1109/BigDataService.2015.30

Henebry, K. L. (1996). Do cash flow variables improve the predictive accuracy of a Cox proportional hazards model for bank failure? *The Quarterly Review of Economics and Finance*, 36(3), 395–409. DOI: 10.1016/S1062-9769(96)90023-X

Jones, S. (2016). A Cash Flow Based Model of Corporate Bankruptcy in Australia. *JAMAR, 14*(1).

Jones, S., & Hensher, D. A. (2004). Predicting Firm Financial Distress: A Mixed Logit Model. *The Accounting Review*, 79(4), 1011–1038. DOI: 10.2308/accr.2004.79.4.1011

Kang, C.-M., Wang, M.-C., & Lin, L. (2022). Financial Distress Prediction of Cooperative Financial Institutions—Evidence for Taiwan Credit Unions. In *International Journal of Financial Studies* (Vol. 10, Issue 2, p. 30). DOI: 10.3390/ijfs10020030

Karas, M., & Režňáková, M. (2015). Are the bankruptcy predictors specific for a given time or branch? the evidence from the czech manufacturing and construction companies. *Finance and Performance of Firms in Science, Education and Practice.* researchgate. Retrieved from https://www.researchgate.net/publication/275353210

Kashyap, S., & Bansal, R. (2019). Modeling Financial Distress Prediction of Indian Companies. *International Journal of Recent Technology and Engineering (IJRTE), 8*(1C2).

Khademolqorani, S., Hamadani, A. Z., & Rafiei, F. M. (2015). A Hybrid Analysis Approach to Improve Financial Distress Forecasting: Empirical Evidence from Iran. In *Mathematical Problems in Engineering* (Vol. 2015, pp. 1–9). DOI: 10.1155/2015/178197

Khoja, L., Chipulu, M., & Jayasekera, R. (2019). Analysis of financial distress cross countries: Using macroeconomic, industrial indicators and accounting data. *International Review of Financial Analysis*, 66(101379), 101379. DOI: 10.1016/j.irfa.2019.101379

Li, J., & Du, W. (2011). An Empirical Study on the Corporate Financial Distress Prediction Based on Logistic Model: Evidence from China's Manufacturing Industry. *International Journal of Digital Content Technology and its Applications, 5*(6). DOI: 10.4156/jdcta.vol5.issue6.44

Li, S., Shi, W., Wang, J., & Zhou, H. (2021). A Deep Learning-Based Approach to Constructing a Domain Sentiment Lexicon: a Case Study in Financial Distress Prediction. In *Information Processing & Management* (Vol. 58, Issue 5, p. 102673). DOI: 10.1016/j.ipm.2021.102673

Liang, C., Wu, Z., Huang, W., & Giles, C. (2015). Measuring Prerequisite Relations Among Concepts. *InProceedings of the 2015 conference on empirical methods in natural language processing*, (pp. 1668-1674). DOI: 10.18653/v1/D15-1193

Maheshwari, Y., & Agrawal, K. (2016). Predicting financial distress: Revisiting the option based model. *South Asian Journal of Global Business Research.*

Malakauskas, A., & Lakštutienė, A. (2021). Financial distress prediction for small and medium enterprises using machine learning techniques. *The Engineering Economist*, 32(1), 4–14. DOI: 10.5755/j01.ee.32.1.27382

Mamatzakis, E., & Koutsomanoli-Filippaki, A. (2009, November). Performance and Merton-type default risk of listed banks in the EU: A panel VAR approach. *Journal of Banking & Finance*, 33(11), 2050–2061. DOI: 10.1016/j.jbankfin.2009.05.009

Martin, D. (1977). Early warning of bank failure: A logit regression approach. *Journal of Banking & Finance*, 1(3), 249–276. DOI: 10.1016/0378-4266(77)90022-X

Mays, N., Roberts, E., & Popay, J. (2001). Synthesising research evidence. In Fulop, N., Allen, P., Clarke, A., & Black, N. (Eds.), *Studying the organisation and delivery of health services: research methods*. Routledge.

Merton, R. C. (1974, December 28-30). On the pricing of corporate debt: The risk structure of interest rates. *The Journal of Finance*, 449-470.

Monti, N. E., & Garcia, R. M. (2010). A Statistical Analysis to Predict Financial Distress. In *Journal of Service Science and Management* (Vol. 03, Issue 03, pp. 309–335). DOI: 10.4236/jssm.2010.33038

Nair, J., & Sachdeva, J. (2018, March). Financial distress indicators of indian manufacturing companies–A factor analysis approach. *Global Journal for Research Analysis*, 7(3), 66–69.

Niemann, M., Schmidt, J. H., & Neukirchen, M. (2008). Improving performance of corporate rating prediction models by reducing financial ratio heterogeneity. *Journal of Banking & Finance*, 32(3), 434–446. DOI: 10.1016/j.jbankfin.2007.05.015

Ninan, N., Roy, J. C., & Thomas, M. R. (2019). Benefits of Cross - Training: Scale Development and Validity. In *Prabandhan: Indian Journal of Management* (Vol. 12, Issue 6, p. 41). DOI: 10.17010/pijom/2019/v12i6/144935

Noronha, M. R., & Mehta, P. U. (2012). Corporate governance and investment decisions. *Prabandhan: Indian Journal of Management*, 5(5), 41. DOI: 10.17010/pijom/2012/v5i5/60202

Nyitrai, T. (2019). Dynamization of bankruptcy models via indicator variables. *Benchmarking*, 26(1), 317–332. Advance online publication. DOI: 10.1108/BIJ-03-2017-0052

Ohlson, J. A. (1980). Financial Ratios and the Probabilistic Prediction of Bankruptcy. *Journal of Accounting Research*, 18(1), 109–131. https://www.jstor.org/stable/2490395. DOI: 10.2307/2490395

Oz, I. O., & Yelkenci, T. (2017). A theoretical approach to financial distress prediction modeling. *Managerial Finance*, 43(2), 212–230. DOI: 10.1108/MF-03-2016-0084

Perry, L. G., Henderson, G. V.Jr, & Cronan, T. P. (1984). Multivariate analysis of corporate bond ratings and industry classifications. *Journal of Financial Research*, 7(1), 27–36. DOI: 10.1111/j.1475-6803.1984.tb00351.x

Pham, M. T., Rajic, A., Greig, J. D., Sargeant, J. M., Papadopoulos, A., & McEwen, S. A. (2014). A scoping review of scoping reviews: Advancing the approach and enhancing the consistency. *Research Synthesis Methods*, 5(4), 371–385. DOI: 10.1002/jrsm.1123 PMID: 26052958

Psillaki, M., Tsolas, I., & Margaritis, D. (2010). Evaluation of credit risk based on firm performance. *European Journal of Operational Research*, 201(3), 873–881. DOI: 10.1016/j.ejor.2009.03.032

Rajpurohit, P. D., & Rijwani, P. R. (2020). Financial reporting practices and internal corporate governance mechanisms in emerging markets. *Prabandhan: Indian Journal of Management*, 13(8-9), 8. DOI: 10.17010/pijom/2020/v13i8-9/155230

Rao, N. V., Atmanathan, G., Shankar, M., & Ramesh, S. (2013). Analysis of bankruptcy prediction models and their effectiveness: An indian perspective. *Great Lakes Herald*, 7(2), 3–17.

Ray, S. (2011). Assessing Corporate Financial Distress in Automobile Industry of India: An Application of Altman's Model. *Research Journal of Finance and Accounting, 2*(3).

Ritchie, J., & Spencer, L. (1994). Qualitative data analysis for applied policy research. In Bryman, A., & Burgess, R. G. (Eds.), *Analysing qualitative data*. Routledge. DOI: 10.4324/9780203413081_chapter_9

Sam, C. Y. (2011). Corporate governance and financial crises: Lessons from Singapore's public sector. *Prabandhan: Indian Journal of Management*, 4(4), 3. DOI: 10.17010/pijom/2011/v4i4/62403

Senbet, L. W., & Wang, T. Y. (2012). Corporate fnancial distress and bankruptcy: A survey. *Found Trends Finance*, 5(4), 243–335. DOI: 10.1561/0500000009

Sharma, N., Arahna, V. P., Saxena, S., Ateef, M., & Samuel, A. J. (2022). Scoping out the scope of scoping reviews in neonatal and pediatric pain management: A scoping review methodological framework. In *Journal of Pediatric Surgery* (Vol. 57, Issue 8, pp. 1599–1608). DOI: 10.1016/j.jpedsurg.2021.11.012

Shilpa, N., & Amulya, M. (2017). Corporate Financial Distress: Analysis of Indian Automobile Industry. *SDMIMD Journal of Management*, 8(1), 85. Advance online publication. DOI: 10.18311/sdmimd/2017/15726

Shivdas, A., & Chandrasekhar, J. (2016). Sustainability through frugal innovations: An application of Indian spiritual wisdom. *Prabandhan: Indian Journal of Management*, 9(5), 7–23. DOI: 10.17010/pijom/2016/v9i5/92567

Shumway, T. (2001, January). Forecasting Bankruptcy More Accurately: A Simple Hazard Model. *The Journal of Business*, 74(1), 101–124. DOI: 10.1086/209665

Srivastava, S., & Bagga, T. (2014). A Comparative Study on the Usage of HRIS in the IT/ITES, Services, and Manufacturing Sectors in the Indian Scenario. In *Prabandhan: Indian Journal of Management* (Vol. 7, Issue 6, p. 21). DOI: 10.17010/pijom/2014/v7i6/59325

Sucharew, H., & Macaluso, M. (2019). Methods for research evidence synthesis: The scoping review approach. *Journal of Hospital Medicine*, 14(7), 416–418. DOI: 10.12788/jhm.3248 PMID: 31251164

Sun, B., & Li, J. (2008). The Recognition of new and old banknotes Based on SVM. *Second International Symposium on Intelligent Information Technology Application, 2*, 95-98. DOI: 10.1109/IITA.2008.157

Sun, J., Fujita, H., Zheng, Y., & Ai, W. (2021). Multi-class financial distress prediction based on support vector machines integrated with the decomposition and fusion methods. In *Information Sciences* (Vol. 559, pp. 153–170). DOI: 10.1016/j.ins.2021.01.059

Sun, J., Li, H., Huang, Q.-H., & He, K.-Y. (2013). Predicting financial distress and corporate failure: A review from the state-of-the-art definitions, modeling, sampling, and featuring approaches. *Knowledge-Based Systems*.

Tinoco, H. M., & Wilson, N. (2013, February 15). Financial distress and bankruptcy prediction among listed companies using accounting, market and macroeconomic variables. *International Review of Financial Analysis*, 30, 394–419. Advance online publication. DOI: 10.1016/j.irfa.2013.02.013

Trujillo-Ponce, A., Samaniego-Medina, R., & Cardone-Riportella, C. (n.d.). Examining What Best Explains Corporate Credit Risk: Accounting-Based versus Market-Based Models. In SSRN *Electronic Journal*. DOI: 10.2139/ssrn.2072176

Tseng, F.-M., & Hu, Y.-C. (2010). Comparing four bankruptcy prediction models: Logit, quadratic interval logit, neural and fuzzy neural networks. *Expert Systems with Applications*, 37(3), 1846–1853. DOI: 10.1016/j.eswa.2009.07.081

Wang, G. (2016). Predicting financial distress for privately held firms in the European Union. *University of Sydney Business School*, 240.

Waqas, H., & Md-Rus, R. (2018). Predicting fnancial distress: Importance of accounting and frm-specifc market variables for Pakistan's listed frms. *Cogent Economics & Finance*, 6(1), 1545739. DOI: 10.1080/23322039.2018.1545739

Xiao, Z., Yang, X., Pang, Y., & Dang, X. (2012). The prediction for listed companies' financial distress by using multiple prediction methods with rough set and Dempster–Shafer evidence theory. *Knowledge-Based Systems*, 26, 196–206. DOI: 10.1016/j.knosys.2011.08.001

Xu, M., & Zhang, C. (2009). Bankruptcy prediction: The case of Japanese listed companies. *Review of Accounting Studies*, 14(4), 534–558. DOI: 10.1007/s11142-008-9080-5

Zhang, R., Zhang, Z., Wang, D., & Du, M. (2022). Financial Distress Prediction with a Novel Diversity-Considered GA-MLP Ensemble Algorithm. In *Neural Processing Letters* (Vol. 54, Issue 2, pp. 1175–1194). DOI: 10.1007/s11063-021-10674-9

Zmijewski, M. E. (1984). Methodological Issues Related to the Estimation of Financial Distress Prediction Models. *Journal of Accounting Research*, 22, 59–82. DOI: 10.2307/2490859

Chapter 9
Review of Citizen Engagement in Public Service Delivery via Volunteered Geographic Information

Munir Ahmad
https://orcid.org/0000-0003-4836-6151
Survey of Pakistan, Pakistan

ABSTRACT

Citizen participation enhances public services by involving individuals in their design, implementation, and evaluation, fostering transparency and accountability. Volunteered geographic information (VGI) leverages technology, such as smartphones and GPS, for efficient, large-scale geospatial data collection, benefiting urban planning, disaster response, and environmental sustainability. Institutional support through open data policies, financial backing, and collaborative partnerships is crucial for VGI's success. OpenStreetMap exemplifies VGI's transformative potential across various sectors. However, challenges like data quality, privacy concerns, and inclusivity need addressing through robust validation systems, training, and standardized methods. Sustaining volunteer engagement and ensuring seamless data integration are essential for maximizing VGI's impact.

DOI: 10.4018/979-8-3693-9246-1.ch009

INTRODUCTION

Citizen participation means the various modalities and possibilities through which people can directly participate in the design, implementation, and evaluation of public services (Jarke, 2021; Nabatchi et al., 2017; Waddington et al., 2019). It must be noted that this participation is very important in delivering improved public services, improving efficiency, and increasing the effectiveness of those services. Public participation is a range of activities and key levels of interaction where people can provide feedback on services, attend meetings or any events related to the community, or participate in delivering and designing the services.

Citizen participation is most informative because it allows the masses to oversee and scrutinize governmental organizations. Engagement initiatives involve people so that they can become active participants in issues that touch on their lives and emotions, enabling them to champion the voices of the population. With these, it is easier to measure the extent of congruency between the public services that are provided and the needs of the people, increasing the likelihood of the public services delivering what is expected. In addition, it cultivates increased interdependence between people and authorities, which increases social cohesion and confidence in the community to overcome obstacles or achieve objectives (Castelnovo, 2016; Fung, 2015; Ponti & Craglia, 2020).

Volunteered geographic information (VGI) refers to geographic information generated and collected by the public without force being applied to accomplish the same (Goodchild, 2007). More specifically, VGI is based on the availability of smartphones, GPS-enabled tools, and internet connections so that everyone with a certain degree of knowledge can participate in geographic mapping and data collection. VGI can be considered valuable because a large amount of geospatial data can be collected and hence can be applied in many sectors, including urban planning, emergency responses, and environmental sustainability, among others (Basiouka et al., 2015; Liu et al., 2020; Olteanu-Raimond et al., 2020). In this regard, VGI methodologies can also help engage the public in data gathering to increase their participation and responsibility. It generates a better and more informed society, which can bring important knowledge and tips related to the local setting that might not be available in other cases. The workflow of the VGI is illustrated in Figure 1.

Furthermore, compared to other common approaches to data collection, working with VGI is likely to be much more efficient. Thus, it can be of particular interest to public authorities without sufficiently large budgets. The advantages derivable from the use of VGI can enable these organizations to source other concerns or serve multiple goals by saving cost and or time. Besides, VGI presents timely information, which can be very useful, especially in emergency cases such as disasters and calamities. The inclusion and provision of proper and early information can go a long way in increasing response effectiveness, often leading to fewer fatalities and losses. Making the data real-time limits the time taken to ensure that the information for decision-making and responding is as up-to-date as possible (Tzavella et al., 2022).

Given the above context, the overall objective of this chapter is to review citizen engagement in public services, with a special focus on Volunteered Geographic Information (VGI). To achieve this objective, the chapter is organized as follows: The next section discusses the enabling technological and institutional frameworks that support VGI. The third section highlights a case study of the use of OpenStreetMap in public services. The fourth section explores the challenges associated with VGI and presents strategies to mitigate these challenges. The final section concludes the chapter, summarizing key insights and findings.

Figure 1. VGI Workflow

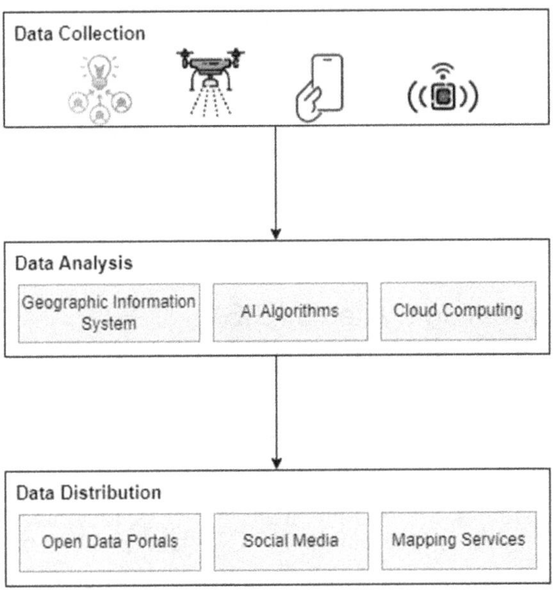

TECHNOLOGICAL AND INSTITUTIONAL PERSPECTIVES

Technology plays an important role in the collection, analysis, and distribution of VGI. These technologies enable people to contribute to the geographic data and ensure that the data collected is utilized appropriately.

Technological Perspective

Smartphones and Tablets

Global Positioning Systems (GPS) enabled in handheld devices such as smartphones and tablets are core to VGI (Antoniou, 2017). These devices make it possible for any common person to become involved, in one way or another, in the collection and sharing of geospatial data, thus reducing the gap between expert geographers and users. These applications are purposely made to be as user-friendly as possible so that users with little statistical background can come up with valuable data. The users can conveniently make annotations, trace lines, and share photos or other relevant information, which are then linked to the general map. This does not only increase the coverage of data but also ensures that the data is updated in a manner that reflects real-world changes.

In addition, smartphones and tablets make it possible to collect data and share VGI in real time. This capability is especially useful in emergencies or other major occurrences because the latest available information can be quickly relayed to the responders and the community. For example, when disasters occur, VGI applications can be used to locate disaster zones and follow the movements of people and points in need of help, thereby increasing the efficiency of the response (Ahmad, 2023; Tzavella et al., 2022).

Geographic Information Systems (GIS)

GIS software like ArcGIS and QGIS can help compile and analyze the geospatial data captured by the VGI. These complex software applications are meant to manage, analyze, and present large amounts of spatial data, converting data into valuable information for decision-making (Giuffrida et al., 2019). Data from various sources like VGI, satellite images, and other GIS databases can be compiled in GIS platforms to capture a holistic view of geographical information. Furthermore, GIS can be used to capture data in map formats, charts, and graphs, which makes complex data convenient for users to understand. Advanced techniques in GIS enable users to make simple queries, measure distances, perform relational operations, and identify patterns in the data. In addition, the GIS platforms can be of great assistance when it comes to decision-making as informed by spatial analysis of available data. This is well illustrated in sectors like urban and rural development, calamity response, and ecology.

By integrating GIS software, large volumes of data gathered through VGI can easily be sorted, filtered, and summarized to allow for the identification of their core and significant aspects. This opens up the use of VGI data for different purposes, including planning and tracking environmental changes and emergencies. The ability to translate the raw data into useful information through maps and analysis enables the stakeholders to make informed decisions and make the right strategies in their geographical information (Giuffrida et al., 2019; González-Ramiro et al., 2016; Verplanke et al., 2016).

Cloud Computing

Cloud computing technologies form the basis of handling large data, such as geospatial information derived from VGI. Such technologies provide immense advantages that facilitate scalability and flexibility and improve collaboration in VGI projects. Cloud platforms offer flexible storage solutions; VGI projects will not run into space constraints while storing data and information. This is particularly significant given that volunteers could submit immense data through GPS gadgets. Cloud services provide a flexible usage of computing resources that can be extended or reduced depending on the demand of the given job. This enables simple applications, where there are repeated data processing updates, as well as complex analysis tasks, to be processed seamlessly (Ahmad & Ali, 2023; Kolak et al., 2020; Tripathi et al., 2020).

Cloud computing allows information to be updated on the go, meaning that any changes made or new information obtained is immediately available. This is especially important for use in systems such as disaster response, where efficient data could greatly improve the emergency assistance process. Data stored centrally also tends to be more accessible since it is stored in the cloud, so participants do not have to be in physical proximity to each other to bring value or participate in a conversation.

Multi-user collaboration is inherent in cloud environments as it enables one or more individuals to work on the same data simultaneously. This increases participation from volunteers, researchers, and policymakers, among other stakeholders. This flexibility permits users to be contained with their privileges in a project and allows project managers/administrators to allocate rights and duties in a project to different users appropriately. Most cloud providers provide a set of tools and services that can be used to support VGI projects, such as data storage, computing facilities, machine learning algorithms, and many others. Cloud platforms can be easily connected to other software and services, including GIS and data visualization tools, so different services become additional modules that help to improve VGI activities.

By utilizing cloud services, VGI projects can avoid the costs associated with maintaining physical servers and other hardware infrastructure (Kolak et al., 2020; Wu et al., 2020). Cloud providers often offer pay-as-you-go pricing models, allowing projects to only pay for the resources they use. This can significantly reduce costs, especially for smaller organizations or community-based initiatives. Cloud services are maintained by the providers, who ensure that the software and hardware are regularly updated. This reduces the burden on VGI project managers to keep up with the latest technology and security practices. Cloud platforms are designed for high reliability and uptime, ensuring that data is available and protected against loss or downtime.

Web Mapping Services

Web mapping services such as Google and Mapbox offer APIs and resources that can help implement VGI into various applications. These services provide a set of features that enable people to visualize, analyze, and share information, such as geospatial data gained through the use of VGI, making it easier and more beneficial to acquire such information. For instance, the Google Maps API offers end users a simple method to incorporate VGI data into their own individualized maps. It can allow the creation of flexible applications that can show both geographic information provided by the end-user and the data offered by Google. By leveraging integrated geolocation services, it becomes much easier to feed location-related information in real-time. This is particularly useful when the users are active participants in the application, for instance, when porting environmental changes or charting community assets (Sathianarayanan et al., 2024).

Furthermore, Google Maps features various visual representations, which can enable anyone to develop dynamic and innovative maps. It can also show the different features of location-based information, such as basic pins and sophisticated heat maps and overlays, which make the analysis of VGI data easier. Additionally, users can insert additional layers of personal information in Google Maps, which in return can provide the means of depicting a special kind of VGI data set. This feature allows the integration of imagery data with points, lines, or polygons on the map. These tools can enable the users to have an insight into the spatial patterns within the data set. This can be useful in decision-making, especially in areas such as urban planning or management of environmental resources/ hazards.

Social Media Networks

Social networks are appropriate in dynamic and emergent environments due to the ability to share information with a large number of participants and gather information quickly. These platforms enable information to be gathered in a short span of time and similarly disseminated. To achieve this, users can share updates in real time, and the best summaries are displayed to ensure that users have the most updated information. These platforms have broad coverage of the user base, which makes it possible to gather details from a large and cosmopolitan pool of participants (Sathianarayanan et al., 2024).

These tools allow participants from the general public to participate by ensuring that everyone can be heard and that data and inputs are gathered from different sources. This makes users feel that they are important and they are also part of the exercise of providing the right data, especially during emergency situations where users' data is vital for emergency response. Text and multimedia content can be created and submitted by users, and these data might include text, photographs, videos, or posts with geographic locations; the availability of such a diverse range of information allows for a deeper understanding of events.

For instance, Twitter users can post live status updates, such as geolocated tweets, that can be particularly helpful in crises or during major occurrences. For instance, a user can convey about traffic situations, floods, strikes, or any other calamities happening in a particular area and can be aware of the situation immediately. Hashtagging provides an optimal way to gather and collect data in the context of events or specific topics with the goal of constructing a geospatial database of such phenomena (Florath et al., 2024a, 2024b).

Like it, Facebook also allows users to share updates, photos, and videos with place check-in, which can be used to build a complete geography of events as they occur. These could be interest-based groups or groups defined by geographical location, hence aiding the process of sample collection and dissemination. Facebook's Crisis Response feature enables users to indicate if they are safe, find ways to help, or share information during emergencies. This feature especially comes in handy when it is needed to spread important information with the help of a large number of users of the platform.

Institutional Perspective

It is essential to note that Volunteered Geographic Information is not completely proactive and that governments are an important factor that either facilitates or hinders the operation of VGI initiatives. All these forms of support relate to various aspects of VGI projects and their implementation with safety and efficiency.

Policy Support

Policy support entails coming up with and enacting policies, practices, and rules that encourage the use of VGI. This means that there are measures such as the open data policies that compel the government to release geospatial data it acquires to the general public (Gao et al., 2023). Such policies help promote the values of open data and also awaken the original data from the citizens. This way, citizens can take a starting point in the form of government data and augment the advanced quality and range of VGI initiatives. Moreover, governments can establish best practices for opening the data, which can include the creation of common guidelines for gathering, storing, and disseminating the data. It is recommended that all VGI data is useful, reliable, and qualitative, which can enable easy blending with official data sets. Standardization can also enhance harmonization and interoperability across various VGI efforts coordination by preventing the creation of divergent standards.

Subsequently, guidelines for data protection and security remain some of the most important requirements that need to be met. Preserving the anonymity of the people who participate in VGI enables a lot of contributors to venture into the project without fearing for their details since they will not be freely exposed on different platforms. Such frameworks should explain how personal data is managed, processed, and shared to retain the public's trust. Furthermore, specific legal frameworks aimed at assigning responsibility for particular projects reduce risk, as all involved parties know their levels of exposure.

Financial Support

Another form of governmental support that is equally vital is funding that is offered to VGI projects. Such support can include financial and award budgets focused on supporting VGI endeavors. Some governments can also support costs such as costs associated with technology infrastructure or community mobilization and training of personnel. Such funding can help VGI projects secure the fundamental requirement that enables a project to work as planned and as planned. Another critical setup through which governments can financially encourage VGI initiatives is through innovation incentives. In particular, tax exemptions, research funding, and public-private partnerships can serve as state stimuli that would contribute to the advancement of platforms and tools for functioning and optimizing VGI data collection, analysis, and application. These incentives help to achieve innovation while ensuring that all VGI projects are advanced (Khosravi Kazazi et al., 2024).

Technological support is also another significant financial commitment. The government can make the necessary financial investments to build cloud storage, faster internet connections, and efficient data analysis. Such investments empower VGI projects in that they scale the capacity to accommodate large chunks of data and process data as they flow in, as well as allow for the integration of participants in real time. In addition, awareness-raising initiatives and development education are vital to prepare all possible actors in VGI, ranging from government employees to laymen, for their participation in the project and its management. Hence, introducing these educational tools can help governments improve the quality of the VGI and the latter's outcomes.

Legislative Support

Legislation can promote openness and citizen participation in the sharing of data, and if these legislations are implemented, the effectiveness of VGI projects will definitely improve. Participation instruments, such as the Freedom of Information laws, can ensure that members of the public have been granted the right to access information concerning government activities, following the legal formalities that have been put in place to encourage the two. These laws also can ensure that the necessary data is there and provided for citizens to use and contribute to VGI projects, which makes them more engaged.

Possible legal measures that can promote and enhance VGI include citizen engagement laws that seek to allow the public to be involved in decision-making processes. These laws can help embrace citizen contribution and shape their contribution towards formulation of public policies, hence promoting the integration of VGI data in the delivery of public services, thus increasing the responsiveness of government actions. Furthermore, national legislation that supports the acquisition, dissemination, diffusion, and application of geospatial information and knowledge in geographical space, more specifically, Geospatial Data Acts, can support the application of VGI in public administration and disaster management. These acts also work to ensure that geospatial data activities across the different departments are integrated to make sure that the data generated is in VGI form.

Partnerships and Collaborations

VGI initiatives widely rely on partnerships and collaborations. These partnerships are cardinal in that they assist in bringing in the specific type of knowledge and assistance that is required to enhance the efficiency and stability of VGI initiatives. For example, the incorporation of Public Private Partnerships (PPP) can lead to the merging of efforts from both the public and private sectors and hence escalate the competence of VGI projects. Some governmental stakeholders may provide regulatory services, public data, and even financial support. On the other hand, the private sector, particularly the companies that focus on mapping services and technologies, may offer efficient products, scientific knowledge, and creative approaches. These collaborations ensure that all future and current VGI projects incorporate the most optimal technological innovations and methodologies regarding data gathering, management, and sharing (Wang et al., 2021).

Nonprofit organizations also have a significant place in VGI because they unite groups of people, increase concern, and stand up for citizens and their activities. Such organizations have surely close links with the targeted communities and thus play a vital role in motivating a large number of individuals and involving them in the VGI initiatives. They can also participate in developing training and capacity-building programs necessary to put the participants in a position to contribute high-quality data. Also, nonprofits can assist in mediating relationships between external and internal constituents, both at the government and business levels, as well as the public (Urrea & Yoo, 2023; Yang et al., 2024).

Besides the government and private companies, academic institutions can play a crucial role in VGI projects through scientific knowledge, analysis, and educational facilities. University institutions and research centers can then fund research projects that seek to verify or extend the techniques used to gather VGI to ensure the quality of the resulting data. They can also create innovations appropriate to VGI projects and develop equipment that will generate the best results. Additionally, organizations receive a source of manpower and ideas from the various student and researcher populations that may be accessible through many academic institutions. It also means that incorporating VGI projects in their respective fields can help develop a new generation of geospatial workers and hobbyists.

Mapping companies bring a wealth of experience and technological capabilities to VGI projects. These companies, specializing in geospatial data and mapping technologies, can offer state-of-the-art software, hardware, and data collection, processing, and visualization platforms. Their expertise in creating and managing geospatial databases is invaluable for ensuring the robustness and scalability of VGI projects. Mapping companies can also facilitate the integration of VGI data with other geospatial datasets, enhancing the overall utility and impact of the information collected (Yang et al., 2024).

Training and Capacity Building

Education and training in the area of humanitarian causes can improve the skills and efficiency of those people and organizations participating in VGI projects. Through the training programs and acquisition of skills development, stakeholders can be assured of participants' performance since these individuals can be equipped with vital skills to enable them to undertake VGI activities. This preparation not only enhances the caliber of the data gathered but also the involvement, enthusiasm, and expertise of the community of contributors (Kitsios & Kamariotou, 2023).

Training activities should be adopted because they are effective in preparing participants with the knowledge and skills needed in VGI projects. These programs generally encompass various areas of study, such as the application of geographic information systems, how data is collected, data accuracy and integrity issues, and legal concerns. By orienting participants to these dimensions, training programs can ensure that the data provided is meaningful, accurate, and relevant to different uses. The importance of extending training and capacity building as part of the several objectives of VGI projects depends on the following: They can assume the responsibility of making participants knowledgeable about these procedures and hence improve the overall performance of VGI initiatives. This ultimately results in better outcomes in terms of data, projects, or even citizens' involvement in projects around their environment.

THE CASE OF OPENSTREETMAP

OpenStreetMap (OSM) is a project that intends to map the world to the extent that it can be amended by anyone, and it collects geographic data from many people (openstreetmap, 2023). This has made OSM to be widely incorporated into the humanitarian effort response activities. For example, when the Haiti earthquake occurred in 2010, people who volunteered coordinated the areas that required help from some relief agencies (Anderson-Tarver, 2015; Mikel, 2011). This enabled the real-time mapping of the affected areas, thus facilitating faster and more efficient rescue operations and the management of resources and supplies in the probable catastrophe. It also showed the significance of VGI in managing such situations.

OpenStreetMap has provided detailed maps and geographic information to the general public, especially those in the developing and inaccessible areas where such information was previously a preserve of a few individuals with expensive equipment and resources. The concerns about the data quality of OSM (Bai et al., 2024; Schott et al., 2024) can be minimized by establishing strong data validation systems to enhance the reliability of the generated VGI data. This may include consistency checks with licensed data and the application of automated diagnostic programs. OpenStreetMap involves a nonprofit mapping model in which senior mappers approve or reject contributions based on their quality. OSM is being broadly utilized in public service delivery over different spaces due to its collaborative and open nature, wealthy geospatial information, and availability.

Figure 2. OpenStreetMap

Emergency Response and Disaster Management

Amid natural disasters such as earthquakes, floods, and tropical storms, OpenStreetMap serves as a basic instrument for mapping influenced areas in real time. Volunteer mappers can quickly update the map with information on damaged infrastructure, road closures, evacuation routes, and relief distribution centers. This up-to-date data makes a difference; crisis responders can arrange and facilitate their endeavors more successfully. Organizations like the Humanitarian OpenStreetMap Team (HOT) collaborate with humanitarian agencies to map vulnerable localities around the world (HOTOSM, 2023). These maps can be utilized for catastrophe readiness, reaction, and recuperation endeavors, guaranteeing that aid reaches those in need expeditiously and effectively (HOTOSM, 2023a).

Urban Planning and Infrastructure Development

OpenStreetMap can empower communities to map their neighborhoods, documenting local amenities, transportation networks, and points of interest (Yang et al., 2024). This community-generated data can illuminate urban planning decisions, such as the location of schools, waste disposal routes, rescue centers, healthcare facilities, and public transportation routes. Moreover, OpenStreetMap data can be used by transportation authorities and app developers to create navigational tools, route planners, and public transit maps. This helps commuters navigate cities more efficiently and promotes the use of sustainable transportation options (Hoffmann et al., 2024).

Environmental Conservation and Management

Environmental organizations can utilize OpenStreetMap to map and monitor natural resources such as woodlands, wetlands, and biodiversity hotspots (Chen, 2024). This data can aid in conservation efforts, habitat restoration, and sustainable land management practices. Moreover, researchers can track animal movements and habitat usage by overlaying GPS data onto OpenStreetMap. This information can assist in wildlife conservation efforts, species protection, and the identification of critical migration corridors.

Public Health and Social Services

OpenStreetMap data can be used to evaluate healthcare accessibility and healthcare facility distribution, particularly in underserved zones (Bartzokas-Tsiompras, 2022; Minghini et al., 2022). This information can help policymakers to close the gaps in healthcare services and allocate assets more evenhandedly. Furthermore, during public health emergencies such as disease outbreaks or pandemics, OpenStreetMap can be utilized to map healthcare facilities, quarantine zones, and testing centers. This supports emergency response efforts and facilitates the dissemination of critical information to the public.

Civic Engagement and Advocacy

OpenStreetMap can empower communities to advocate for improved public services and infrastructure by mapping areas lacking essential amenities such as clean water sources, sanitation facilities, and electricity access. This grassroots mapping effort raises awareness of socioeconomic disparities and mobilizes support for targeted interventions. OpenStreetMap can also be used to map accessibility features such as wheelchair ramps, pedestrian crossings, and tactile paving. This data promotes inclusivity and supports advocacy efforts for individuals with disabilities, ensuring equal access to public spaces and services (Hasselwander et al., 2022).

CHALLENGES AND LIMITATIONS

Decision-makers should note that VGI-driven citizen engagement initiatives are not without several challenges, as presented below.

Data Quality

The data gathered through VGI may be less accurate than the data gathered through a more professional approach because of the variation in skills among the volunteers (Antoniou, 2023; Tavra et al., 2024). It can be useful to record how inconsistent data accuracy, data completeness, and data reliability affect the application of VGI in decision-making. Volunteers usually do not have training or experience in geo-location, which results in errors in the gathered data. Data quality can be attributed to the fact that different approaches and tools can be used in data collection, which leads to inconsistency. Moreover, the validity of the VGI data requires time and a lot of resources to ensure they are accurate and satisfactory.

To this end, awareness and development of training programs and accessories aimed at the volunteers can enhance the quality of collected data. Moreover, upgraded data validation and verification can improve data quality by cross-checking the reports with official statistics or utilizing error control and checking mechanisms. Furthermore, standardized data collection can be encouraged through the promotion of guidelines that will make the quality of VGI data more consistent.

Privacy Concerns

It is imperative to understand that the collection and dissemination of geospatial data can present acute privacy concerns. If not well managed, personal information and location data provided through VGI projects can be used maliciously to infringe on individual rights and lead to a loss of confidence in crowdsourced activities. In the process of collecting geospatial data, it is easy to come across other sensitive information, such as people's homes or their movement patterns. Moreover, data collected from the sample can be accessed and misused when appropriate security measures are not adopted. Therefore, it becomes critical to guarantee that volunteers understand how the data they provide will be used and even seek their permission.

To tackle privacy concerns, anonymization and data masking solutions can be applied while preserving the data's usefulness. In addition, incorporating features like passwords or keys for data encryption can help discourage forgery and unauthorized access to the VGI data. Moreover, making privacy policies clear and gaining informed consent from the participants can help in gaining their trust and avoid compromising ethics in data collection.

Inclusivity

There are several issues that can make the involvement of the public in VGI projects a difficult process. Technological access barriers due to socioeconomic differences may contribute to constraints on the ability to participate in VGI projects. A limitation to VGI may be based on the low representation of rural and remote areas, as these may record low participation. To address the problem of lack of access to assistive technology among under-represented groups, the provision of technology along with training in its use may help overcome the lack of participation. Moreover, incorporating effective and user-friendly data capture tools may go a long way in creating an enabling environment for collecting data from multiple people and settings, especially those who may not necessarily have the technical know-how.

Sustainability

One of the major challenges that various organizations experience is how to retain their volunteers for long-term service. The volunteers may start to work with a low level of commitment, so they may get bored and stop volunteering, reducing data collection. Volunteers can disengage and stop volunteering after a while, be it due to a lack of visible contribution or appeal in the engagement activities. Moreover, the personnel or volunteers may have other personal and professional engagements, which is why they cannot dedicate a long time to the cause.

Compensation encourages volunteers by offering them something in return, such as appreciation/reward money or small grants. Moreover, training and development can be provided to the volunteers and staff to help them gain more experience in the project and increase their commitment level. To maintain interest and continued volunteer participation, volunteers must feel involved and valued as members of a community, which is achieved through frequent progress reports, end-user feedback, and organized community activities.

One of the areas of difficulty within VGI is funding, where interested parties can sometimes struggle to secure steady funding for the continuation of the project. It may be difficult to sustain appropriate technology, training, and data obtainment due to poor funding. The dependence on employees in public agencies and organizations might be a problem where there is no available and experienced staff to administrate and analyze the data from VGI projects. In order for technological platforms used in VGI to operate continuously and be effective, constant upgrades and maintenance must be done. This means there has to be continuous funding and application of technical know-how. While adopting a certain technology, VGI projects need to have their working means and tools updated frequently since new ones may develop frequently or for various reasons.

To foster VGI initiatives, one has to ensure that such investigations are owned and endorsed by community members through the formation of trust relationships. Otherwise, volunteers may lose their motivation to support causes or organizations. Information transfer among the organizers and volunteers should be on a daily or weekly basis, depending on the number of events in a week, since it is important to keep the participants informed on all occasions.

Grant collaborations and crowdfunding can be used to expand the funding framework for reliable revenue. Moreover, volunteers can be further encouraged by incentives like rewards, recognition, or even small financial grants that they can be willing to give in constant volunteerism. To monitor and get volunteers reactivated, more feedback and updates should regularly be made available to the volunteers regarding the extent to which they are helping to improve the cause.

Data Integration

There are a number of challenges associated with including VGI data with existing official data sets and systems; for example, the format, standards, and quality of these sources can vary significantly. This can negatively affect the likelihood and efficiency of VGI in enhancing public service delivery.

VGI data may be in various forms; thus, the data may not be well-suited to fit key government and GIS databases. This incompatibility can cause issues that manifest in a lack of integration or an inability to effectively employ the associated modules. Since the commonly used app, platform, and device's VGI data originate from these sources, they may also inherently possess different structures and lack uniformity to be merged directly with the standardized dataset. A stronger emphasis could also be placed on advocating for the utilization of data collection techniques that are standard and compatible with current systems. The integration of VGI data with other data sources to which it is to be merged requires the use of APIs, middleware, and other integration means.

As the volunteers may have different levels of expertise and can use a range of approaches to collect VGI data, the quality of these data may vary. Methods for verifying the accuracy of the gathered data are crucial to incorporating the information into the database. The observations using VGI data can also have volunteer bias in the selection of area or population of interest, resulting in the data not being generalizable and accurate in large-scale applications. Implementing good data validation and data quality checks leads to improved quality of VGI data, therefore making it more acceptable to users and for inclusion in other higher-level systems. Moreover, it is crucial that VGI initiatives, governments, and stakeholders collaborate to promote the sharing of experiences and strategies along with the underlying technology solutions for the integration of data.

While data sharing of traditional GIS is challenge-free since the standards are well-defined, VGI data lacks clear guidelines for integration with the official data. Implementation of standards is another crucial factor that enhances interoperability. To support the application based on the VGI data and integrate such a system with the existing applications, it is essential to have APIs and other tools that can help in such integration; this means that integration work also requires technical work, time, and staff. Ensuring VGI data works seamlessly with existing structures may be achieved by implementing interoperability standards, including implementations of the Open Geospatial Consortium (OGC). Some public agencies may be reluctant to use and incorporate VGI data into their platforms since the data might not be accurate and reliable, or the integration process may cause a shift away from the customary organizational working patterns. Moreover, some measures of regulating and using information sharing and data may become a limitation in integrating VGI data.

CONCLUSION

Citizen participation involves various ways through which individuals can engage in the design, implementation, and evaluation of public services, enhancing their efficiency and effectiveness. This participation allows people to provide feedback, attend community events, and take part in service delivery, fostering transparency and accountability in governmental organizations. VGI is a form of citizen participation where individuals collect and share geographic data using smartphones, GPS tools, and the internet. VGI is valuable for collecting large amounts of geospatial data efficiently and cost-effectively, benefiting urban planning, emergency responses, and environmental sustainability. Engaging the public in VGI not only enhances data quality and relevance but also fosters a more informed and cohesive society.

Technology plays a crucial role in the collection, analysis, and distribution of VGI. Smartphones and tablets with GPS capabilities enable widespread public participation in geospatial data collection by allowing users to easily annotate, trace, and share relevant information. These devices facilitate real-time data collection, which is particularly useful in emergencies, enhancing the efficiency of disaster response efforts. GIS can help compile and analyze this data, converting it into valuable information for decision-making in urban planning, disaster response, and environmental management. Cloud computing further enhances VGI projects by providing scalable and flexible storage solutions, enabling real-time data updates and multi-user collaboration. Additionally, web mapping services like Google Maps and social media networks enable users to visualize, analyze, and share VGI, making data collection and dissemination more efficient and accessible.

Institutional support is essential for the success of VGI initiatives, with governments playing a critical role in facilitating or hindering these projects. Policy support through open data policies and guidelines for data protection ensures the reliability and interoperability of VGI data. Financial support, including funding for technology infrastructure and community mobilization, as well as innovation incentives, can significantly boost VGI projects. Legislative measures such as Freedom of Information laws and citizen engagement laws promote transparency and public participation in VGI. Partnerships and collaborations with public-private sectors, nonprofit organizations, academic institutions, and mapping companies enhance the efficiency and sustainability of VGI projects. These collaborations bring together regulatory services, financial support, technological expertise, and community engagement, ensuring that VGI initiatives are well-supported and impactful. Training and capacity-building programs further empower participants by equipping them with the necessary skills and knowledge to contribute effectively to VGI projects, resulting in higher-quality data and greater community involvement.

OpenStreetMap has proven to be a transformative tool in various sectors by leveraging Volunteered Geographic Information. Its open and collaborative nature has enabled the creation of detailed maps that are invaluable, especially in areas where geographic data was previously inaccessible. In humanitarian response, OSM's impact is profound, as seen in the aftermath of the 2010 Haiti earthquake, where it facilitated real-time mapping, efficient resource management, and effective rescue operations. OSM's application extends beyond disaster management to urban planning, environmental conservation, public health, and civic engagement, underscoring its versatility and utility.

Despite its benefits, the quality of data in OSM remains a concern. Variability in volunteer skills and the inconsistency of collected data pose challenges. Implementing robust data validation systems, training programs, and standardizing data collection methods can significantly enhance the reliability of OSM. Addressing privacy concerns through anonymization and encryption and ensuring informed consent is crucial to maintaining public trust in VGI initiatives. Inclusivity is another challenge, as technological and socioeconomic barriers can limit participation. Providing access to technology and user-friendly data capture tools can help bridge this gap. The sustainability of volunteer engagement is vital for long-term success, which can be supported through recognition, training, and regular feedback. Integrating VGI data with existing systems requires overcoming issues related to data compatibility and quality. Standardizing data collection techniques and promoting interoperability through tools like APIs and middleware can facilitate smoother integration. Collaboration among VGI initiatives, governments, and stakeholders is essential to share strategies and enhance data integration efforts.

VGI offers vast potential for improvement in future citizen engagement crusades as defined by emergent technologies and progressive tactics. A new generation of mobile applications needs to be designed that are easier to interact with and reach a broader audience within VGI projects. Moreover, risk communication should be improved through the use of social media in emergencies to increase the opportunities for the rapid collection and sharing of data. Adapting gamification features to VGI systems to increase the level, duration, and quality of volunteers' engagement in data collection tasks through positive reinforcement based on intrinsic rewards obtained from game-like tasks is another future direction that can be taken up. AI can efficiently analyze and process VGI data and then conduct a comprehensive analysis of the extent and volume of information in a shorter time than humans can. Consequently, AI algorithms in VGI practice can be applied to confirm and improve the quality of crowdsourced data because the former can identify irregularities and amend errors. As such, AI can utilize VGI analytics data to drive predictive actions on issues that are likely to precipitate in order to enable early public service provision.

REFERENCES

Ahmad, M. (2023). Connecting People and Places: How Citizen Diplomacy and VGI Are Strengthening Disaster Response and Community Development. In Kavoğlu, S., & Köksoy, E. (Eds.), *Global Perspectives on the Emerging Trends in Public Diplomacy* (pp. 195–226). IGI Global. DOI:10.4018/978-1-6684-9161-4.ch009

Ahmad, M., & Ali, A. (2023). Mapping the future of sustainable development through cloud-based solutions: A case study of openstreetmap. In *Promoting Sustainable Management Through Technological Innovation* (pp. 153–176). IGI Global. DOI:10.4018/978-1-6684-9979-5.ch011

Anderson-Tarver, C. (2015). Crisis Mapping the 2010 Earthquake in OpenStreetMap Haiti. In *ProQuest Dissertations and Theses*. jn8sf5hk5v.search.serialssolutions.com/?ctx_ver=Z39.88-2004&ctx_enc=info:ofi/enc:UTF-8&rfr_id=info:sid/ProQuest+Dissertations+%26+Theses+A%26I&rft_val_fmt=info:ofi/fmt:kev:mtx:dissertati

Antoniou, V. (2017). Sources of VGI for Mapping. In *Mapping and the Citizen Sensor*. DOI:10.5334/bbf.b

Antoniou, V. (2023). On volunteered geographic information quality: a framework for sharing data quality information. In *Geoinformatics for Geosciences: Advanced Geospatial Analysis using RS*. GIS and Soft Computing. DOI:10.1016/B978-0-323-98983-1.00009-0

Bai, A., Satarpour, M., Mohebbi, F., & Forati, A. M. (2024). Digital Crowdsourcing and VGI: impact on information quality and business intelligence. *Spatial Information Research*, 1–9.

Bartzokas-Tsiompras, A. (2022). Utilizing OpenStreetMap data to measure and compare pedestrian street lengths in 992 cities around the world. *European Journal of Geography*, 13(2), 127–141. Advance online publication. DOI:10.48088/ejg.a.bar.13.2.127.138

Basiouka, S., Potsiou, C., & Bakogiannis, E. (2015). OpenStreetMap for cadastral purposes: An application using VGI for official processes in urban areas. *Survey Review*, 47(344), 333–341. Advance online publication. DOI:10.1179/1752270615Y.0000000011

Castelnovo, W. (2016). *Co-production makes cities smarter: Citizens' participation in smart city initiatives*. SpringerBriefs in Applied Sciences and Technology. DOI:10.1007/978-3-319-30558-5_7

Chen, Y. (2024). Urban Green Space Maps Based on GeoAI. In *Handbook of Geospatial Approaches to Sustainable Cities* (pp. 75–93). CRC Press. DOI:10.1201/9781003244561-7

Florath, J., Chanussot, J., & Keller, S. (2024a). Rapid natural hazard extent estimation from twitter data: Investigation for hurricane impact areas. *Natural Hazards*, 120(7), 6775–6796. Advance online publication. DOI:10.1007/s11069-024-06488-2

Florath, J., Chanussot, J., & Keller, S. (2024b). Utilizing Volunteered Geographic Information for Real-Time Analysis of Fire Hazards: Investigating the Potential of Twitter Data in Assessing the Impacted Areas. *Fire (Basel, Switzerland)*, 7(1), 6. Advance online publication. DOI:10.3390/fire7010006

Fung, A. (2015). Putting the Public Back into Governance: The Challenges of Citizen Participation and Its Future. *Public Administration Review*, 75(4), 513–522. Advance online publication. DOI:10.1111/puar.12361

Gao, Y., Janssen, M., & Zhang, C. (2023). Understanding the evolution of open government data research: Towards open data sustainability and smartness. *International Review of Administrative Sciences*, 89(1), 59–75. Advance online publication. DOI:10.1177/00208523211009955

Giuffrida, N., Le Pira, M., Inturri, G., & Ignaccolo, M. (2019). Mapping with stakeholders: An overview of public participatory GIS and VGI in transport decision-making. *ISPRS International Journal of Geo-Information*, 8(4), 198. Advance online publication. DOI:10.3390/ijgi8040198

González-Ramiro, A., Gonçalves, G., Sánchez-Ríos, A., & Jeong, J. S. (2016). *Using a VGI and GIS-based multicriteria approach for assessing the potential of rural tourism in Extremadura (Spain)*. Sustainability., DOI:10.3390/su8111144

Goodchild, M. F. (2007). Citizens as sensors: The world of volunteered geography. In *GeoJournal*. DOI:10.1007/s10708-007-9111-y

Hasselwander, M., Kiko, M., & Johnson, T. (2022). Digital civic engagement, open data, and the informal sector: A think piece. *Transportation Research Interdisciplinary Perspectives*, 16, 100700. Advance online publication. DOI:10.1016/j.trip.2022.100700

Hoffmann, M. T., Ostapowicz, K., Bartoń, K., Ibisch, P. L., & Selva, N. (2024). Mapping roadless areas in regions with contrasting human footprint. *Scientific Reports*, 14(1), 4722. Advance online publication. DOI:10.1038/s41598-024-55283-3 PMID:38413813

HOTOSM. (2023a). *Hot's Response To The Turkey & Syria Earthquake*. https://www.hotosm.org/projects/join-the-turkey-and-syria-earthquake-response/

HOTOSM. (2023b). *What We Do | Humanitarian OpenStreetMap Team*. https://www.hotosm.org/what-we-do

Jarke, J. (2021). Co-creating digital public services. In *Public Administration and Information Technology* (Vol. 6). DOI:10.1007/978-3-030-52873-7_3

Khosravi Kazazi, A., Rahmani, Y., Amiri, F., & Hosseinali, F. (2024). A School of Thought on VGI Challenges: A Literature Review. *Papers in Applied Geography*, 10(1), 53–68. Advance online publication. DOI:10.1080/23754931.2023.2256344

Kitsios, F., & Kamariotou, M. (2023). Digital innovation and entrepreneurship transformation through open data hackathons: Design strategies for successful start-up settings. *International Journal of Information Management*, 69, 102472. Advance online publication. DOI:10.1016/j.ijinfomgt.2022.102472

Kolak, M., Steptoe, M., Manprisio, H., Azu-Popow, L., Hinchy, M., Malana, G., & Maciejewski, R. (2020). *Extending Volunteered Geographic Information (VGI) with Geospatial Software as a Service: Participatory Asset Mapping Infrastructures for Urban Health*. DOI:10.1007/978-3-030-19573-1_11

Liu, X., Meng, X., Chen, X., Su, W., Zhang, X., Li, L., & Su, X. (2020). The Research on Assistance Information Acquisition for Earthquake Emergency Response by Aggregating VGI Data and GF-2 Images. *Remote Sensing Technology and Application*, 34(6), 1286–1295.

Mikel. (2011). Haiti OSM Mapping for Japan and Libya. In *Humanitarian OpenStreetMap Team*. https://hot.openstreetmap.org/weblog/2011/03/haiti-osm-mapping-for-japan-and-libya/\nhttp://hot.openstreetmap.org/weblog

Minghini, M., Sarretta, A., & Napolitano, M. (2022). OpenStreetMap Contribution to Local Data Ecosystems in COVID-19 Times: Experiences and Reflections from the Italian Case. *Data*, 7(4), 39. Advance online publication. DOI:10.3390/data7040039

Nabatchi, T., Sancino, A., & Sicilia, M. (2017). Varieties of Participation in Public Services: The Who, When, and What of Coproduction. *Public Administration Review*, 77(5), 766–776. Advance online publication. DOI:10.1111/puar.12765

Olteanu-Raimond, A. M., See, L., Schultz, M., Foody, G., Riffler, M., Gasber, T., Jolivet, L., le Bris, A., Meneroux, Y., Liu, L., Poupée, M., & Gombert, M. (2020). Use of automated change detection and VGI sources for identifying and validating urban land use change. *Remote Sensing (Basel)*, 12(7), 1186. Advance online publication. DOI:10.3390/rs12071186 openstreetmap. (2023). *Why use OpenStreetMap?*https://welcome.openstreetmap.org/why-openstreetmap/

Ponti, M., & Craglia, M. (2020). *Citizen-generated data for public policy*. European Commission.

Sathianarayanan, M., Hsu, P. H., & Chang, C. C. (2024). Extracting disaster location identification from social media images using deep learning. *International Journal of Disaster Risk Reduction*, 104, 104352. Advance online publication. DOI:10.1016/j.ijdrr.2024.104352

Schott, M., Zell, A., Lautenbach, S., Sumbul, G., Schultz, M., Zipf, A., & Demir, B. (2024). Analyzing and Improving the Quality and Fitness for Purpose of OpenStreetMap as Labels in Remote Sensing Applications. In *Volunteered Geographic Information*. DOI:10.1007/978-3-031-35374-1_2

Tavra, M., Lisec, A., Galešić Divić, M., & Cetl, V. (2024). Unpacking the role of volunteered geographic information in disaster management: focus on data quality. In *Geomatics, Natural Hazards and Risk* (Vol. 15, Issue 1). DOI:10.1080/19475705.2023.2300825

Tripathi, A. K., Agrawal, S., & Gupta, R. D. (2020). Cloud enabled SDI architecture: a review. In *Earth Science Informatics* (Vol. 13, Issue 2). DOI:10.1007/s12145-020-00446-9

Tzavella, K., Skopeliti, A., & Fekete, A. (2022). Volunteered geographic information use in crisis, emergency and disaster management: A scoping review and a web atlas. *Geo-Spatial Information Science*. Advance online publication. DOI:10.1080/10095020.2022.2139642

Urrea, G., & Yoo, E. (2023). The role of volunteer experience on performance on online volunteering platforms. *Production and Operations Management*, 32(2), 416–433. Advance online publication. DOI:10.1111/poms.13879

Verplanke, J., McCall, M. K., Uberhuaga, C., Rambaldi, G., & Haklay, M. (2016). A Shared Perspective for PGIS and VGI. *The Cartographic Journal*, 53(4), 308–317. Advance online publication. DOI:10.1080/00087041.2016.1227552

Waddington, H., Sonnenfeld, A., Finetti, J., Gaarder, M., John, D., & Stevenson, J. (2019). Citizen engagement in public services in low- and middle-income countries: A mixed-methods systematic review of participation, inclusion, transparency and accountability (PITA) initiatives. In *Campbell Systematic Reviews* (Vol. 15, Issues 1–2). DOI:10.1002/cl2.1025

Wang, H., Zhao, Y., Gao, X., & Gao, B. (2021). Collaborative decision-making for urban regeneration: A literature review and bibliometric analysis. *Land Use Policy*, 107, 105479. Advance online publication. DOI:10.1016/j.landusepol.2021.105479

Wu, B., Tian, F., Zhang, M., Zeng, H., & Zeng, Y. (2020). Cloud services with big data provide a solution for monitoring and tracking sustainable development goals. *Geography and Sustainability*, 1(1), 25–32. Advance online publication. DOI:10.1016/j.geosus.2020.03.006

Yang, A., Fan, H., Jia, Q., Ma, M., Zhong, Z., Li, J., & Jing, N. (2024). How do contributions of organizations impact data inequality in OpenStreetMap? *Computers, Environment and Urban Systems*, 109, 102077. Advance online publication. DOI:10.1016/j.compenvurbsys.2024.102077

Chapter 10
Exploring AI-Driven Solutions for Fake News:
A Qualitative Investigation Through Interviews of Industry Experts

Harshwardhani Sharma
https://orcid.org/0000-0003-3485-1463
Manav Rachna International Institute of Research and Studies, India

Akshay Upmanyu
Sharda University, India

Disha Batra
https://orcid.org/0000-0001-8372-5290
Manav Rachna International Institute of Research and Studies, India

Saket Kumar Bhardwaj
https://orcid.org/0009-0001-5906-2965
Hemwati Nandan Bahuguna Garhwal University, India

Bala Lakhendra
Banaras Hindu University, India

ABSTRACT

The proliferation of fake news challenges information integrity and societal well-being in the digital age. Despite AI advancements, detecting misinformation, including deep fakes, remains difficult. However, AI offers potential solutions. This study explores AI's role in fake news detection and the impact of public policy on

DOI: 10.4018/979-8-3693-9246-1.ch010

fact-checking through interviews with experts—Sandeep Amar, Nikhil Pahwa, and Pratyush Ranjan. The first objective examines AI's efficacy in detecting fake news, with Sandeep Amar discussing AI models. The second objective focuses on enhancing fact-checking through AI, with Pratyush Ranjan emphasising the importance of critical thinking alongside AI tools. The third objective explores government policy's role, with Nikhil Pahwa analysing Indian regulations and their impact on AI deployment for fact-checking. The study contributes to debates on addressing fake news, offering insights for policymakers, researchers, and practitioners. It highlights the need for a hybrid approach integrating AI's capabilities with human judgment to ensure information integrity.

INTRODUCTION

Fake news, a form of deliberate misinformation, has emerged as a significant challenge in today's digital age, posing threats to information integrity and societal well-being (Sadiku et al., 2018). The proliferation of fake news, primarily through social media, complicates discerning the truth, a growing concern for individuals and institutions (Devi, 2019). This issue has been particularly prominent in political contexts, such as the 2016 US presidential election, highlighting misinformation's widespread impact and potential dangers (Thielen, 2018).

The historical roots of fake news are deep and complex, with its origins traceable to the Byzantine Empire and its proliferation noted in 18th-century London and Paris (Adams & Darnton, 2017). The term "fake news" has evolved and is now commonly associated with disinformation and the intentional spread of false information for harmful purposes. The rise of social media has exacerbated the issue, enabled the rapid dissemination of dubious content, and undermined mainstream media's credibility (Rogers & Niederer, 2020).

Despite fake news's longstanding presence, some scholars argue that the panic surrounding it in recent years may be overstated. They suggest that misinformation has always been a part of journalism, with the primary change being its increased visibility and spread due to digital platforms (Gorbach, 2018). Nonetheless, the impact of fake news on public perceptions and trust in media is undeniable, necessitating enhanced media literacy and effective interventions to combat misinformation (Collier & Van Duyn, 2022) (Alalawi et al., 2023).

Various studies have explored the potential of Artificial Intelligence (AI) in detecting and mitigating misinformation to address the complex issue of fake news (Sharma & Bhardwaj, 2024). Researchers have proposed novel AI-based approaches for identifying fake news, highlighting the capabilities and limitations of these technologies. For instance, Rao et al. (2020) suggest innovative AI techniques for fake news detection, while Cybenko & Cybenko (2018) discuss AI's potential to counteract human susceptibilities that facilitate the spread of misinformation. Botha and Pieterse (2020) emphasize the dangers of AI-created deep fakes and the techniques to detect them, and Velentzas et al. (2020) underscores the role of semantics in AI's ability to identify false information.

Despite the promise of AI in combating fake news, challenges remain, such as the need for more accurate and comprehensive tools. Studies by Goksu and Cavus (2019) and Paredes (2023) highlight the success rates of AI tools, while Školkay and Filin (2019) raise concerns about their accuracy and comprehensiveness. Additionally, Shae and Tsai (2019) propose an AI blockchain platform to enhance collaboration between researchers and journalists, providing reliable sources of information and improving the overall fact-checking process.

The integration of AI in fact-checking practices has been further explored through the lens of industry experts. This chapter aims to delve into the efficacy of AI-driven solutions for fake news detection, the influence of public policy on these practices, and the enhancement of fact-checking accuracy. The study provides valuable insights into AI's current landscape and future directions in combating misinformation through in-depth interviews with Sandeep Amar, Pratyush Ranjan, and Nikhil Pahwa.

In the subsequent sections, this second section will explore the efficacy of AI in detecting fake news, drawing on insights from Sandeep Amar, Founder of PD Lab, who discusses AI models and techniques used in the fight against misinformation. The third section focuses on enhancing fact-checking practices through AI-driven techniques, featuring insights from Pratyush Ranjan, Head of Digital Services at Press Trust of India, on the identification of fake news and the role of critical thinking. The fourth section will examine the intersection of government policy and AI in fact-checking, with Nikhil Pahwa, Founder of Medianama, providing an analysis of Indian technology policy within the framework of the IT Act. The final section concludes the chapter with a summary of key findings and their implications.

By exploring these dimensions, this chapter aims to contribute to the ongoing debate on fake news detection and mitigation, providing valuable perspectives for policymakers, researchers, and practitioners seeking to advance the role of AI in ensuring information integrity in society.

AI'S EFFICACY IN DETECTING FAKE NEWS

Many previous studies have explored AI's efficacy in detecting fake news, with promising results. Paredes (2023) and Shetty et al. (2023) highlight AI's potential in identifying and classifying fake news, particularly machine learning and natural language processing. Goksu (2019) further supports this, reporting success rates of over 90% in using AI tools for fake news detection. Zhang et al. (2018) introduce a specific AI model, FAKEDETECTOR, which uses a deep diffusive neural network to simultaneously learn representations of news articles, creators, and subjects, showing promising results in detecting fake news. These studies collectively underscore the potential of AI in effectively detecting fake news. In one interview with Mr. Sandeep Amar, founder of PD Lab, he describes AI as more of a contributor to creating fake news than a savior in preventing it. He explains that AI can generate and spread misinformation at a large scale, and its algorithms can be manipulated to promote false narratives. While AI can also detect and flag fake news through models like sentiment analysis and content verification, he emphasizes that by the time fake news is detected or fact-checked, it has often already done significant damage. He explains that human fact-checking is more objective and accurate than AI models for detecting fake news. He mentions that human fact-checkers can analyze information critically, connect the dots, and verify the accuracy of news sources. Human fact-checkers also possess a deeper understanding of context and can consider factors like bias, intention, and motivation behind the creation of fake news.

Amar also mentions that there are limitations to current AI models when detecting fake news but acknowledges their potential for development and improvement in the future. He suggests that combining AI data analysis and human fact-checking can lead to more comprehensive and effective detection systems for fake news. However, he also acknowledges that AI models have the advantage of processing large amounts of data quickly, which can aid in identifying patterns or anomalies that may indicate misinformation. Natural Language Processing (NLP) models, Sentiment Analysis, Content Verification, and Automated Fact-Checking Tools are potential ways AI can help detect and combat fake news. AI models can analyze the language used in news articles, social media posts, and other sources to identify patterns and determine the likelihood of fake news. They were hugging Face Transformers, a popular library that provides pre-trained models for various natural language processing tasks, including sentiment analysis. Models like BERT, RoBERTa, and DistilBERT can be fine-tuned for sentiment analysis and potentially adapted for fake news detection.

DistilRoBERTa Model

The DistilRoBERTa propaganda class 2 model represents a significant advancement in the field. This model, a distilled version of RoBERTa, leverages the efficiency of knowledge distillation to maintain robust performance while operating with reduced computational overhead. By fine-tuning datasets specifically annotated for propaganda detection, the DistilRoBERTa model adeptly identifies nuanced patterns and rhetorical techniques characteristic of propaganda, a critical component of fake news. The expert highlighted the model's utility in real-time applications due to its accelerated processing capabilities, making it particularly advantageous for media organisations and fact-checking bodies that require timely intervention.

Moreover, based on the Transformer framework, the model's architecture allows for sophisticated contextual analysis, enabling it to discern subtle manipulative tactics within the text. However, the expert also underscored the model's dependency on the diversity and quality of training data, which can influence its generalizability and effectiveness across different contexts. Overall, deploying the DistilRoBERTa model in fake news detection exemplifies the intersection of advanced machine learning techniques and practical applications, offering a potent tool for mitigating the spread of misinformation. Table 1.0 illustrates a systematic approach to integrating the DistilRoBERTa propaganda class 2 model into a fake news detection pipeline, highlighting its role in identifying and mitigating the spread of misinformation.

Table 1. Usage of DistilRoBERTa Model in Fake News Detection

S. No.	Steps	Description
1.	Data Collection	Tweets are collected via the Twitter API based on specific keywords, hashtags, or user accounts.
2.	Pre-processing	- Tokenisation: Splitting tweets into individual tokens (words). - Cleaning: Removing URLs, mentions, hashtags, and special characters. - Normalisation: Standardising text (e.g., lowercasing).
3.	Indexing	Tweets are organized into an indexed structure for efficient retrieval and analysis.
4.	Initial Analysis	The indexed tweets are initially analyzed for essential features like frequency of terms, sentiment, and metadata (e.g., user information, timestamp).
5.	LLM Processing	- The indexed and pre-processed tweets are fed into the Large Language Model (LLM) for deeper analysis - The model uses NLU techniques to understand the context and detect linguistic patterns indicative of misinformation.
6.	Verification	The LLM cross-references claim in tweets against a reliable knowledge base or fact-checking databases to verify the information.
7.	Conversational Interaction	Users can interact with the conversational AI to query specific tweets or claims, prompting the model to retrieve and provide detailed analysis or verification results.
8.	Flagging and Reporting	Tweets identified as potentially fake news are flagged. - Detailed reports, including the analysis and evidence supporting the classification, are generated.
9.	Continuous Learning	The model is updated with new data and feedback, improving its accuracy and ability to detect evolving misinformation tactics.

Explanation of Each Step:

1. Data Collection: Tweets are gathered using the Twitter API, focusing on relevant keywords, hashtags, or specific user accounts that may be involved in spreading misinformation.
2. Pre-processing: The collected tweets undergo pre-processing to prepare the text for analysis. It includes tokenisation, cleaning (removing URLs, mentions, hashtags, and special characters), and normalisation (standardising the text).
3. Indexing: Tweets are organised into an indexed structure, allowing efficient retrieval and analysis. This indexing is crucial for handling large volumes of data and ensuring quick access during analysis.
4. Initial Analysis: Basic analysis of the indexed tweets includes examining term frequency, sentiment analysis, and metadata analysis (such as user information and timestamps).
5. LLM Processing: The pre-processed and indexed tweets are fed into the Large Language Model (LLM) for deeper contextual analysis. The LLM uses natural language understanding (NLU) techniques to identify linguistic patterns and contextual cues that may indicate misinformation.
6. Verification: The LLM verifies the claims made in the tweets by cross-referencing them against a reliable knowledge base or fact-checking databases. This step helps determine the integrity of the information.
7. Conversational Interaction: Users can query specific tweets or claims using conversational AI. The model can retrieve and provide detailed analysis or verification results, enhancing user engagement and understanding.
8. Flagging and Reporting: Tweets identified as potentially containing fake news are flagged. Detailed reports are generated to document the analysis and evidence supporting the classification, which can be used for further action.
9. Continuous Learning: The model is continuously updated with new data and feedback, improving its ability to detect evolving misinformation tactics and enhancing its overall accuracy.

LlamaIndex Framework

AI models equipped with sentiment analysis capabilities can analyze public opinion about specific news topics or events and identify potential misinformation based on the overall sentiment expressed in social media posts or comments. Microsoft Azure's Text Analytics service offers sentiment analysis, key phrase extraction, and language detection, which can help analyze news articles' sentiments and identify potentially misleading content. Also, The LlamaIndex Framework, leveraging conversational AI, represents a cutting-edge approach in the architecture for fake news detection. The framework's architecture is designed to integrate large language models (LLMs) with an indexing system that enhances the model's ability to understand, retrieve, and analyze vast amounts of text data efficiently. At its core, the LlamaIndex Framework utilizes a multi-layered indexing mechanism that organizes information into a structured format, allowing the LLM to perform rapid and contextually aware retrieval of relevant data. This architecture is compelling for fake news detection as it enables the conversational AI to dynamically interact with and interrogate data sources, extracting nuanced insights and verifying information against a reliable knowledge base.

In this framework, the LLM is pre-trained on diverse datasets to develop a broad understanding of language and then fine-tuned on specific datasets related to news and misinformation. This dual training process ensures the model can discern subtle linguistic patterns and contextual cues indicative of fake news. When a news article is input into the system, the LlamaIndex Framework pre-processes the text, breaking it into indexed segments. These segments are then fed into the LLM, which processes the information using advanced natural language understanding (NLU) techniques to detect inconsistencies, biases, or typical markers of fake news.

The conversational AI aspect allows for an interactive querying process where users can engage in a dialogue with the model to probe deeper into specific claims or elements of the article. This interactive capability is facilitated by the framework's indexing system, which ensures that responses are grounded in accurately retrieved information. The LlamaIndex Framework provides a scalable and efficient real-time fake news detection solution by combining robust indexing with conversational AI's adaptive and contextual strengths. This architecture enhances the model's accuracy and response time. It enables continuous learning and adaptation as new data and misinformation tactics emerge, making it a potent tool in the ongoing battle against fake news.

Table 2. Usage of LlamaIndex Framework in Fake News Detection with Twitter Data

S. No.	Steps	Description
1.	Input Article	The process begins with an input news article that needs to be evaluated for fake news.
2.	Pre-processing	- Tokenisation: Splitting the text into individual tokens (words). - Stop Word Removal: Eliminating common words that do not contribute to the detection process - Normalisation: Standardising text (e.g., lowercasing, removing punctuation).

continued on following page

Table 2. Continued

S. No.	Steps	Description
3.	DistilRoBERTa Model	- Feature Extraction: The model extracts relevant features from the pre-processed text. - Context Analysis: Utilising the Transformer's self-attention mechanism to understand the context and relationships within the text. - Propaganda Detection: Identifying and classifying propaganda techniques present in the text.
4.	Model Inference	- Probability Scores: The model outputs probabilities indicating the likelihood of various propaganda techniques. - Classification: Based on the probabilities, the article is classified as containing propaganda or not, and if so, which type.
5.	Post-processing	- Result Interpretation: Interpreting the model's output to determine the presence and type of propaganda. - Flagging Propaganda: Marking sections of the article or the entire article as potentially containing propaganda.
6.	Further Action	- Fact-Checking: Articles flagged by the model are subjected to additional fact-checking processes to verify the accuracy. - Report Generation: Generating reports on the findings for stakeholders, such as journalists, editors, or automated systems for further action.

Explanation of Each Step:

1. Input Article: The process begins with an input news article that needs to be evaluated for fake news.
2. Pre-processing: The article undergoes pre-processing steps including:
    a. Tokenisation: Splitting the text into individual tokens (words).
    b. Stop Word Removal: Eliminating common words that do not contribute to the detection process.
    c. Normalisation: Standardising text (e.g., lowercasing, removing punctuation).
3. DistilRoBERTa Model:
    a. Feature Extraction: The model extracts relevant features from the pre-processed text.
    b. Context Analysis: Utilising the Transformer's self-attention mechanism to understand the context and relationships within the text.
    c. Propaganda Detection: Identifying and classifying propaganda techniques present in the text.
4. Model Inference:
    a. Probability Scores: The model outputs probabilities indicating the likelihood of various propaganda techniques.
    b. Classification: Based on the probabilities, the article is classified as containing propaganda or not, and if so, which type.
5. Post-processing:

   a. Result Interpretation: Interpreting the model's output to determine the presence and type of propaganda.
   b. Flagging Propaganda: Marking sections of the article or the entire article as potentially containing propaganda.
6. Further Action:
   a. Fact-Checking: Articles flagged by the model are subjected to additional fact-checking processes to verify their accuracy.
   b. Report Generation: Generating reports on the findings for stakeholders, such as journalists, editors, or automated systems for further action.

AI-DRIVEN TECHNIQUES FOR FACT-CHECKING

In recent years, many studies have made significant strides in automating fact-checking through AI-driven techniques. Guo et al. (2022) provide a comprehensive overview of the field, highlighting the use of natural language processing, machine learning, and knowledge representation in predicting the veracity of claims. Mahmood et al. (2023) introduce a method for fact-checking AI-generated reports, using image-sentence associations to differentiate between real and fake findings. Alhindi et al. (2018) extend a fact-checking dataset by incorporating justifications, significantly improving machine learning models' accuracy in classifying claims. These studies collectively underscore the potential of AI in enhancing fact-checking practices. The interview with Mr. Pratyush Ranjan, Head of Digital Services at Press Trust of India, emphasizes the importance of critical thinking and vigilance in the digital age. He advises individuals to scrutinize the source of information, considering the credibility and reliability of the author or the originating platform. Evaluating the currency of the information is crucial; outdated content might be irrelevant or misleading in the current context. Ranjan also underscores the necessity of cross-referencing the information with multiple reputable sources to verify its authenticity and accuracy. This triangulation process helps identify inconsistencies and ensure a broader consensus within the credible media landscape that supports the information.

Tips for Fact-Checking Information Before Sharing

Pratyush Ranjan emphasizes the necessity of a rigorous approach to fact-checking information before individuals disseminate it on social media. This section explores his recommended strategies, encompassing source verification, cross-referencing information, and evaluating content currency.

a. Source Verification

As Ranjan highlighted, the first step in effective fact-checking is verifying the information source. It involves assessing the credibility and reliability of the author or the platform where the information originates. In the era of digital media, where content can be created and shared by anyone, the credibility of the source becomes paramount. Ranjan suggests that individuals should always consider whether the source has a history of providing accurate and unbiased information. Established and reputable news organisations, academic institutions, and official government websites are typically reliable sources. Conversely, an individual should approach sources with a known history of spreading misinformation or lacking transparency about their authorship and editorial processes should be approached with scepticism.

b. Cross-Referencing Information

Another critical tip Ranjan provides is the importance of cross-referencing information with multiple credible sources. This practice helps in verifying the authenticity and accuracy of the content. By comparing the data against other reputable sources, individuals can identify inconsistencies and ensure that a broader consensus within the credible media landscape corroborates the content. Cross-referencing involves checking different sources, such as news articles, academic journals, and official reports. This triangulation process helps paint a more accurate picture and guards against the reliance on a single, potentially biased source.

c. Evaluating Content Currency

Evaluating the currency of the information is another essential aspect of fact-checking. Outdated content may no longer be relevant or have been superseded by new information. Ranjan advises that individuals should always check the publication date and ensure that the information still applies in the current context. It is imperative in fast-moving news environments where situations can change rapidly. For instance, health guidelines during a pandemic or political developments can evolve quickly, and relying on outdated information can lead to misunderstandings and misinformed decisions.

Reverse Image Search Tools

Ranjan highlights the utility of reverse image search tools, such as Google Reverse Image Search or TinEye, in verifying the authenticity of photos and videos circulating on social media. These tools allow users to trace the origin of an image by searching the web for instances where it appears. Users can determine if the image has been previously published, modified, or misrepresented. This process involves uploading the image or providing its URL to the search tool and retrieving instances of the image from various sources. This method can uncover whether the image is used in its original context or has been repurposed with misleading or false information. Such verification helps identify manipulated content and distinguish authentic visuals from deceptive ones.

Ranjan discusses the evolving role of AI in enhancing fact-checking practices. AI-driven techniques, including deep learning and natural language processing, can automate the detection of false information and identify patterns indicative of misinformation. AI tools can quickly analyze vast amounts of data, cross-referencing content across multiple databases to identify discrepancies. Moreover, AI can assess the credibility of sources by evaluating their historical accuracy and reputation. Implementing AI in fact-checking can also include real-time monitoring of social media platforms to flag suspicious content promptly. This proactive approach enables faster response times in countering misinformation, safeguarding public discourse from the detrimental effects of false information.

Reverse Image Search Tools to Verify Authenticity

Ranjan also highlights the importance of verifying the authenticity of photos and videos, which are potent forms of media that can easily be manipulated. Reverse image search tools, such as Google Reverse Image Search and TinEye, are practical. This section explores how individuals can use these tools to trace an image's origin and uncover manipulations.

Reverse image search tools allow users to upload an image or provide its URL to search the web for instances where it appears. These tools then retrieve instances of the image from various sources, helping users determine if the image has been previously published, modified, or misrepresented. By tracing the image back to its original context, users can identify whether it has been used accurately or manipulated to convey false information.

To conduct a reverse image search, users can follow these steps:

a. Upload the Image: Use a reverse image search tool to upload the image in question or provide its URL.
b. Analyse Search Results: The tool will generate a list of instances where the image appears online. Users should carefully analyse these results to identify the source and context of the image.

c. Verify Authenticity: By examining the source, users can determine whether the image has been altered or used out of context. It will help distinguish authentic visuals from deceptive ones.

Reverse image search tools are handy in debunking viral images that claim to depict recent events but are old or taken out of context. For example, In India, an old picture of a natural disaster was circulated during a new incident to amplify its impact, even though it was from a previous event. However, by conducting a reverse image search, individuals can reveal the true origin of the image and prevent the spread of misinformation.

Ranjan discusses AI's transformative role in enhancing fact-checking practices. AI-driven techniques, including deep learning and natural language processing, can automate the detection of false information and identify patterns indicative of misinformation. This section explores how AI can be utilized in fact-checking, including its capabilities and challenges.

AI technologies have advanced significantly, enabling the development of sophisticated fact-checking tools. Deep learning algorithms can quickly analyze vast amounts of data, identifying patterns and anomalies that may indicate false information. Natural language processing (NLP) can assess text credibility by evaluating factors such as language complexity, sentiment, and historical accuracy of the source.

One significant advantage of AI in fact-checking is its real-time monitoring of social media platforms. AI algorithms can continuously scan content, flagging suspicious posts for further review. This proactive approach enables faster response times in countering misinformation, safeguarding public discourse from the detrimental effects of false information. Automated detection systems can identify common misinformation patterns, such as fake news headlines, doctored images, or misleading claims, allowing fact-checkers to focus on the most pressing cases.

Despite its potential, AI in fact-checking faces several challenges. AI systems rely on large datasets to learn and make accurate predictions, and the quality of these datasets is crucial. The AI system may produce inaccurate results if the data is biased or incomplete. Additionally, AI algorithms may struggle with nuanced content, such as satire or context-dependent information, which requires human judgment to interpret correctly. There is also the risk of over-reliance on AI, where the human element of critical thinking and contextual understanding is diminished. Hence, a balanced approach that combines AI capabilities with human expertise is essential for effective fact-checking.

Significance of Media Literacy and Critical Thinking

Beyond the technical aspects, Ranjan underscores the importance of media literacy and critical thinking in combating misinformation. This section explores how individuals can develop these skills to become more discerning information consumers.

Developing Media Literacy

Media literacy involves understanding how media content is created, distributed, and consumed. It encompasses critically evaluating media messages and recognizing biases, stereotypes, and propaganda. Ranjan advocates for media literacy education, which can empower individuals to navigate the complex information landscape. This education can be integrated into school curricula, community workshops, and online resources, helping people of all ages develop the skills to identify and counter misinformation.

Cultivating Critical Thinking

Critical thinking is the ability to analyze information objectively and make reasoned judgments. It involves questioning assumptions, evaluating evidence, and considering alternative perspectives. Ranjan emphasizes the need for individuals to cultivate critical thinking skills, which can help them discern the credibility of information and resist the influence of false narratives. By fostering a culture of critical inquiry, society can build resilience against misinformation.

Encouraging Scepticism and Verification

Encouraging scepticism and verification is crucial in the fight against misinformation. Ranjan advises individuals to adopt a sceptical mindset, questioning the validity of information before accepting it as accurate. It involves verifying facts, checking the credibility of sources, and seeking out multiple perspectives. By making verification habitual, individuals can reduce the spread of false information and contribute to a more informed public discourse.

PRACTICAL EXAMPLES

Ranjan provides several case studies and practical examples to illustrate the effectiveness of fact-checking practices. This section explores these examples, highlighting the real-world impact of misinformation and the importance of vigilant fact-checking. Ranjan discusses instances where misleading social media posts have caused public panic or influenced political opinions. For example, false claims about candidates can spread rapidly during elections, shaping voter perceptions and potentially altering election outcomes. By conducting thorough fact-checking and debunking these claims, fact-checkers can mitigate the impact of misinformation and promote a more informed electorate.

Health misinformation, particularly during the COVID-19 pandemic, has had significant real-world consequences. Ranjan highlights efforts to combat false health claims, such as fake cures and conspiracy theories, through rigorous fact-checking. These efforts have involved collaboration between fact-checkers, health organisations, and social media platforms to identify and remove harmful misinformation, ensuring that accurate information reaches the public.

Ranjan's insights provide a comprehensive framework for enhancing fact-checking practices through critical thinking, media literacy, and AI-driven techniques. Individuals can become more discerning information consumers by verifying sources, cross-referencing information, and utilizing reverse image search tools. AI technologies offer powerful tools for automated detection and real-time monitoring, though they must be complemented by human judgment. Fostering a culture of media literacy and critical thinking is essential for building resilience against misinformation and promoting a more informed public discourse. As the digital landscape continues to evolve, these practices will remain vital in ensuring the integrity and accuracy of the information that shapes our world.

GOVERNMENT POLICY AND AI IN FACT-CHECKING

The intersection of government policy and AI in fact-checking is complex and evolving. Nguyen et al. (2018) emphasize the need for transparent AI models to facilitate effective human interaction, while Guo et al. (2022) survey the use of natural language processing and machine learning in automated fact-checking, highlighting the challenges for future research. Thorne and Vlachos (2018) unify task formulations and methodologies in automated fact-checking research, emphasizing evidence as crucial. These studies underscore the importance of a human-AI partnership, the need for transparent AI models, and the use of evidence in automated fact-checking.

The interview with Mr Nikhil Pahwa, founder of MediaNama, sheds light on the subjectivity in regulating fake news, particularly under the IT Act. The subjectivity involved in controlling phoney news enforcement stems from various factors. Firstly, determining what constitutes "fake news" can be subjective as it assesses the accuracy and truthfulness of information. Different individuals or authorities may have different interpretations or biases when evaluating content.

Secondly, enforcing regulations against fake news requires distinguishing between harmless misinformation and deliberate disinformation with malicious intent. This determination can be challenging due to variations in context, purpose, and interpretation. Additionally, the IT Act may allow platforms to remove unlawful content. However, what is deemed "unlawful" can vary across jurisdictions and legal frameworks. This subjectivity creates challenges in consistent enforcement globally.

Furthermore, determining liability for fake news poses a challenge as platforms often act as intermediaries rather than publishers of content. Determining whether platforms should be held accountable for user-generated content raises questions about their role in moderating and regulating speech online. According to Pahwa, the subjectivity involved in the IT Act in the context of fake news is related to determining what constitutes misinformation or unlawful content. The interpretation of what is considered fake news can vary, and this subjectivity raises challenges in enforcing regulations and taking action against such content.

For example, if a video containing fake information is shared on a platform, it may be challenging to ascertain whether it falls under misinformation or lawful speech. Individuals may have different perspectives on whether the content is misleading or harmful. This subjectivity makes it challenging for platforms to determine what content should be removed or flagged as fake news. Another example he gave relates to identifying misinformation about geopolitical issues. For instance, if there's a dispute between two countries regarding territorial claims over a particular region, one country might consider any information favouring their claim as factual. In contrast, another country may label it as false propaganda. The subjectivity involved in determining which side has accurate information can complicate efforts to tackle fake news.

He emphasises that subjective interpretation and differing perspectives make it difficult for platforms and authorities to uniformly address fake news through existing legal frameworks like the IT Act. It highlights the need for more straightforward guidelines and policies to help define standards for identifying and addressing misinformation more effectively within legal frameworks. Determining responsibility for the output generated by an AI bot is a complex issue. It can involve multiple parties, including the AI system's developers, operators, and users. The level of responsibility may depend on various factors, such as the purpose and design of the AI system, legal frameworks in place, and ethical considerations.

In some cases, developers might be held responsible if they have not adequately trained or tested their AI systems or if there are flaws in their algorithms. Operators or platform providers could also share responsibility for the content generated by their AI bots. However, it is essential to note that assigning legal liability to an AI bot is still a relatively new and evolving field. Laws and regulations around this issue are still being developed in many jurisdictions worldwide.

According to him, public policy plays a crucial role in shaping the deployment of AI technologies for fact-checking purposes. Public policy can set guidelines and regulations that define how AI systems are developed, trained, and used for fact-checking. One aspect highlighted is the need for policies that ensure transparency and accountability in AI systems. It includes requirements for explainability, where AI models can provide insights into their decision-making process. Policy frameworks can also address issues related to bias and fairness in AI algorithms, ensuring that they do not perpetuate or amplify existing biases during the fact-checking process.

Another area where public policy influences the deployment of AI technologies is data protection and privacy laws. Fact-checking often involves processing large amounts of data, including personal information. Policies focused on data protection can help establish safeguards to ensure user privacy while using these technologies. Public policy can also play a role in establishing standards and certifications for AI-based fact-checking tools. By setting standards for accuracy, reliability, and effectiveness, policymakers can guide organisations in choosing trustworthy tools while promoting competition within this field.

Additionally, public policies may address liability concerns related to the use of AI technologies in fact-checking. Determining who holds responsibility when erroneous or misleading information is propagated through an automated system requires clear guidelines outlined by policymakers. Overall, the public policy provides a framework within which organizations develop and deploy AI technologies for fact-checking purposes while addressing important ethical considerations such as transparency, fairness, privacy protection, and liability attribution.

CONCLUSION

In conclusion, exploring AI-driven solutions for fake news detection reveals a complex interplay of technology, human judgment, and policy. As the digital age progresses, the proliferation of fake news seriously threatens information integrity and societal well-being. Artificial Intelligence, with its powerful data processing capabilities, offers a promising approach to combating this issue. Advanced AI models like DistilRoBERTa and frameworks like LlamaIndex have shown significant potential in detecting misinformation by swiftly and accurately analyzing vast amounts of data. However, these technologies are not without their limitations. The efficacy of AI in identifying fake news is greatly enhanced when combined with human critical thinking and contextual understanding. Human fact-checkers are crucial in providing nuanced insights and deeper analysis that AI alone cannot achieve.

Public policy also plays a vital role in deploying and regulating AI-driven fact-checking tools. Chu-Ke and Dong (2024), in their recent article, state that

In the age of generative AI, misinformation and AI literacy targeting misinformation should be reconsidered. We must embark on ethical AI development and ethical AI use to combat the rising tide of misinformation. The call to action should and has been increasingly made for AI practitioners and lawmakers to embrace ethical guidelines and regulations.

Nikhil Pahwa's analysis of Indian technology policy within the framework of the IT Act accentuates the importance of clear and adaptable regulations that support the ethical use of AI in combating misinformation. Effective public policies can help ensure that AI technologies are used responsibly, protecting both freedom of speech and the need for accurate information.

Continuous improvement and adaptation are essential for AI models to remain effective in the ever-evolving landscape of fake news. Developing new algorithms and incorporating feedback and new data will enable AI systems to detect and counteract sophisticated misinformation tactics better. Additionally, integrating Business Intelligence tools can enhance the effectiveness of AI-driven solutions by providing deeper data insights and advanced analytics, further refining the fact-checking process.

Educating the public on media literacy and the importance of verifying information is another critical component in the fight against fake news. While AI tools can flag potential misinformation, individuals must be equipped to assess and verify the information they encounter critically. This combined approach of technological innovation, human oversight, supportive policy frameworks, and public education can significantly mitigate the impact of fake news.

The insights from industry experts Sandeep Amar, Nikhil Pahwa, and Pratyush Ranjan highlight the multifaceted nature of this issue and the need for a comprehensive strategy. By leveraging the strengths of AI and human intelligence, supported by robust public policies and enhanced media literacy, we can advance the fight against misinformation. This holistic approach addresses the immediate challenge of fake news and promotes long-term information integrity and trust in the digital age.

REFERENCES

Adams, P.D., & Darnton, RC (2017). The true history of fake news.

Alalawi, S., Baalfaqih, S., Almeqbaali, M., & Masud, M. M. (2023). Social Media Misinformation Propagation and Detection. *2023 15th International Conference on Innovations in Information Technology (IIT)*, 240-245.

Alhindi, T., Petridis, S., & Muresan, S. (2018). Where is Your Evidence: Improving Fact-checking by Justification Modeling.

Botha, J.G., & Pieterse, H. (2020). Fake news and deepfakes: A dangerous threat for 21st century information security.

Chu-Ke, C., & Dong, Y. (2024). *Misinformation and Literacies in the Era of Generative Artificial Intelligence: A Brief Overview and a Call for Future Research.* Emerging Media. DOI: 10.1177/27523543241240285

Collier, J. R., & Van Duyn, E. (2022). Fake news by any other name: Phrases for false content and effects on public perceptions of US news media. *Journal of Applied Communication Research*, 51(4), 424–443. DOI: 10.1080/00909882.2022.2148487

Cybenko, A. K., & Cybenko, G. V. (2018). AI and Fake News. *IEEE Intelligent Systems*, 33(5), 1–5. DOI: 10.1109/MIS.2018.2877280

Goksu, M., & Cavus, N. (2019). Fake News Detection on Social Networks with Artificial Intelligence Tools: Systematic Literature Review.

Gorbach, J. (2018). Not Your Grandpa's Hoax: A Comparative History of Fake News. *American Journalism*, 35(2), 236–249. DOI: 10.1080/08821127.2018.1457915

Guo, Z., Schlichtkrull, M., & Vlachos, A. (2022). A survey on automated fact-checking. *Transactions of the Association for Computational Linguistics*, 10, 178–206. DOI: 10.1162/tacl_a_00454

Mahmood, R., Wang, G., Kalra, M., & Yan, P. (2023). Fact-Checking of AI-Generated Reports. *ArXiv, abs/2307.14634.*

Nguyen, A. T., Kharosekar, A., Krishnan, S., Krishnan, S., Tate, E., Wallace, B. C., & Lease, M. (2018). Believe it or not: Designing a Human-AI Partnership for Mixed-Initiative Fact-Checking. *Proceedings of the 31st Annual ACM Symposium on User Interface Software and Technology*. DOI: 10.1145/3242587.3242666

Paredes, D. G. (2023). *Strategies based on artificial intelligence for the detection of fake news*. AWARI. DOI: 10.47909/awari.58

Rao, A., Shetty, A., Uphade, A., Thawani, P., & Rl, P. (2020). A Proposal for a novel approach to analyse and detect the fake news using AI techniques. *2020 3rd International Conference on Intelligent Sustainable Systems (ICISS)*, 582-589.

Rogers, R., & Niederer, S. (2020). The Politics of Social Media Manipulation. *The Politics of Social Media Manipulation.*

Sadiku, M.N., Eze, T.P., & Musa, S.M. (2018). Fake news and misinformation.

Shae, Z., & Tsai, J. J. (2019). AI Blockchain Platform for Trusting News. *2019 IEEE 39th International Conference on Distributed Computing Systems (ICDCS)*, 1610-1619.

Sharma, H., & Bhardwaj, S. (2024). *Ethical Explorations: AI's Role in Shaping the Journalism Ecosystem Across Asia.* New Global Studies., DOI: 10.1515/ngs-2024-0002

Shetty, M., & Shetty, A. (2023). Automated Method for Fake News Detection Using Machine Learning. *2023 International Conference on Network, Multimedia and Information Technology (NMITCON)*, 1-6. DOI: 10.1109/NMITCON58196.2023.10275880

Školkay, A., & Filin, J. (2019). *A Comparison of Fake News Detecting and Fact-Checking AI Based Solutions*. Studia Medioznawcze. DOI: 10.33077/uw.24511617.ms.2019.4.187

Thielen, A. (2018). Fake News – Två ord, två betydelser: En statsvetenskaplig begreppsstudie på traditionella medieartiklar.

Thorne, J., & Vlachos, A. (2018). Automated Fact Checking: Task Formulations, Methods and Future Directions. *ArXiv, abs/1806.07687.*

Velentzas, J., Broni, G., & Professor, A. (2020). *Fake News and the Role of Semantics and AI (Artificial Intelligence).* How to Recognise the Falsity of the Information.

Zhang, J., Dong, B., & Yu, P. S. (2018). FakeDetector: Effective Fake News Detection with Deep Diffusive Neural Network. *2020 IEEE 36th International Conference on Data Engineering (ICDE)*, 1826-1829.

Chapter 11
Word Prediction Using Fuzzy Sets and Computational Intelligence

D. Rajalakshmi
https://orcid.org/0000-0003-4851-7685
SASTRA University, India

G. Revathy
https://orcid.org/0000-0002-0691-1687
SASTRA University, India

V. Prakash
https://orcid.org/0000-0002-6087-0839
SASTRA University, India

R. Bhavani
https://orcid.org/0000-0002-9062-4348
SASTRA University, India

ABSTRACT

Fuzzy soft sets find an extensive assortment of applications in many decision-making problems. Fuzzy sets are applied to overcome uncertainty, and soft sets came into existence as an extension. It is called soft because the boundary depends on the parameters. Soft sets are the generalization of fuzzy sets. Parameter reduction in soft sets shows an energetic role in reducing the complexity of any decision-making problems. Fuzzy soft sets are applied to identify a word. The set of alphabets is taken as the universal set, and the words formed using these alphabets give the parameter

DOI: 10.4018/979-8-3693-9246-1.ch011

set. The length of the word is first calculated. The first three letters of the word to be identified are given as the input. The parameter set consists of words of the same length, and the first three letters are generated using NLG. The word is identified using machine learning of word classification and prediction. The results are compared, and the final accuracy is calculated.

INTRODUCTION

Fuzzy soft sets have various applications in decision-making scenarios. They address uncertainty and were developed as an extension of the Fuzzy Set (FS). The term "soft" is used because parameters determine the boundaries. Soft sets are a broader version of FS. Simplifying parameters in soft sets is crucial for easing the intricacy of decision-making problems (Wang et al., 2023).

In this study, we utilize fuzzy soft sets to determine a word. The universal set is demarcated as the collection of alphabets, and words created from these alphabets make up the parameter set. Initially, the length of the word is determined, followed by inputting the first three letters of the word to be identified. Through Natural Language Generation (NLG), a parameter set consisting of words of equal length and beginning with the specified three letters is generated. The word is then identified using similarity and matching algorithms (Al-Shami, 2023). A comparison of the outcomes from both algorithms allows for the successful detection of the word.

FUZZY INTRODUCTION

FS theory helps decision-making by providing a structure for effectively handling imprecise and subjective information (Zadeh et al., 1996). It is credited with creating FSs that have been instrumental in finding solutions for unclear situations. Conventional decision-making models often face confrontations with ambiguity and uncertainty in practical situations. FSs offer a solution for managing uncertain data.

When faced with uncertainty and rough data, FS theory provides a way for decision-makers to describe the situation accurately. By utilizing graded membership FSs, uncertainty can be visually represented more intuitively (Garg et al., 2018). FSs introduce the idea of partial membership, where objects can fit into a set to varying degrees between 0 and 1 (Ejegwa, 2022). This membership level is determined by a membership function that assigns a value based on how well the element fits within the set. Incorporating Fuzzy Logic (FL) allows decision-makers to navigate situations where category distinctions are not clear-cut and objects cannot easily be quantified or categorized as binary entities (Zhou et al., 2023).

FS allows for a more adaptable and refined way to characterize uncertainty in decision-making, leading to a more authentic simulation of complex real-world issues. They offer a unique depiction of the inherent uncertainty and vagueness observed in several decision-making scenarios by illustrating the sluggish shift between belonging and now not belonging (Thiruvenkatasuresh et al., 2020). Compared to other techniques, utilizing a fuzzy selection approach offers numerous blessings. Here they may be:

- Handling uncertainty: FL is first rate at handling uncertainty and imprecision when making decisions. Unlike ordinary binary good judgment, which requires clean differences among companies, FL allows sluggish shifts and levels of belonging, making it best for conditions with doubtful limitations and unsuitable records.
- Flexibility: FL is first-rate for dealing with complicated international issues. It enables selection makers to use language and FSs to handle qualitative records quantitatively.
- Incorporating subjectivity: FL allows decision-makers to remember subjective exams and expert reviews while selecting. By incorporating human knowledge and domain expertise, FL advises selection-makers to combine imprecise and subjective records in a selection-making manner.
- Robustness: FL models are useful for dealing with imperfect enter, as they can handle noise and lack statistics. The ability of FL to tolerate imprecision and uncertainty makes it reliable when facing fluctuations and uncertainties in entering records, leading to more reliable choice outcomes.
- One of the key advantages of FL models is their interpretability. Unlike complex mathematical models, FL models are simpler to comprehend. The language regulations and FSs utilized in FL structures are easy to understand and interpret, making it easier for decision-makers to grasp the reasoning behind the system's findings.
- FL systems are not static; they can adapt and improve over time. By considering feedback and new records, FL systems can regulate membership capabilities and rules, increasing their performance and accuracy as they learn. This adaptability is a key feature of FL systems, making them more effective in real-world decision-making scenarios.

Below are key steps for applying FSs to practical decision-making issues:

- When tackling a decision-making problem, clearly outlining the objectives, restrictions, and variables involved is crucial. It is also essential to recognize the important parameters and criteria that will be key in guiding the decision-making process.

- Describe qualitative factors and expert opinions connected to decision criteria using language phrases such as "high," "medium," and "low" to indicate the near pleasure or importance of each criterion.
- When working with membership functions, we assign them to each linguistic term to measure how much an element belongs to an FS. These functions help us understand the connection between language terms and the numerical quantities of the variables.
- Using membership functions, develop FSs for every choice criterion. These FSs outline the uncertain boundaries of each criterion and allow for the representation of ambiguity and vagueness in the decision-making process.
- One way to design systems is by creating rules that connect input variables to output decisions. These rules typically involve using IF-THEN statements to determine the logic behind decision-making using FSs and language concepts.
- Utilise Fuzzy inference methods such as Mamdani or Sugeno to assess fuzzy guidelines and input data, achieving accurate output values. The technique uses FL operations like fuzzy AND, OR, and fuzzy implication to make the last selection.
- Defuzzification techniques such as the centroid and maximization techniques are usually applied to deduce a clear numeric price from the vague outputs of a fuzzy system. This allows trustworthy interpretation and choice of action.
- Both validating the outputs against acknowledged records and acquiring expert viewpoints on the version behavior are important for comparing accuracy. Sensitivity evaluation, furthermore, allows for the estimation of ways fluctuations in input variables will affect results, aiding with enhancement.
- Software-primarily based implementation facilitates choice-makers exploring scenarios and weighing the FL consequences for well-informed judgments. Integrating the model into a utility streamlines the evaluation and choice of alternatives, similar to its indeterminate reasoning.

By following those methods and using FS principles, selection-makers can effectively rent FL to handle real-world decision-making issues in a dependent and methodical way. FL's flexibility, interpretability, and robustness make it a green tool for coping with various disciplines' complex and ambiguous selection situations.

The advantages of using a fuzzy technique in selection-making encompass the capability to deal with uncertainty, flexibility in modeling complicated conditions, subjectivity inclusion, resilience in noisy information, model interpretability, and model-to-converting conditions. These benefits make FL an efficient tool for dealing with several decision-making issues in various disciplines.

Maji et al. (2022) first suggested the theory of soft sets in 1999 as a generalization of FSs. Soft sets have a parameter set that offers extra statistics about the situation. The smooth set's border is 0 to 1. Because of their potential to address uncertainties and vague records, smooth units play a critical role in selection-making. Traditional formal modeling and computation technology are often crisp and deterministic, making them unsuitable for complex problems with ambiguous inputs. Soft set theory affords a clear framework that allows for approximate descriptions of factors without requiring particular answers. Because it helps many parameterizations consisting of words, phrases, actual numbers, capabilities, and mappings, it is very convenient and easy to apply in practice. Soft sets may be used to explain preferences, barriers, and uncertainties in more intuitive and adaptive decision-making settings. By establishing soft sets over unique universe sets with parameters, choice-makers can verify and compare numerous opportunities based on their appropriateness with distinctive selection factors.

Soft units can be used to express the splendor of houses, consumer options, or any other decision standards in a manner that displays the facts' approximate nature. In the context of gentle sets, weighted preference fee refers to giving a couple of preference parameters weights so that judgments can be made based totally on their relative relevance. Individuals frequently have options or standards that vary in importance while making choices. Decision-makers can better prioritize and analyze options by giving weights to these criteria (Ejegwa et al., 2023).

The weighted choice-value in soft sets is computed by multiplying the weight of each parameter by its corresponding value for every object being considered. This computation aids in assessing the overall desirability or suitability of an object based on the weighted criteria.

For example, suppose a decision maker chooses a house based on factors such as beauty, material, surroundings, and repair status. In that case, they can assign weights to each parameter according to its importance. The decision maker can determine the optimal choice that best suits their preferences by computing the weighted choice value for each house using the supplied weights (Zhou et al., 2023). The weighted choice price offers an established technique for selection-making by factoring within the relative significance of many variables during the evaluation procedure. It allows decision-makers to make informed selections representing their possibilities and priorities, resulting in more effective selection outcomes in complicated choice-making troubles with using soft units. Overall, the importance of fuzzy gentle units in decision-making stems from their capability to efficiently control obscure and unsure data, making them an invaluable tool for modeling and evaluating complex preference troubles. FSs offer a wide variety of programs, including actual-time selection-making. Soft units have penetrated the literature as a generalization of FSs.

Definition: A pair (A, P) is a tender set over A if and handiest if f is a mapping from P to the energy set of A. A gentle set is a parameterized family of subsets of A.

Example: Let A denote the set of motors under consideration, and P denote the set of parameters. Each parameter is a word or a sentence. P = {beautiful, cheap, expensive, Petrol fuelled, diesel fuelled}. A soft set means pointing out beautiful cars, expensive cars, etc. The soft set (A, P) describes the "attractiveness of the car" a person will purchase.

Let us assume that there are three cars in the universe F, given by {C1, C2, C3} and P = {Q1,Q2, Q3, Q4}, Where

Q1 stands for a beautiful car.
Q2 stands for expensive cars.
Q3 stands for petrol-fuelled cars and
Q4 stands for diesel-fuelled cars.

Let

F(Q1) = { beautiful cars =C1, C2},
F(Q2) = { Expensive cars = C2},
F(Q3) = { Petrol Fuelled cars = { C1, C2},
F(Q4) = { Diesel fuelled cars = C3}

We can view (F, P) as a soft set.

Al-Shami and Mhemdi (2023) proposed an immediate application for soft sets. They discussed the application by displaying soft sets in tabular form and reached a judgment. This remarkably impacted the literature, resulting in a new range of decision-making difficulties. The soft set is represented in tabular format as follows.

Cars	Q1	Q2	Q3	Q4
C1	1	0	1	0
C2	1	1	1	0
C3	0	0	0	1

To reach a judgment, the row total is added, and the maximum sum is considered.

The car with the most value is declared as the best purchase choice. In the event of a tie in the total, we can assign appropriate weights to each parameter and repeat the procedure. Many applications have evolved in all sectors, and in this paper, we look at a decision-making problem for predicting a term.

WORD FORMATION

Various NLP models and techniques can be employed to generate the remaining alphabet sets based on the original three letters of a word. Here are several approaches, along with corresponding NLP models, that could be utilized:

Character-Level Language Models

Recurrent Neural Networks (RNNs) and Long Short-Term Memory Networks (LSTMs)

These models may be trained to anticipate the following character in a sequence based on the typescripts that originated beforehand. Recurrent Neural Networks (RNNs) and Long Short-Term Memory Networks (LSTMs) are effective designs for natural language processing (NLP) applications involving sequential input, such as text creation, machine translation, and sentiment analysis. Look at their architectures and see how they work.

Recurrent Neural Networks (RNNs)

Recurrent Neural Networks (RNNs) are a category of neural networks intended to handle sequential input by keeping a hidden state representing the network's memory of the sequence it has seen thus far. RNdNs process input sequences one at a time while keeping a hidden state encompassing evidence from previous items in the order.

Training RNNs

- RNNs are trained via backpropagation complete time (BPTT), a version of backpropagation that considers sequential data.
- Gradient updates are computed by unrolling the network through time and propagating errors back through the network's sequential connections.

Long Short-Term Memory Networks (LSTMs)

Long-short-term memory Networks (LSTMs) are a type of RNN architecture designed to address the vanishing gradient problem. This problem can occur during traditional RNN training when gradients decrease exponentially over time, making it problematic for the network to absorb long-range dependencies in sequential data.

Key Components

- Cell State: The cell state acts as a conveyor belt that conveys evidence crosswise time stages without being modified by the recurrent connections.
- Gates: LSTMs incorporate three types of gates—the forget gate, input gate, and output gate—which adjust the current of data entering and leaving the cell state. These gates are controlled by sigmoid and tanh functions that selectively allow the network to reminisce or overlook material over time.

Training LSTMs

- LSTMs are trained using the same principles as traditional RNNs, with backpropagation through time (BPTT) used to compute inclines and apprise model parameters.
- Because of its more complicated design, LSTMs can better retain long-term addictions in succeeding data than regular RNNs, making them idyllic for claims requiring memory across extended periods.

Applications in Predicting Next Characters

RNNs and LSTMs are both good at forecasting the ensuing charm in a sequence based on the types that arose earlier than them, creating ideal options for errands like language modeling and text production.

- Training Objective: The network is qualified to forecast the probability distribution over the following characters in the order assumed by the prior types.
- Loss Function: Cross-entropy loss is classically used to measure the incongruity between the forecast chance distribution and the true distribution of the next character.

In conclusion, RNNs and LSTMs are strong architectures for processing sequential data. They are especially successful for applications where the order of items in the sequence is critical, such as predicting the next letter in a text string. LSTMs, in particular, have been extensively used in NLP applications because they can detain long-term relationships and alleviate the endangered gradient problem in typical RNNs.

Gated Recurrent Unit (GRU): Similar to LSTMs, GRUs are another recurrent neural network that can be cast for character-level sequence prediction tasks.

GRUs are a variant of RNNs that, like LSTMs, address the vanishing gradient problem and capture long-range dependencies in sequential data. They achieve this with a simplified architecture compared to LSTMs, consisting of two gates: the update gate and the reset gate.

- **Update Gate:** Controls the evidence passed from the preceding hidden state to the current one.
- **Reset Gate:** Controls how much earlier evidence should be forgotten.

GRUs combine the benefits of RNNs with a more streamlined architecture, making them computationally more efficient and easier to train than LSTMs. They have fewer parameters and require less memory, which can be advantageous in scenarios with limited computational resources.

Comparison

- LSTMs: They have a more complex architecture with three gates and a memory cell, allowing them to capture more intricate patterns in the data. They are predominantly convenient for tasks requiring modeling of long-range dependencies and handling long sequences.
- GRUs: The suggestion is a simpler architecture with two gates, making them computationally extra competent and easier to train than LSTMs. They are apposite for errands where computational capitals are inadequate or when a simpler model is preferred.

LSTMs and GRUs have remained widely used in numerous NLP tasks, language modeling, text generation, sentiment analysis, machine translation, and sequence prediction tasks like next character prediction. The choice between LSTMs and GRUs depends on the detailed requirements of the task, the available computational resources, and the trade-offs between model complexity and performance.

Word-Level Language Models

N-Gram Language Models: These models envisage the subsequent term in a classification founded on the previous N-1 words. They can be proficient in a corpus of text to seize word-level dependencies.

When predicting the succeeding expression in a categorization built on the preceding N-1 words, N-gram language Models are geometric mockups frequently employed in natural language processing (NLP). Based on conditional probability, they calculate the likelihood of witnessing a word given its context—that is, the arguments that originate previously from it (N–1)—using a training corpus of text data.

Key Components of N-Gram Language Models

1. N-Gram: An N-gram is a contiguous sequence of N items from a text. In language modeling, an N-gram refers to an arrangement of N words. For example, a bigram (2-gram) contains two words, and a trigram (3-gram) consists of three words.
2. Conditional Probability: The core idea behind N-gram language Models is to guess the conditional probability of observing the succeeding term in a categorization given the preceding N-1 words. This probability is computed based on the occurrence frequency of N-grams in the training corpus.
3. Smoothing: To handle unseen N-grams or N-grams with zero frequency in the training data, smoothing practices such as Laplace (add-one) smoothing or Good-Turing smoothing may be employed. These techniques adjust the probability estimates to ensure all N-grams have non-zero probabilities.

Training Process

- Corpus Collection: N-gram language Models are qualified on a quantity of text data, which could be a collection of documents, books, or other textual sources.
- Tokenization: The text corpus is tokenized into individual words or tokens. Punctuation marks, special characters, and whitespace may also be treated as separate tokens, depending on the tokenization scheme.
- N-Gram Extraction: From the tokenized corpus, all possible N-grams are mined laterally with their frequencies of occurrence.
- Probability Estimation: The conditional probability of observing the next word, given the context of the preceding N-1 words, is estimated for each N-gram. This is classically finished using maximum likelihood estimation (MLE) or other smoothing techniques.

Inference Process

1. Context Formation: Given a sequence of words, the last N-1 words are used as context to envisage the next word.
2. N-Gram Lookup: The N-gram conforming to the context is in the model's probability table.

3. Next Word Prediction: Based on the prospects associated with the candidate's next disputes in the N-gram, the word with the maximum likelihood is selected as the predicted next word.

Example

Let us consider a simple example using a bigram (2-gram) language model trained on the following sentence: "The quick brown fox jumps over the lazy dog."

- Bigrams:
- "The quick"
- "quick brown"
- "brown fox"
- "fox jumps"
- "jumps over"
- "over the"
- "the lazy"
- "lazy dog"
- Probability Estimation:
- P("quick" | "The") = count("The quick") / count("The") = 1 / 1 = 1
- P("brown" | "quick") = count("quick brown") / count("quick") = 1 / 1 = 1
- P("fox" | "brown") = count("brown fox") / count("brown") = 1 / 1 = 1
- ...
- Inference:
- Given the context "quick brown", the next word prediction would be "fox" with a probability of 1.
- N-Gram Language Models are frequently used in NLP applications such as text creation, machine translation, spelling correction, and speech recognition because they capture word-level relationships and can create coherent, contextually appropriate text based on the input context. However, they have limits in capturing long-range relationships and may have sparsity difficulties when occupied with big languages or rare N-grams.

Transformer Models (e.g., GPT, BERT)

Pre-trained transformer-based models, such as GPT (Generative Pre-trained Transformer) and BERT (Bidirectional Encoder Representations from Transformers), may create text grounded on context and condition. Transformer models, such as GPT (Generative Pre-trained Transformer) and BERT (Bidirectional Encoder Representations from Transformers), have transformed natural language processing (NLP) by achieving cutting-edge performance in a variety of tasks such as text generation, machine translation, inquiry responding, and sentiment analysis. These models are based on the transformer architecture, which uses self-attention techniques to detect long-range relationships in sequential data.

Key Components of Transformer Models

- Self-Attention Mechanism: Transformer models' main novelty is the self-attention mechanism, which permits the model to weigh the value of various phrases in the input arrangement while producing predictions. This method allows transformers to efficiently capture long-range relationships by focusing on key bits of the input sequence.
- Multi-Head Attention: Transformer models use multi-head attention to describe self-attention by applying the attention mechanism in parallel with varied input linear projections. The model may focus on many segments of the input sequence at once, capturing various aspects of context.
- Positional Encoding: Transformer models cannot represent the order of arguments in the input sequence. Therefore, positional encoding is added to the input embeddings. This allows the model to identify words created on their location in the sequence.
- Feedforward Neural Networks: Transformer models often comprise feedforward neural networks with linear transformations and non-linear activation functions, totaling self-attention layers. Feedforward networks let the model capture complicated relationships between words in the input sequence.

GPT (Generative Pre-trained Transformer)

GPT is a modernized language model established by OpenAI. It is trained using a generative pre-training approach, where the model is proficient in forecasting the subsequent word in a classification given the preceding context. GPT is unidirectional, meaning it generates text from left to right based on the context provided. It has stayed broadly used for text generation tasks, including story generation, dialogue generation, and summarization.

BERT (Bidirectional Encoder Representations from Transformers)

Google created another transformer-based language paradigm, BERT. Unlike GPT, BERT is bidirectional, making predictions based on the left and right contexts. BERT is trained with a masked language modeling (MLM) goal, which involves masking random tokens in the input categorization and training the model to predict the masked tokens based on the surrounding context. BERT has demonstrated cutting-edge performance across various NLP tasks, including question answering, text categorization, and named entity identification.

Text Generation with Transformer Models

Transformer models may be fine-tuned for text production tasks using a prompt or conditioning text to generate future tokens based on the learned language model. The model creates text token by token, sampling from the probability distribution across the vocabulary at each step. By conditioning on diverse prompts or circumstances, transformer models may create coherent and contextually appropriate text for the variability of submissions, such as story production, conversation systems, and content creation.

Sequence-to-Sequence Models

Sequence-to-Sequence (Seq2Seq) with Attention

Seq2Seq models, often equipped with attention mechanisms, can be trained to generate sequences of characters or words given an input sequence. They are frequently cast off for errands such as machine translation and text summarization.

Sequence-to-sequence (Seq2Seq) models with attention mechanisms are robust neural network designs widely utilized in natural language processing (NLP) for sequence-generating tasks. These models have become popular for tasks such as machine translation and text summarization because they can properly manage variable-length input and output sequences.

Components of Seq2Seq Models with Attention

Encoder-Decoder Architecture

- Seq2Seq models include two major components: an encoder and a decoder.

- The encoder converts the input classification into a fixed-size context vector, which is then used by the decoder to construct the output sequence.

Recurrent Neural Networks (RNNs) or Transformers

- Traditionally, RNNs such as Long Short-Term Memory (LSTM) or Gated Recurrent Unit (GRU) cells are cast off in Seq2Seq models.
- More recently, transformer-based architectures, which do not rely on recurrence, have become prevalent due to their capability to more effectively detain long-range dependencies.

Attention Mechanism

- The attention method enables the decoder to engender output tokens by focusing on different sections of the input sequence.
- It computes attention weights, which define the significance of each encoder output at each decoding step.
- The devoutness weights are utilized to create a context vector, the weighted amount of encoder outputs.

Training Process

Encoder Processing

- The encoder converts the input classification into hidden states, each representing an instant of the sequence up to that moment.

Attention Calculation

- At each decoding step, the attention mechanism analyses consideration scores between the decoder's hidden state and each encoder's hidden state.
- These attention scores are standardized to obtain attention weights.

Context Vector Calculation

- The context vector is totaled as a prejudiced totality of the encoder hidden states, where the attention weights give the weights.

Decoder Generation

- The decoder receives the situation vector and its prior hidden state as input to create the subsequent token in the output sequence.
- This process is frequent until an end-of-sequence token is generated or a maximum length is reached.

Inference Process

During inference, the trained model creates output sequences specified input sequences. The inference process is comparable to the training process but with a few differences:

Greedy Decoding or Beam Search

- Greedy decoding selects the token with the highest probability at each step.
- Beam search explores multiple hypotheses and selects the most likely sequence overall.

Sampling

- Instead of selecting the token with the highest probability, the model can sample from the probability distribution over the vocabulary at each step.

CONDITIONAL TEXT GENERATION MODELS:

Conditional Variational Autoencoders (CVAEs)

CVAEs can generate text conditioned on some input, such as the initial letters of a word. They learn a latent representation of the input and generate text conditioned on this representation.

Conditional Variational Autoencoders (CVAEs) are a variant of variational autoencoder (VAE) models that generate text conditioned on some input. Unlike traditional VAEs, which generate data from a random noise vector, CVAEs can generate data based on specific conditions or inputs. This makes them predominantly convenient for errands such as text generation, where the generated text must be conditioned on certain attributes or contexts.

Key Components of Conditional Variational Autoencoders (CVAEs)

Encoder Network

- The encoder network receipts the input, such as the preliminary letters of a word, and plots it to a latent space representation.
- In the case of text data, the input may be encoded using techniques like one-hot encoding or word embeddings.
- Latent Space
- CVAEs learn a probabilistic latent space depiction of the input data.
- This latent space is typically represented by a mean vector and a variance vector, which encode the mean and variance of a Gaussian distribution, respectively.
- Sampling
- During training, CVAEs sample latent vectors from the scholarly latent space distribution.
- This sampling process introduces stochasticity into the model, allowing it to produce diverse outputs for a given input.
- Decoder Network
- The decoder network takes the sampled latent vector and any additional conditioning information and generates the output data.
- In the case of text generation, the decoder may consist of recurrent neural networks (RNNs) or transformer-based architectures.
- Objective Function
- CVAEs are trained using a variational inference approach that maximizes a lower bound on the log-likelihood of the data.
- This objective function consists of two terms: a reconstruction loss, which procedures how well the generated data matches the input data, and a regularization term, which heartens the latent space to follow a prior distribution (usually a unit Gaussian).

Training Process

- **Encoder Training:**
- The encoder network is trained to map the input data to a distribution in the latent space that matches the prior distribution.
- This is typically done by minimizing the KL divergence between the learned distribution and the prior distribution.

- **Decoder Training:**
- The decoder network is trained to recreate the input data from the sampled latent vectors.
- This is achieved by minimizing the reconstruction loss, which actions the variance between the generated data and the input data.

Generation Process

- **Input Conditioning:**
- During inference, the model is provided with conditioning information, such as the initial letters of a word.
- This conditioning information is used to guide the generation process.
- **Latent Vector Sampling:**
- The model samples a latent vector from the learned latent space distribution.
- **Text Generation:**
- The sampled latent vector, along with the conditioning information, is agreed over the decoder network to generate the output text.
- The decoder generates one token at a time, conditioning on the previously generated tokens.

Applications

CVAEs have been applied to various text generation tasks, including:

- Text Completion: Generating text conditioned on partial input text.
- Text Style Transfer: Generating text in a particular style or tone.
- Text Summarization: Generating concise summaries of longer texts.
- Dialogue Generation: Generating responses in conversational agents.

CONDITIONAL GENERATIVE ADVERSARIAL NETWORKS (cGANS)

cGANs can generate text conditioned on a given input. By training on pairs of input-output sequences, they can absorb and generate text sequences created on the input.

Conditional Generative Adversarial Networks (cGANs) are a variant of Generative Adversarial Networks (GANs) that incorporate conditional information into the generation process. In the context of text generation, cGANs can generate text sequences conditioned on specific attributes, input sequences, or other forms of conditioning information. CGANs learn to generate text sequences based on the provided input by training on pairs of input-output sequences.

Key Components of Conditional Generative Adversarial Networks (cGANs)

Generator Network

- The generator takes random noise vectors as input, along with the conditioning information.
- It learns to generate text sequences conditioned on the provided input.

Discriminator Network

- The discriminator distinguishes between real text sequences and generated text sequences.
- It takes the generated text sequences and the conditioning information as input and outputs a probability score indicating the likelihood that the input sequence is real.

Conditional Information

- The conditioning evidence can yield several systems, such as input sequences, attributes, or class labels.
- It is provided as input to the generator and discriminator networks to guide the generation process.

Objective Function

- cGANs are trained using a min-max objective function, where the generator tries to generate realistic text sequences that fool the discriminator. In contrast, the discriminator tries to discriminate between real and generated text sequences.
- The generator aims to minimize the log probability of the discriminator making a correct classification, while the discriminator aims to maximize it.

Training Process

Data Preparation

- Training data consists of pairs of input-output sequences, where the input sequences are used as conditioning information.
- Both the generator and discriminator are trained on these paired sequences.
  Generator Training
- The generator takes random noise vectors as input, along with the conditioning information, and generates text sequences.
- It aims to produce text sequences that are indistinguishable from real text sequences, given the conditioning information.

Discriminator Training

- The discriminator takes real and generated text sequences and the conditioning information and classifies them as real or generated.
- It aims to correctly classify real text sequences as real and generate text sequences based on the provided conditioning information.

Inference Process

- The trained cGAN model can generate text sequences conditioned on new or unseen conditioning information during inference. The inference process involves feeding the desired conditioning information to the generator network and sampling from the generated text sequences.

Applications

cGANs have various applications in text generation tasks, including:

- Conditional Text Generation: Generating text sequences conditioned on specific attributes, input sequences, or class labels.
- Text Style Transfer: Modifying the style or attributes of a given text while preserving its content.
- Content Generation: Generating text sequences with desired attributes specified by conditioning information.

RULE-BASED SYSTEMS

Finite State Transducers (FSTs)

FSTs can define rules for generating text based on input sequences. They are commonly used in tasks such as spelling correction and morphological analysis.

Finite State Transducers (FSTs) are computational models used in natural language processing and other fields to represent and manipulate sequences of symbols. By associating input symbols with output symbols or sequences, FSTs can define rules for generating text based on input sequences. They are commonly used in spelling correction, morphological analysis, and text generation tasks.

Key Components of Finite State Transducers (FSTs)

- States: FSTs consist of a set of states, each representing a particular configuration or position in the transducer.
- Transitions: Transitions connect states and are labeled with input and output symbols. They define the mapping from input symbols to output symbols.
- Start State: The start state is the transducer's initial state, where the generation process begins.
- Final States: Final states indicate the end of the generation process. When the transducer reaches a final state, it outputs the generated text.

Types of Finite State Transducers

- Sequential FSTs (SFSTs): SFSTs define a one-to-one mapping between input and output symbols. They are commonly used in tasks such as spelling correction, where each input symbol corresponds to exactly one output symbol.
- Weighted FSTs (WFSTs): WFSTs associate weights with transitions, allowing for probabilistic modeling. They are used in tasks such as speech recognition and machine translation, where the likelihood of different output sequences may vary.

Applications of Finite State Transducers (FSTs)

- Spelling Correction: FSTs can be used to generate corrections for misspelled words by mapping them to their correct spellings based on predefined rules.
- Morphological Analysis: FSTs are commonly used to analyze the morphological structure of words in languages with complex inflectional or derivational morphology.

- Text Generation: FSTs can define rules for generating text based on input sequences, such as generating inflected forms of words or expanding abbreviations.
- Information Retrieval: FSTs can be used in search engines and information retrieval systems to process queries and match them to relevant documents or web pages.

Example

Let's consider an example of a spelling correction system implemented using an SFST. Suppose we have the following rules:

If the input is "helo," the output should be "hello."
If the input is "writ," the output should be "write."
If the input is "cat," the output should remain "cat."

In this case, the SFST would consist of states representing the input sequences "h," "e," "l," "o," "w," "r," "i," "t," "c," "a," and "t," with transitions connecting these states according to the defined rules.

When presented with an input sequence, the SFST would traverse the states based on the input symbols, following the defined transitions, and output the corresponding generated text.

Overall, FSTs provide a flexible and efficient framework for defining rules for generating text based on input sequences. They are widely used in various NLP tasks and continue to be an important tool in computational linguistics and related fields.

HYBRID APPROACHES

Combination of Models

Hybrid approaches that combine multiple models, such as using a character-level RNN for initial predictions and a word-level transformer for refinement, can sometimes yield better results.

Combining multiple models through hybrid approaches is a powerful strategy in machine learning and natural language processing, as it allows leveraging the strengths of different models to achieve improved performance. One hybrid approach combines a character-level Recurrent Neural Network (RNN) for initial predictions with a word-level transformer for refinement. This approach can often yield better results than using either model in isolation.

Key Components of the Hybrid Approach

- Character-Level RNN
- The character-level RNN is used for initial predictions. It processes input text character by character, capturing fine-grained patterns and dependencies at the character level.
- RNNs, such as LSTM or GRU, are commonly used for this purpose due to their ability to capture sequential dependencies.
- Word-Level Transformer
- The word-level transformer is employed for refinement. It operates at the word level and can more effectively capture long-range dependencies and contextual information in the input text.
- Transformers like the GPT or BERT models are well-suited for this task due to their self-attention mechanism and large-scale pre-training.
- Combination Strategy
- The predictions from the character-level RNN serve as initial inputs or embeddings for the word-level transformer.
- The transformer refines these initial predictions, incorporating contextual information and making more accurate predictions based on the global context of the input text.

Training and Inference Process

- Training
- The hybrid model is trained using a combination of supervised learning and possibly transfer learning techniques.
- The character-level RNN and word-level transformer components are trained separately or jointly, depending on the specific architecture and training objectives.
- Inference
- During inference, the input text is first processed by the character-level RNN to generate initial predictions.
- These initial predictions are then fed into the word-level transformer, which refines them and generates the final output sequence.

Benefits of Hybrid Approaches

- Complementary Strengths

- Combining models operating at different levels of abstraction (character level vs. word level), hybrid approaches can capture fine-grained and global patterns in the input text, leading to more robust and accurate predictions.
- Synergistic Effects
- Hybrid models can exploit synergies between different components, resulting in improved performance compared to individual models in isolation.
- Flexibility
- Hybrid architectures offer flexibility in model design and training, allowing practitioners to tailor the model to specific tasks and datasets.

Applications

Hybrid approaches combining character-level RNNs and word-level transformers have been successfully applied in various NLP tasks, including:

- Text Generation: Generating coherent and contextually relevant text sequences.
- Machine Translation: Translating text from one language to another with improved accuracy and fluency.
- Text Summarization: Generating concise summaries of longer texts while preserving key information.

Overall, hybrid approaches combining multiple models offer a promising avenue for advancing state-of-the-art natural language processing and addressing the complex challenges posed by real-world text data. Each of these NLP models and techniques has its strengths and weaknesses, and the choice of model depends on factors such as the complexity of the task, the availability of training data, and computational resources. Experimentation with different approaches and model architectures is often necessary to determine the most effective solution for a given problem.

NATURAL LANGUAGE GENERATION

Natural Language Generation (NLG) using fuzzy soft sets involves a hybrid approach that combines fuzzy set theory with soft set theory to generate natural language text. This approach leverages fuzzy sets' and soft sets' flexibility and uncertainty handling capabilities to model and generate text based on imprecise or uncertain information.

Components of NLG using Fuzzy Soft Sets

- Fuzzy Sets
- Fuzzy set theory extends classical set theory by allowing elements to have degrees of membership between 0 and 1, representing the degree of belongingness of an element to a set.
- Fuzzy sets represent linguistic variables and fuzzy linguistic terms, capturing the vagueness and uncertainty inherent in natural language.
- Soft Sets
- Soft set theory provides a framework for dealing with uncertainty and ambiguity by representing sets using a parameterized family of subsets.
- Soft sets allow for the representation of indeterminate and incomplete information, making them suitable for modeling imprecise or uncertain linguistic data.
- NLG Process
- The NLG process using fuzzy soft sets involves several steps:
- Input Representation: Input data, such as structured data or knowledge bases, are represented using fuzzy sets and soft sets to capture uncertainty and imprecision.
- Knowledge Extraction: Fuzzy soft-set operations are applied to extract relevant knowledge or patterns from the input data.
- Text Generation: The extracted knowledge is used to generate natural language text, where fuzzy soft set operations help incorporate uncertainty and vagueness into the generated text.
- Fuzzy Soft Set Operations
- Fuzzy soft set operations, such as union, intersection, complement, and aggregation, are used to manipulate and combine fuzzy and soft sets during the NLG process.
- These operations help handle uncertainty, imprecision, and ambiguity in the input data and the generated text.

Applications of NLG Using Fuzzy Soft Sets

- Summarization and Paraphrasing
- NLG systems can summarize or paraphrase input text by extracting key information using fuzzy soft set operations and generating concise and coherent summaries or paraphrases.
- Data-to-Text Generation

- NLG systems can convert structured or semi-structured data into natural language text by incorporating fuzzy soft-set operations to handle uncertainty and ambiguity in the input data.
- Dialogue Systems
- NLG systems in dialogue systems can generate natural language responses based on fuzzy soft-set representations of user queries and system knowledge, allowing for flexible and contextually appropriate responses.
- Personalization and Adaptation
- NLG systems can personalize generated text based on user preferences or adapt the generated text to different contexts or domains using fuzzy soft-set representations of user-profiles and contextual information.

Benefits of NLG using Fuzzy Soft Sets

- Flexibility and Expressiveness
- Fuzzy soft sets provide a flexible and expressive framework for representing and manipulating uncertain and imprecise linguistic data, allowing NLG systems to capture complex linguistic phenomena.
- Robustness to Uncertainty
- NLG systems using fuzzy soft sets can handle uncertainty, vagueness, and ambiguity in both the input data and the generated text, resulting in more robust and reliable NLG performance.
- Interpretability
- Fuzzy soft sets offer an interpretable representation of linguistic data, making it easier to understand and interpret the generated text and the underlying NLG process.

DECISION MAKING

Decision-making with Natural Language Generation (NLG) involves using NLG techniques to generate natural language explanations, summaries, or recommendations to support the decision-making process. NLG can help make complex data and analysis more understandable and accessible to decision-makers by presenting key insights, trends, and implications in a human-readable format. Here's how NLG can be applied in decision-making:

Data Analysis and Interpretation

- Data Summarization: NLG can summarize large datasets, providing decision-makers with concise overviews of the most relevant information.
- Insight Generation: NLG can analyze data and generate natural language explanations of patterns, trends, and correlations, helping decision-makers understand the underlying insights.

Risk Assessment and Management

- Risk Communication: NLG can articulate risks associated with different options or scenarios, helping decision-makers evaluate potential outcomes.
- Scenario Analysis: NLG can describe the implications of various scenarios or decision paths, allowing decision-makers to assess risks and make informed choices.

Financial Analysis and Planning

- Financial Reporting: NLG can generate natural language summaries of financial reports, highlighting key metrics, performance indicators, and financial trends.
- Budgeting and Forecasting: NLG can explain budgeting decisions, forecast outcomes, and provide a rationale for financial planning choices.

Policy Analysis and Evaluation

- Policy Recommendations: NLG can generate natural language explanations of policy options, outlining the pros and cons of each option and providing recommendations based on analysis.
- Impact Assessment: NLG can assess the potential impact of policies or decisions on various stakeholders, communities, or the environment, helping decision-makers understand the consequences of their choices.

Business Intelligence and Strategy

- Competitive Analysis: NLG can analyze competitors' strategies and market trends, providing decision-makers with insights into competitive dynamics and opportunities.

- Strategic Planning: NLG can explain strategic plans, goals, and objectives, helping stakeholders understand the rationale behind strategic decisions and aligning them with organizational goals.

Benefits of NLG in Decision-Making

- Clarity and Understanding: NLG can transform complex data and analysis into easy-to-understand natural language explanations, facilitating comprehension and decision-making.
- Efficiency: NLG automates the generation of textual summaries and insights, saving time and effort for decision-makers.
- Consistency: NLG ensures consistency in reporting and communication by standardizing the presentation of information.
- Transparency: NLG provides transparent explanations of decisions, enabling stakeholders to understand the reasoning behind them.
- Personalization: NLG can tailor explanations and recommendations to the specific needs and preferences of individual decision-makers.

CONCLUSION

The methodology presented in this study underscores the effectiveness of fuzzy soft sets in enhancing word prediction and identification. By determining the length of a given word and generating a parameter set of potential matches through Natural Language Generation, we have demonstrated how fuzzy soft sets can streamline the decision-making process. The exact word is identified with high precision using machine learning algorithms, effectively reducing the complexity and time required for word prediction tasks. This approach optimizes the accuracy of word identification and contributes to the development of more efficient computational models. Integrating fuzzy soft sets with computational intelligence proves to be a robust framework, offering significant improvements in the speed and reliability of word prediction systems. This research provides a foundation for future advancements in natural language processing and decision-making applications, paving the way for more sophisticated and efficient methodologies.

REFERENCES

Al-shami, T. M. (2023). (2, 1)-Fuzzy sets: Properties, weighted aggregated operators and their applications to multi-criteria decision-making methods. *Complex & Intelligent Systems*, 9(2), 1687–1705. DOI: 10.1007/s40747-022-00878-4

Al-shami, T. M., & Mhemdi, A. (2023). Generalized frame for orthopair fuzzy sets:(m, n)-fuzzy sets and their applications to multi-criteria decision-making methods. *Information (Basel)*, 14(1), 56. DOI: 10.3390/info14010056

Ejegwa, P. A., Onyeke, I. C., Kausar, N., & Kattel, P. (2023). A new partial correlation coefficient technique based on intuitionistic fuzzy information and its pattern recognition application. *International Journal of Intelligent Systems*, 2023, 1–14. Advance online publication. DOI: 10.1155/2023/5540085

Ejegwa, P. A., Onyeke, I. C., Terhemen, B. T., Onoja, M. P., Ogiji, A., & Opeh, C. U. (2022). Modified Szmidt and Kacprzyk's intuitionistic fuzzy distances and their applications in decision-making. *J. Niger. Soc. Phys. Sci.*, 4, 175–182. DOI: 10.46481/jnsps.2022.530

Garg, H., & Kumar, K. (2018). A novel correlation coefficient of intuitionistic fuzzy sets based on the connection number of set pair analysis and its application. *Scientia Iranica*, 25(4), 2373–2388.

Maji, P. K., Roy, A. R., & Biswas, R. (2002). An application of soft sets in a decision making problem. *Computers & Mathematics with Applications (Oxford, England)*, 44(8-9), 1077–1083. DOI: 10.1016/S0898-1221(02)00216-X

Thiruvenkatasuresh, M. P., Venkatachalam, V., & Revathy, G. (2020). Brain Tumor Detection in MRI Images, using Fuzzy C-Means and Cuckoo Search Algorithm. *Solid State Technology*, 63(5), 6759–6769.

Wang, T., Wu, X., Garg, H., Liu, Q., & Chen, G. (2023). A prospect theory-based MABAC algorithm with novel similarity measures and interactional operations for picture fuzzy sets and its applications. *Engineering Applications of Artificial Intelligence*, 126, 106787. DOI: 10.1016/j.engappai.2023.106787

Zadeh, L. A. (1996). On fuzzy algorithms. In *fuzzy sets, fuzzy logic, and fuzzy systems: selected papers By Lotfi A Zadeh* (pp. 127-147).

Zhou, Y., Ejegwa, P. A., & Johnny, S. E. (2023). Generalized similarity operator for intuitionistic fuzzy sets and its applications based on recognition principle and multiple criteria decision making technique. *International Journal of Computational Intelligence Systems*, 16(1), 85. DOI: 10.1007/s44196-023-00245-2

Chapter 12
Natural Language Processing for Legal Document Classification

Tomin Joseph
Christ University, India

A. Vijayalakshmi
Christ University, India

ABSTRACT

The legal domain generates an immense volume of textual data daily, ranging from court opinions and statutes to contracts and legal briefs. As this volume grows, so does the need for effective methods to organize, categorize, and extract insights from these documents. Natural language processing (NLP) has emerged as a powerful tool to address these challenges by automating the analysis of legal texts. This chapter aims to explore the application of NLP techniques in legal document classification, delving into its methods, challenges, and real-world applications.

ARTIFICIAL INTELLIGENCE AND LAW

One of the most legitimate arguments for artificial intelligence is its ability to increase efficiency and productivity. AI-powered tools can analyze a wide range of legal documents, regulatory documents, and regulations in a fraction of the time it takes a human lawyer. This allows legal professionals to focus on the more negative aspects of their job, such as client advice, negotiations, and courtroom arguments. For example, AI-powered contract review software can save lawyers hours of quality review by detecting risks and conflicts in legal documents. Additionally, AI can

DOI: 10.4018/979-8-3693-9246-1.ch012

aid legal research by scanning large volumes of legal documents and identifying important cases according to specific standards. This can reduce the time it takes for lawyers to build a strong case for their clients. Artificial intelligence has the ability to be fair as well as efficient. For many individuals and small businesses, legal services are often prohibitively expensive. AI-powered legal chatbots can provide legal advice and answer frequently asked questions, especially in areas such as family law, consumer rights, and landlord bankruptcy compliance. These services allow people to resolve their legal problems on their own, even if they do not have the resources to hire a lawyer. Additionally, AI-powered document automation tools can help people create legal documents, such as wills or contracts, at a lower cost than legal services. By reducing financial barriers to entry, Artificial Intelligence can play a greater part in ensuring that justice is served. AI-powered algorithms can analyze big data to identify cases in the judicial system, which can improve the relationship between the decision and the outcome of the case.

Additionally, AI can be used to develop risk assessment tools to help judges decide on pretrial detention or incarceration. Legal and ethical procedures must be put in place to ensure that these tools are bias-free and that judges still have the final say. Artificial intelligence algorithms are only as good as the data they learn from. For example, if an AI system used in criminal justice analysis is trained on data related to existing injustices in the criminal justice system, the system could lead to racial discrimination. Additionally, human monitoring is important to ensure that algorithm results are reviewed and do not discriminate. A clear definition of liability is important to ensure that there is a repeatable process by which AI-driven legal systems become just or unjust. This becomes particularly important when considering the potential of human intelligence in law. Although such a process is far from reality, this possibility raises ethical questions about the role of human decision-making and the meaning of justice in the AI legal environment. And it's a fascinating rivalry. A legal framework must be established to address issues such as injustice, accountability, and the role of human judgment. Only through careful consideration of these ethical issues and ideas can the power of wisdom be used to create a more just and beneficial society for all (Surden et al., 2019).

FUNDAMENTALS OF NATURAL LANGUAGE PROCESSING (NLP)

Introduction to NLP

Natural language processing (NLP) stands at the crossroads of computer science, linguistics, and artificial intelligence (AI). It delves into the intricate world of human language, aiming to bridge the gap between how humans communicate and how machines interpret information. This essay delves into the fascinating realm of NLP, exploring its core functionalities, the techniques employed, its diverse applications, and the exciting possibilities it holds for the future. At its heart, NLP helps us to make the computer understand, analyze, and manipulate human language. This encompasses a wide range of tasks, from the seemingly simple act of identifying parts of speech to the highly complex process of extracting meaning from a nuanced conversation. NLP algorithms can help us understand the statistical relationships that govern human language.

One of the fundamental techniques employed in NLP is machine learning. For instance, an algorithm trained for sentiment analysis might be exposed to sentences labeled as "positive," "negative," or "neutral." By analyzing the patterns within these labeled examples, the algorithm learns to identify similar patterns in unseen text, ultimately enabling it to predict the sentiment of a new sentence (Walters et al.,2021). Another cornerstone of NLP is statistical language modeling. This technique involves analyzing the statistical probabilities of word sequences. By understanding the likelihood of one word appearing after another, NLP systems can perform tasks like sentence completion, text prediction, and even machine translation. Imagine an NLP system trained on a vast corpus of news articles. It can analyze the statistical probability of words appearing together, like "earthquake," often following "devastating." This allows the system to not only understand the meaning of individual words but also grasp the relationships between them, forming a more comprehensive understanding of the text.

Beyond these core techniques, NLP encompasses a variety of specialized tasks. One crucial area is named entity recognition (NER). NER plays a pivotal role in identifying and classifying entities within text, such as people, organizations, locations, and dates. Imagine an NLP system analyzing a news article about a company announcing a new product launch. NER would identify the company name, the product name, and the location of the company headquarters. This information extraction capability is vital for tasks like information retrieval and knowledge base population.

Another crucial task is text summarization. Here, NLP algorithms condense large amounts of text into a concise and informative summary. This proves invaluable for tasks like summarizing news articles, research papers, or lengthy legal documents.

NLP systems achieve this by identifying key sentences and phrases, discarding irrelevant information, and generating a concise representation of the original text. Sentiment analysis, another essential task within NLP, delves into the emotional undercurrents of text. NLP algorithms analyze the language used to determine the sentiment expressed by the author, be it positive, negative, or neutral. This has vast applications in social media monitoring, customer review analysis, and even market research, allowing businesses to gauge public opinion and make informed decisions (McCarthy et al., 2021).

The applications of NLP extend far beyond the realm of academia. It plays a vital role in the development of intelligent virtual assistants, enabling them to respond to human queries. Machine translation, the cornerstone of applications like Google Translate, leverages NLP to bridge the communication gap between languages. Additionally, NLP is revolutionizing the field of chatbots, allowing them to engage in more natural and engaging conversations with users. As NLP continues to evolve, the possibilities it unlocks are truly astounding. In healthcare, NLP systems can understand medical records and research papers to identify potential drug interactions or predict the course of a disease. NLP also holds promise for personalized education, where intelligent tutoring systems can tailor their approach to each student's learning style and needs.

However, the burgeoning field of NLP is not without its challenges. One major hurdle is the inherent ambiguity of human language. Unlike computer code, human language is rife with sarcasm, metaphors, and cultural references that can make it difficult for NLP systems to comprehend. The sheer diversity of languages across the globe necessitates the development of NLP tools that can handle the unique nuances of each language.

Another challenge lies in the potential for bias within NLP systems. Like any machine learning algorithm, NLP systems are likely to be biased based on the training data. This raises ethical concerns, particularly when NLP systems are used in areas like loan approvals or criminal justice. Mitigating bias necessitates careful selection and curation of training data sets, ensuring they are representative and diverse (Alarie et al., 2021). Looking towards the future, NLP research delves into exciting new frontiers. One area of exploration is the development of explainable AI (XAI) for NLP. Explainable AI enables transparency, allowing humans to understand the reasoning behind an NLP system's decisions. This is crucial for building trust and ensuring that data is used ethically and responsibly.

Role of NLP in Legal Document Analysis

The legal domain thrives on precision and clarity. Mountains of contracts, regulations, and case law form the bedrock of the legal system, and navigating this complex landscape can be a daunting task. Enter Natural Language Processing (NLP), a powerful AI technology that is transforming the way legal professionals analyze and interpret legal documents. This essay explores the multifaceted role of NLP in legal document analysis, highlighting its potential to streamline workflows, enhance accuracy, and democratize access to legal information.

The most compelling application of NLP in legal document analysis lies in its ability to automate repetitive tasks. Legal professionals often spend countless hours reviewing voluminous documents to identify key information, such as specific clauses, parties involved, or deadlines. NLP-powered tools can automate this process by leveraging techniques like named entity recognition (NER) and keyword extraction. NER allows NLP systems to identify and classify legal entities within a document, such as names of individuals, organizations, locations, and dates. Keyword extraction focuses on pinpointing specific terms or phrases of legal significance. By automating these tasks (Robaldo et al., 2021).

Beyond automation, NLP empowers legal professionals with advanced search capabilities. Imagine searching through a vast database of legal documents to find specific precedents related to a unique case. NLP-powered search engines can crawl through legal text and identify relevant documents based on semantic similarity, not just exact keyword matches. This allows lawyers to delve deeper into the legal landscape, uncovering potentially game-changing precedents that might have been missed through traditional keyword-based searches.

Furthermore, NLP holds immense potential for risk assessment and compliance purposes. Legal documents, particularly contracts, often contain clauses with hidden risks or potential for non-compliance. NLP systems can be trained to identify these clauses based on predefined criteria, flagging potential issues for legal review. This proactive approach allows businesses to mitigate legal risks before they escalate, minimizing the potential for costly lawsuits or regulatory fines.

The impact of NLP extends beyond efficiency gains. NLP-powered tools can be used to analyze legal documents for sentiment and tone. By analyzing the language used within a contract or legal opinion, NLP systems can identify areas of potential disagreement or ambiguity. This information can be invaluable for lawyers during negotiations or when drafting legal documents, allowing them to craft clear and concise language that minimizes the risk of future disputes.

Another potential of NLP applications is to provide free access to legal information. Legal documents are often shrouded in complex legal jargon, making them inaccessible to the average person. NLP-powered tools can be used to summarize legal documents, which allows individuals to understand the essence of a contract or legal ruling in the absence of extensive legal expertise. This can empower individuals to navigate legal issues more confidently and make informed decisions regarding their rights.

However, the integration of NLP into the legal domain necessitates careful consideration of potential drawbacks. Legal documents, particularly contracts, are nuanced and often hinge on subtle details. NLP systems, while powerful, are still under development and susceptible to errors. It's crucial to emphasize that NLP tools should be viewed as assistants, not replacements, for human legal judgment.

Additionally, the legal field is rife with complex legal concepts and established legal precedents. NLP systems require substantial training data sets that include these nuances. Potential biases within training data can lead to biased outputs by NLP systems. This necessitates careful selection and curation of training data, ensuring they are comprehensive and representative of the legal landscape (Calister et al.,2020).

In conclusion, NLP is revolutionizing the way legal documents are analyzed and interpreted. By automating tasks, enhancing search capabilities, and identifying potential risks, NLP empowers legal professionals to work more efficiently and strategically. Furthermore, NLP holds promise for making legal information more accessible to the public. However, it's vital to acknowledge the limitations of this technology and ensure that human oversight remains paramount in the legal decision-making process. As NLP continues to evolve, a symbiotic relationship between technology and human expertise will undoubtedly pave the way for a more streamlined, accurate, and inclusive legal system.

Preprocessing Techniques for Legal Text

NLP algorithms can work their magic only if the raw data is well prepared. This crucial stage is known as preprocessing, and it plays a vital role in ensuring the accuracy and efficiency of NLP tasks. This essay delves into the world of NLP preprocessing, exploring the key techniques, their benefits, and the considerations for effective application. Feeding unprocessed text data into an NLP system can lead to unreliable and misleading results. Preprocessing aims to clean, structure, and standardize the raw text, transforming it into a format that NLP algorithms can readily understand and analyze (Bommarrito et al., 2021).

Text Cleaning involves removing unwanted elements from the text that don't contribute to the meaning or can introduce noise into the data. Examples of elements removed during text cleaning include punctuation marks, special characters, HTML tags, and URLs. Removal of stop words – the most common words in a language (e.g., "the," "a," "an") - which carry little semantic meaning can play a pivotal role in reducing the dimensionality. While seemingly insignificant, removing these elements can significantly improve the performance of NLP models, as they focus on the words that truly matter.

Another crucial technique is tokenization. Imagine breaking down a sentence into individual words. Tokenization achieves precisely that. It segments the text data into smaller units, typically words but sometimes sentences or phrases. This process allows NLP algorithms to analyze the individual building blocks of language and understand the relationships between them. Techniques like stemming and lemmatization come into play after tokenization (Tsarapatsanis et al., 2021).

Stemming is a crucial step that aims at reducing words to their base form, often referred to as the stem. While stemming can be effective in reducing data sparsity, it can sometimes lead to a loss of meaning. Consider the words "teach" and "taught." Stemming would reduce both to "teach," potentially confusing in the context of a sentence. Lemmatization takes a more nuanced approach in reducing the word to its lemma form. In this case, "running," "runs," and "ran" would all be reduced to the lemma "run." This approach preserves the core meaning of the word while ensuring consistency in the data. The choice between lemmatization and stemming can be made based on their need and the level of analysis required.

Normalization is another essential preprocessing technique. It encompasses a range of methods to ensure consistency in the text data. This can involve tasks like lowercase conversion – converting all characters to lowercase – or handling abbreviations and acronyms. Normalization reduces ambiguity and allows NLP algorithms to focus on the meaning of words rather than their capitalization or variations. (Lehr et al., 2017)

Text normalization can also involve correcting spelling errors. Typos and grammatical mistakes are common in user-generated text, and they can significantly impact the performance of NLP tasks. Techniques like spell-checking and dictionary-based corrections can be employed to address this issue. However, it's crucial to note that these techniques might not always be perfect, especially for domain-specific vocabulary or slang.

Beyond these core techniques, additional preprocessing steps might be necessary depending on the specific NLP task. For example, sentiment analysis might involve identifying and removing emoticons or sarcasm markers to focus on the core emotional tone of the text. Similarly, topic modeling, which aims to identify thematic clusters within a corpus of documents, might benefit from removing irrelevant information like headers or footers from the documents (Jia et al., 2014).

The benefits of effective preprocessing are numerous. Firstly, it improves the accuracy of NLP tasks. By removing noise and inconsistencies, clean and preprocessed data allows NLP algorithms to focus on the essential information within the text. This leads to more reliable and meaningful results, be it sentiment analysis, text summarization, or machine translation. Secondly, preprocessing enhances the efficiency of NLP algorithms. By structuring the data and reducing its complexity, preprocessing allows algorithms to process information faster and with fewer computational resources. Preprocessing can be particularly effective when dealing with a large, copious amount of data, such as a large legal database.

Finally, preprocessing promotes consistency in NLP models. By applying standardized preprocessing techniques across different datasets, NLP models become more robust and generalizable. This allows them to perform well on unseen data, ensuring their applicability to real-world scenarios. However, implementing effective preprocessing requires careful consideration of various factors. One key aspect is the specific NLP task at hand. The preprocessing steps employed might differ depending on whether the goal is sentiment analysis, machine translation, or topic modeling. Additionally, the chosen techniques need to be tailored to the nature of the text data. For instance, preprocessing techniques for social media text might differ from those used for legal documents due to differences in language style and vocabulary (Bach et al., 2003).

MACHINE LEARNING MODELS FOR LEGAL DOCUMENT CLASSIFICATION

Document classification automates the process of categorizing documents based on their content. Traditionally, this task relied on manual review by humans, a time-consuming and error-prone process. ML models, however, offer an automated and scalable solution. The training process allows the model to learn the patterns and features that distinguish documents belonging to different categories. Once trained, the model can then classify new, unseen documents with a high degree of accuracy (Von et al., 2008).

There are several key machine-learning approaches employed for document classification, each with its strengths and limitations. One popular approach is supervised learning, where the model is explicitly trained on labeled data.

Supervised Learning Techniques

- **Naive Bayes:** This probabilistic classifier analyzes the probability of words appearing in a document belonging to a specific category. It's a simple and efficient algorithm, particularly for well-defined categories with distinct vocabulary. However, Naive Bayes can struggle with complex classification tasks or highly imbalanced datasets.
- **Support Vector Machines (SVMs):** SVMs aim to create a hyperplane that separates documents belonging to different categories in a high-dimensional

feature space. They offer good performance for high-dimensional data that are less susceptible to overfitting compared to some other models. However, SVMs can be computationally expensive to train and require careful parameter tuning.

- **Decision Trees:** These algorithms create a tree-like structure where each node represents a decision based on a specific feature of the document. Documents traverse the tree based on their features, reaching a final leaf node that represents the predicted category. Decision trees are interpretable, allowing for an understanding of the classification logic. However, they can be susceptible to overfitting and might not generalize well to unseen data.

Unsupervised Learning Techniques

While supervised learning requires labeled data, unsupervised learning can be useful when such data is scarce.

- **K-Means Clustering:** This technique groups documents into a predefined number of clusters. While K-Means is efficient and easy to interpret, the number of clusters chosen needs to be determined beforehand and should be optimal. The clusters themselves might not always map neatly to desired categories.
- **Hierarchical Clustering:** Creates a hierarchical structure where data points are grouped in a bottom-up or top-down fashion based on their similarity.
- **Density-Based Spatial Clustering of Applications with Noise (DBSCAN):** Clusters are identified based on their corresponding densities.

The choice of the most suitable ML model for document classification depends on several factors. The nature of the text data, the availability of labeled data, the desired balance between accuracy and interpretability, and the computational resources available are all crucial considerations. Furthermore, feature engineering plays a vital role. This involves extracting relevant features from the text data that can be used by the ML model for classification. Common feature extraction techniques include bag-of-words, which represents documents as histograms of word frequencies, or TF-IDF, which assigns weightage to a word based on its frequency. Selecting the most appropriate features can significantly impact the model's performance.

Once the model is trained, it's crucial to evaluate its performance on a separate test dataset. This evaluation process ensures that the model responds well to general data rather than overfitting the trained data. Metrics like accuracy, precision, recall, and F1-score are used to assess the model's effectiveness in classifying documents correctly. The benefits of employing machine learning models for document classification are numerous. They offer significant efficiency gains compared to manual classification. It frees up valuable human resources for more strategic tasks. Machine Learning models can also handle vast amounts of data with ease, making them ideal for managing large document collections (Palacio-Niño et al., 2009).

Spectral Clustering for Legal Document Classification

Spectral Clustering leverages the power of linear algebra and graph theory to identify clusters within a dataset. It relies on the concept of constructing a similarity graph, where each data point is represented as nodes, and edges connect nodes that exhibit some degree of similarity. This intricate network captures the relationships between data points, allowing spectral Clustering to establish communities within the graph with the help of inherent connectivity.

The key to spectral clustering lies in its utilization of the graph's spectral properties, specifically its eigenvalues and eigenvectors. Imagine a matrix representation of the similarity graph, where each entry reflects the similarity between data points. By applying mathematical operations to this matrix, spectral Clustering extracts a set of eigenvalues and corresponding eigenvectors. These eigenvalues and eigenvectors hold valuable information about the underlying structure of the graph.

The power of spectral clustering lies in the observation that it encodes information about the community structure of the graph. Intuitively, these eigenvectors capture the notion of "smoothness" within the graph. Clusters often exhibit a high degree of internal similarity, meaning data points within a cluster are well connected to each other. This smoothness translates into lower eigenvalues associated with the corresponding eigenvectors.

Spectral clustering algorithms exploit this relationship by analyzing the Fiedler eigenvectors. One of the most common approaches involves the following steps:

1. Construct the Similarity Graph: This stage involves constructing a similarity matrix based on the chosen similarity measure (e.g., Euclidean distance for numerical data, cosine similarity for textual data). Edges are then added between data points based on their similarity values.
2. Compute the Laplacian Matrix: The Laplacian matrix is a representation of the graph structure that captures the connectivity between nodes. It essentially reflects the difference between the node's degree (number of connections) and the sum of similarities to its neighbors.
3. Eigenvalue and Eigenvector Decomposition: Perform eigenvalue decomposition on the Laplacian matrix. This process identifies eigenvalues and their corresponding eigenvectors.

4. Dimensionality Reduction: Project the data points onto a lower-dimensional space using the Fiedler eigenvectors (those corresponding to the smallest eigenvalues). This helps in ensuring that data complexity is reduced while maintaining or preserving the information.
5. Clustering in the Reduced Space: Implement a traditional clustering algorithm (e.g., k-means Clustering) on the projected data points in the reduced space to identify clusters.

This sequential process allows spectral Clustering to efficiently identify clusters within the data, leveraging the inherent structure of the underlying graph. Spectral Clustering takes into account its own set of considerations.

One key factor to consider is the selection of the appropriate similarity measure. As mentioned earlier, different metrics cater to different data types. Choosing the right metric ensures that the edges within the graph accurately represent the relationships between data points.

Furthermore, the choice of the number of clusters can impact the outcome. Spectral Clustering can be used to identify a predefined number of clusters, similar to k-means clustering. However, determining the optimal number of clusters can be a challenge, requiring careful analysis of the data and the eigenvalues.

Another consideration is the computational complexity of spectral Clustering. While it offers advantages over traditional clustering methods for certain types of data, the eigenvalue decomposition step can be computationally expensive for very large datasets. Despite these considerations, spectral Clustering offers several advantages that make it a valuable tool in the data scientist's arsenal. Firstly, it is well-suited for data with complex, non-linear relationships. Unlike k-means clustering, which assumes spherical clusters, spectral Clustering can handle more intricate cluster shapes.

Secondly, spectral Clustering provides a theoretical foundation for understanding the clustering process. By leveraging the properties of the Laplacian matrix and eigenvalues, spectral Clustering offers a more interpretable approach compared to some purely distance-based clustering methods. Finally, spectral Clustering offers a level of flexibility in its application. It can be integrated with other clustering techniques to create hybrid approaches, leveraging the strengths of different algorithms to achieve optimal results.

While the core algorithm involves eigenvalue decomposition and dimensionality reduction, there are different approaches for assigning data points to final clusters in the reduced space. This explanation focuses on two such methods: Discretization and Cluster_QR.

Discretization

This is a straightforward approach often used in conjunction with spectral Clustering. Here's how it works:

1. **Spectral Embedding:** A lower-dimensional space using the Fiedler eigenvectors is created to ensure the preservation of information. This creates a new representation of the data that captures the underlying cluster structure.
2. **Thresholding:** Each dimension in the reduced space is analyzed, and a threshold is chosen (often the median value). This threshold essentially divides the data points along each dimension into two groups: above or below the threshold.
3. **Cluster Assignment:** Based on the binary assignments (above or below threshold) in each dimension, a new binary code is generated for each data point. This code represents the data point's position in the reduced space relative to the chosen thresholds.
4. **Final Clustering:** Traditional clustering algorithms like k-means clustering can then be employed on these binary codes to group data points into final clusters. Discretization essentially transforms the continuous values in the reduced space into discrete categories, allowing k-means to partition the data effectively.

Advantages of Discretization

- **Simple and Efficient:** Discretization is a computationally inexpensive approach that leverages existing clustering algorithms like k-means.
- **Interpretability:** The binary codes generated during Discretization provide a basic level of interpretability regarding the data point's position in the reduced space relative to the chosen thresholds.

Disadvantages of Discretization

- **Information Loss:** Discretization inherently leads to information loss, as continuous values are converted into binary categories. This can potentially impact the accuracy of the final clustering results.
- **Sensitivity to Threshold Selection:** The choice of the threshold can significantly influence the resulting clusters. Finding an optimal threshold might require experimentation.

CLUSTER_QR

Introduced as a recent addition to spectral clustering libraries (like scikit-learn), Cluster_QR offers an alternative approach to assigning data points to clusters directly from the reduced space. Here's a simplified explanation:

1. **Spectral Embedding:** Similar to Discretization, Cluster_QR projects the data points onto a lower-dimensional space using the Fiedler eigenvectors.
2. **QR Decomposition:** This step applies QR decomposition on the projected data points in the reduced space. QR decomposition essentially factors a matrix into an orthogonal matrix (Q) and an upper triangular matrix (R).
3. **Cluster Assignment:** Cluster_QR utilizes the Q matrix directly to assign data points to clusters. The specific logic behind this assignment varies depending on the implementation, but it generally involves analyzing the signs and magnitudes of elements within the Q matrix.

Advantages of Cluster_QR

- **Efficiency:** Cluster_QR is often computationally faster compared to Discretization, as it bypasses the need for k-means clustering in the final step.
- **No Parameter Tuning:** Unlike Discretization, Cluster_QR requires no additional parameter tuning (like choosing thresholds). This simplifies the process and reduces the risk of user bias.
- **Potentially Higher Accuracy:** Studies suggest that Cluster_QR can achieve better clustering accuracy compared to Discretization, particularly for well-separated clusters.

Disadvantages of Cluster_QR

- **Limited Interpretability:** Cluster_QR offers less interpretability compared to Discretization, as it doesn't explicitly generate binary codes for data points.
- **Less Established Technique:** As a relatively new approach, Cluster_QR has less research and practical application than Discretization.

Choosing Between Discretization and Cluster_QR

The choice between these two methods depends on several factors:

- **Data Complexity:** For simpler data with well-separated clusters, Cluster_QR might offer better accuracy and efficiency.
- **Interpretability Needs:** If you require a basic level of interpretability regarding data point location in the reduced space, Discretization might be preferable.
- **Computational Resources:** If computational efficiency is a primary concern, Cluster_QR might be the better choice.

EVALUATION METRICS FOR LEGAL DOCUMENT CLUSTERING

Homogeneity

Homogeneity focuses on the purity of individual clusters. It measures how well each cluster contains data points that belong to the same class (in the case of supervised Clustering with labeled data) or exhibit high similarity based on the chosen features.

Calculating Homogeneity

There are several ways to measure homogeneity, but a common approach is using the Jaccard Similarity Coefficient. This metric helps us compare the number of data points belonging to the same class (or with high similarity) within a cluster to the total number of data points in that cluster.

Interpretation

- A high homogeneity score (close to 1) indicates that a cluster primarily contains data points belonging to the same class (supervised) or exhibiting high similarity (unsupervised).
- A low score (close to 0) suggests that a cluster is heterogeneous, containing data points from multiple classes or with significant dissimilarity.

Limitations

- Homogeneity can be misunderstood when we are dealing with a large heterogeneous dataset. In a dataset with a dominant class, clustering algorithms might create clusters that are mostly homogeneous but don't capture the minority classes effectively.

- Homogeneity focuses on individual clusters in isolation. It doesn't necessarily guarantee that all relevant data points from a class are grouped across all clusters.

Completeness

Completeness complements homogeneity by focusing on how well a clustering algorithm captures all data points belonging to a specific class (supervised) or exhibiting high similarity (unsupervised). It essentially measures the extent to which clusters recover the true class structure of the data.

Calculating Completeness

One way to calculate completeness is by utilizing the Adjusted Rand Index. This metric compares the compatibility between the true class labels (or inherent similarity) and the clustering assignments. It penalizes clusters that miss out on data points from a particular class or group similar data points in separate clusters.

Interpretation

- A high completeness score (close to 1) indicates that the clustering algorithm successfully captures most of the data points belonging to each class or similarity group.
- A low score (close to 0) suggests that the Clustering misses out on significant portions of some classes, leading to incomplete cluster recovery.

Limitations

- Similar to homogeneity, completeness can be affected by imbalanced datasets. In such cases, it might prioritize capturing the majority class while neglecting the minority classes.
- Completeness only considers the overall class structure, not the quality of individual clusters. It's possible to have high completeness, with some clusters still being impure.

V-Measure

The V-metric behaves similarly to the F1 score for Clustering, which combines the power of both homogeneity and completeness into a single metric. It essentially calculates the harmonic mean of the homogeneity and completeness scores, providing a balanced view of cluster quality.

Calculating V-Measure

The V-measure is equal to the harmonic mean of homogeneity (h) and completeness (c):

Interpretation

- A high V-measure score is a good indicator that the Clustering achieves a good balance between homogeneity and completeness, ensuring both the purity of individual clusters and the overall recovery of class structure.
- A low score (close to 0) suggests that the Clustering suffers from either low homogeneity (impure clusters) or low completeness (missing data points from some classes).

Advantages

- V-measure provides a more comprehensive evaluation of cluster quality compared to analyzing homogeneity and completeness separately.
- It avoids the pitfalls of considering only purity or only recovery, offering a balanced perspective.

Limitations

- V-measure inherits some limitations from homogeneity and completeness, such as being sensitive to imbalanced datasets.
- It might not be suitable for all clustering scenarios, particularly when focusing heavily on one aspect (purity or recovery) is necessary.

Rand Index

The Rand Index provides a different perspective on cluster evaluation, focusing on the overall agreement between the clustering assignments and the true class labels (supervised) or inherent similarity structure (unsupervised).

Calculating Rand Index

The Rand Index considers two types of agreements:

- True Positives (TP): Data points assigned to the same cluster in both the Clustering and the true class labels/similarity structure.
- True Negatives (TN): Data points assigned to different clusters in both the Clustering and the true class labels/similarity structure.

About the CUAD Dataset

Contract Understandings Dataset (CUAD) V1 is a collection of 13,000 labels from the commercial legal contract landscape, which are manually labeled by lawyers with a background in Contractual law. Forty-one types of legal clauses concerning corporate transactions were identified, including mergers and acquisitions of various entities.

Figure 1 describes the distribution of documents taken from the CUAD dataset for Clustering. The documents are unevenly distributed across the various types of contracts. This can be particularly difficult given the limited number of documents in the dataset. A highly heterogeneous dataset with 11 different classes makes Clustering particularly difficult.

Figure 1. Various Types of Documents are in the CUAD Dataset

Types of Documents	Total
Collaboration	38
Marketing	30
License	26
Supply	24
Service	20
Development	17
Sponsorship	15
Agency	12
Intellectual Property	11
Manufacturing	10
Maintenance	10
Total	213

METHODOLOGY

Term Frequency-Inverse Document Frequency

TF-IDF (Term Frequency-Inverse Document Frequency) emerges as a powerful tool for this task. This essay delves into the core concepts of TF-IDF, exploring its calculation methods, practical applications, and considerations for effective use.

At its core, TF-IDF assigns a weight to each word within a document, reflecting its significance relative to the entire collection of documents (corpus). This weighting scheme balances two key factors:

1. Term Frequency (TF): This measures how often a particular word appears within a specific document. Intuitively, words that appear frequently within a document are likely to be more relevant to its content.
2. Inverse Document Frequency (IDF): This captures how rare or common a word is across the entire corpus. Words that appear very frequently in many documents become less informative for distinguishing the specific content of a single document.

By combining these two factors, TF-IDF provides a more nuanced weighting scheme compared to simply counting word occurrences. Here's how the calculation works:

TF-IDF (t, d) = TF(t, d) * IDF(t)

- TF(t, d): This represents the frequency of the word 't' within document 'd'. It's often calculated as the number of times 't' appears in 'd' divided by the total number of words in 'd'.
- IDF(t): This represents the inverse document frequency of the word 't' across the corpus.

Latent Semantic Analysis (LSA)

emerges as a powerful lighthouse, illuminating hidden relationships and semantic connections within documents. This essay delves into the core concepts of LSA, exploring its working principles, benefits, and limitations, equipping you with insights into this valuable technique for natural language processing (NLP).

At its heart, LSA bridges the gap between traditional keyword-based approaches and the inherent meaning conveyed by text. It operates on the fundamental principle that words with similar semantic meaning tend to appear in similar contexts. By analyzing the co-occurrence patterns of words across a large corpus of documents, LSA uncovers the latent semantic structure – the hidden relationships that define the true meaning behind the words.

Imagine a vast library – the corpus of documents. LSA meticulously analyzes this library, tracking how often words appear together. It observes that "king" and "queen" often co-occur in documents about royalty, while "car" and "engine" frequently appear together in automotive discussions. Through this analysis, LSA builds a hidden layer of understanding, recognizing the semantic connection between these terms without explicitly being programmed with their definitions.

The technical core of LSA revolves around a mathematical technique called Singular Value Decomposition (SVD). SVD decomposes a document-term matrix, where rows represent documents and columns represent terms, into three key components:

- U matrix: Represents the documents in a reduced-dimensional space, capturing the essential semantic themes present across the corpus.
- Σ matrix: Represents the singular values, which indicate the importance of each dimension (latent semantic concept) identified by SVD.
- V matrix:** Represents the terms in the reduced-dimensionality space, revealing how each term contributes to the identified latent concepts.

RESULTS AND DISCUSSION

The Clustering of Legal Documents involved the process of TF-IDF Vectorization of the documents followed by LSA, which further reduced the dimensionality. We can observe the result in the following table, which details the results obtained.

Table 1. Observation of the Efficiency of the Various Clustering Algorithms

Algorithms	Homogeneity	Completeness	V-Measure	Rand-Index	Silhouette score
Spectral Clustering	0.0188	0.649	0.035	0.004	0.353
Spectral Clustering with LSA	0.316	0.366	0.323	0.084	0.084
Spectral Clustering (LSA+Discretize)	0.396	0.405	0.400	0.171	0.069
Spectral Clustering (LSA+QR Cluster)	0.347	0.350	0.349	0.121	0.069

From Table 1, we notice a considerable change in the results based on the vectorization, dimensionality reduction method applied, and type of Spectral Clustering technique. We observe inconsistent results when Spectral Clustering is applied to the dataset without any form of dimensionality reduction techniques applied. However, we observed a very consistent result when both Discretization and LSA were applied.

REFERENCES

Alarie, B., Niblett, A., & Yoon, A. H. (2018). How artificial intelligence will affect the practice of law. *University of Toronto Law Journal*, 68(supplement 1), 106–124. DOI:10.3138/utlj.2017-0052

Bach, F., & Jordan, M. (2003). Learning spectral Clustering. *Advances in Neural Information Processing Systems*, 16.

Bommarito, M. J.II, Katz, D. M., & Detterman, E. M. (2021). LexNLP: Natural language processing and information extraction for legal and regulatory texts. In *Research handbook on big data law* (pp. 216–227). Edward Elgar Publishing. DOI:10.4337/9781788972826.00017

Callister, P. D. (2020). Law, artificial intelligence, and natural language processing: A funny thing happened on the way to my search results. *Law Library Journal*, 112, 161. DOI:10.31228/osf.io/dw29y

Jia, H., Ding, S., Xu, X., & Nie, R. (2014). The latest research progress on spectral Clustering. *Neural Computing & Applications*, 24(7-8), 1477–1486. DOI:10.1007/s00521-013-1439-2

Lehr, D., & Ohm, P. (2017). Playing with the data: What legal scholars should learn about machine learning. *UCD Law Review*, 51, 653.

McCarty, L. T. (2018). Finding the right balance in artificial intelligence and law. In *Research Handbook on the Law of Artificial Intelligence* (pp. 55–87). Edward Elgar Publishing. DOI:10.4337/9781786439055.00013

Palacio-Niño, J. O., & Berzal, F. (2019). Evaluation metrics for unsupervised learning algorithms. *arXiv preprint arXiv:1905.05667.*

Robaldo, L., Villata, S., Wyner, A., & Grabmair, M. (2019). Introduction for artificial intelligence and law: Special issue "natural language processing for legal texts". *Artificial Intelligence and Law*, 27(2), 113–115. DOI:10.1007/s10506-019-09251-2

Surden, H. (2019). Artificial intelligence and law: An overview. *Georgia State University Law Review*, 35, 19–22.

Surden, H. (2021). Machine learning and law: An overview. *Research Handbook on Big Data Law*, 171-184.

Tsarapatsanis, D., & Aletras, N. (2021). On the ethical limits of natural language processing on legal text. *arXiv preprint arXiv:2105.02751*. DOI:10.18653/v1/2021.findings-acl.314

Von Luxburg, U., Belkin, M., & Bousquet, O. (2008). Consistency of spectral Clustering. *Annals of Statistics*, 555–586.

Walters, R., & Novak, M. (2021). Artificial Intelligence and Law. In *Cyber Security, Artificial Intelligence, Data Protection & the Law* (pp. 39–69). Springer Singapore. DOI:10.1007/978-981-16-1665-5_3

Compilation of References

A future that works: Automation, employment, and productivity. (2017, January 12). McKinsey & Company. https://www.mckinsey.com/featured-insights/digital-disruption/harnessing-automation-for-a-future-that-works/de-DE

A survey of Generative AI Applications. (n.d.). Ar5iv. https://ar5iv.labs.arxiv.org/html/2306.02781

Abbas, A. (2021). Social network analysis using deep learning: Applications and schemes. *Social Network Analysis and Mining*, 11(1), 106. Advance online publication. DOI: 10.1007/s13278-021-00799-z

Acquisti, A., & Gross, R. (2006, June). Imagined communities: Awareness, information sharing, and privacy on the Facebook. In *International workshop on privacy enhancing technologies* (pp. 36-58). Springer Berlin Heidelberg. DOI: 10.1007/11957454_3

Adams, P.D., & Darnton, RC (2017). The true history of fake news.

Afful-Dadzie, E., Nabareseh, S., Oplatková, Z. K., & Klímek, P. (2014, August). Enterprise competitive analysis and consumer sentiments on social media-insights from telecommunication companies. In *International Conference on Data Management Technologies and Applications* (Vol. 2, pp. 22-32). SciTePress. DOI: 10.5220/0004991300220032

Agarwal, S., & Mehta, S. (2022). GNPA: A hybrid model for social influence maximization in dynamic networks. *Multimedia Tools and Applications*, 83(1), 1–28. DOI: 10.1007/s11042-021-11606-4

Aghaee, Z., Ghasemi, M., Ahmadi Beni, H., Bouyer, A., & Fatemi, A. (2021). A survey on meta-heuristic algorithms for the influence maximization problem in social networks. *Computing*, 103(11), 2437–2477. Advance online publication. DOI: 10.1007/s00607-021-00945-7

Ahmad, M. (2023). Connecting People and Places: How Citizen Diplomacy and VGI Are Strengthening Disaster Response and Community Development. In Kavoğlu, S., & Köksoy, E. (Eds.), *Global Perspectives on the Emerging Trends in Public Diplomacy* (pp. 195–226). IGI Global. DOI: 10.4018/978-1-6684-9161-4.ch009

Ahmad, M., & Ali, A. (2023). Mapping the future of sustainable development through cloud-based solutions: A case study of openstreetmap. In *Promoting Sustainable Management Through Technological Innovation* (pp. 153–176). IGI Global. DOI: 10.4018/978-1-6684-9979-5.ch011

Ahuja, V. (2024, June 3). Ethical Considerations for Generative AI. Forbes. https://www.forbes.com/sites/forbestechcouncil/2024/02/14/ethical-considerations-for-generative-ai/

Ajiga, D. I., Ndubuisi, N. L., Asuzu, O. F., Owolabi, O. R., Tubokirifuruar, T. S., & Adeleye, R. A.. (2024). AI-driven predictive analytics in retail: A review of emerging trends and customer engagement strategies. *International Journal of Management & Entrepreneurship Research*, 6(2), 307–321. DOI: 10.51594/ijmer.v6i2.772

Alaka, H., Oyedele, L., Toriola-Coker, O., Owolabi, H., Akinade, O., & Bilal, M. (2015) Methodological approach of construction businesses failure prediction studies: a review. In: Raiden, AB, & Aboagye-Nimo, E (eds) *Proceedings 31st annual ARCOM conference*, 7–9 September 2015. Association of Researchers in Construction Management, Lincoln UK, pp 1291–1300

Alalawi, S., Baalfaqih, S., Almeqbaali, M., & Masud, M. M. (2023). Social Media Misinformation Propagation and Detection. *2023 15th International Conference on Innovations in Information Technology (IIT)*, 240-245.

AlAli, M. S. (2019). Examining the Effect of Altman's Zeta Model Score on the Share Price of Healthcare Companies Listed at Kuwait Stock Exchange. In International Journal of Economics and Finance (Vol. 11, Issue 4, p. 25). DOI: 10.5539/ijef.v11n4p25

Alarie, B., Niblett, A., & Yoon, A. H. (2018). How artificial intelligence will affect the practice of law. *University of Toronto Law Journal*, 68(supplement 1), 106–124. DOI: 10.3138/utlj.2017-0052

Alhindi, T., Petridis, S., & Muresan, S. (2018). Where is Your Evidence: Improving Fact-checking by Justification Modeling.

Alifiah, M. N., & Tahir, M. S. (2019). Macroeconomic Variables and the Prediction of Financial Distress Companies in the Manufacturing Sector in Malaysia. *International Journal of Recent Technology and Engineering*, 8(1S).

Ali, M., & Amir, H. (2024). Understanding consumer herding behavior in online purchases and its implications for online retailers and marketers. *Electronic Commerce Research and Applications*, 64, 101356. DOI: 10.1016/j.elerap.2024.101356

Alonso, C., Kothari, S., & Rehman, S. (2020, December 2). *How artificial intelligence could widen the gap between rich and poor nations*. IMF. https://www.imf.org/en/Blogs/Articles/2020/12/02/blog-how-artificial-intelligence-could-widen-the-gap-between-rich-and-poor-nations

Al-shami, T. M. (2023). (2, 1)-Fuzzy sets: Properties, weighted aggregated operators and their applications to multi-criteria decision-making methods. *Complex & Intelligent Systems*, 9(2), 1687–1705. DOI: 10.1007/s40747-022-00878-4

Al-shami, T. M., & Mhemdi, A. (2023). Generalized frame for orthopair fuzzy sets:(m, n)-fuzzy sets and their applications to multi-criteria decision-making methods. *Information (Basel)*, 14(1), 56. DOI: 10.3390/info14010056

Al-Shayea, Q. K., & El-Refae, G. A. (2012). Evaluation of Banks Insolvency Using Artificial Neural Networks. *In Proceedings of the 11th WSEAS international conference on Artificial Intelligence, Knowledge Engineering and Data Bases (AIKED'12)*, (pp. 22-24). Cambridge.

Altman, E. I. (1968). Financial Ratios, Discriminant Analysis and the Prediction of Corporate Bankruptcy. *The Journal of Finance*, 23(4), 589–609. DOI: 10.1111/j.1540-6261.1968.tb00843.x

Alzubaidi, L., Zhang, J., Humaidi, A. J., Al-Dujaili, A., Duan, Y., Al-Shamma, O., Santamaría, J., Fadhel, M. A., Al-Amidie, M., & Farhan, L. (2021). Review of deep learning: Concepts, CNN architectures, challenges, applications, future directions. *Journal of Big Data*, 8(1), 53. DOI: 10.1186/s40537-021-00444-8 PMID: 33816053

Amazon Titan Text Premier Demo (1:30). (n.d.). [Video]. Amazon Web Services, Inc. https://aws.amazon.com/bedrock/titan/

Amodei, D., & Hernandez, D. (2018, May 16). *AI and compute*. IBM. https://openai.com/index/ai-and-compute/

Anderson-Tarver, C. (2015). Crisis Mapping the 2010 Earthquake in OpenStreetMap Haiti. In *ProQuest Dissertations and Theses*. jn8sf5hk5v.search.serialssolutions.com/?ctx_ver=Z39.88-2004&ctx_enc=info:ofi/enc:UTF-8&rfr_id=info:sid/ProQuest+Dissertations+%26+Theses+A%26I&rft_val_fmt=info:ofi/fmt:kev:mtx:dissertati

Antoniou, V. (2017). Sources of VGI for Mapping. In *Mapping and the Citizen Sensor*. DOI: 10.5334/bbf.b

Antoniou, V. (2023). On volunteered geographic information quality: a framework for sharing data quality information. In *Geoinformatics for Geosciences: Advanced Geospatial Analysis using RS*. GIS and Soft Computing. DOI: 10.1016/B978-0-323-98983-1.00009-0

Arksey, H., & O'Malley, L. (2005). Scoping studies: towards a methodological framework. In International Journal of Social Research Methodology (Vol. 8, Issue 1, pp. 19–32). DOI: 10.1080/1364557032000119616

Asudani, D. S., Nagwani, N. K., & Singh, P. (2023). Impact of word embedding models on text analytics in deep learning environment: A review. *Artificial Intelligence Review*, 56(9), 10345–10425. DOI: 10.1007/s10462-023-10419-1 PMID: 36844886

Attri, R., & Pahwa, M. S. (2012). Learning for public sector oil marketing companies in India to manage business in turbulent times. *Prabandhan & Taqniki. Management Research Journal*, 6, 29–36.

Ausat, A. M. A., Permana, R. M., Angellia, F., Subagja, A. D., & Astutik, W. S. (2023). Utilisation of Social Media in Market Research and Business Decision Analysis. *Jurnal Minfo Polgan*, 12(1), 652–661. DOI: 10.33395/jmp.v12i1.12485

Auti, A., Patil, D., Zagade, O., Bhosale, P., & Ahire, P. (2022). Bitcoin Price Prediction Using Svm. *Int. J. Eng. Appl. Sci. Technol*, 6(11), 226–229. https://www.ijeast.com/papers/226-229,%20Tesma611,IJEAST.pdf

Auto-Regressive models in generative AI | Saturn Cloud. (2023, July 14). https://saturncloud.io/glossary/autoregressive-models-in-generative-ai/

Azeem, M. (2024). Internet of Things (IoT) impact on inventory management. IGI Global. https://www.igi-global.com/chapter/internet-of-things-iot-impact-on-inventory-management/339252

Babatunde, S. O., Odejide, O. A., Edunjobi, T. E., & Ogundipe, D. O. (2024). The role of AI in marketing personalization: A theoretical exploration of consumer engagement strategies. *International Journal of Management & Entrepreneurship Research*, 6(3), 936–949. DOI: 10.51594/ijmer.v6i3.964

Bach, F., & Jordan, M. (2003). Learning spectral Clustering. *Advances in Neural Information Processing Systems*, 16.

Back, B., Laitinen, T., & Sere, K. (1996). Neural Networks and Genetic Algorithms for Bankruptcy Predictions. *Expert Systems with Applications*, 11(4), 407–413. DOI: 10.1016/S0957-4174(96)00055-3

Bai, A., Satarpour, M., Mohebbi, F., & Forati, A. M. (2024). Digital Crowdsourcing and VGI: impact on information quality and business intelligence. *Spatial Information Research*, 1–9.

Baimwera, B., & Muriuki, A. M. (2014). Analysis of corporate financial distress determinants: A survey of non-financial firms listed in the NSE. *International Journal of Current Business and Social Sciences*, 1(2), 58–80.

Bandi, A., Adapa, P. V. S. R., & Kuchi, Y. E. V. P. K. (2023). The power of Generative AI: A review of requirements, models, Input–Output formats, evaluation metrics, and challenges. *Future Internet*, 15(8), 260. DOI: 10.3390/fi15080260

Banerjee, S., Jenamani, M., & Pratihar, D. (2020). A survey on influence maximization in a social network. *Knowledge and Information Systems*, 62(9), 3417–3455. Advance online publication. DOI: 10.1007/s10115-020-01461-4

Bartzokas-Tsiompras, A. (2022). Utilizing OpenStreetMap data to measure and compare pedestrian street lengths in 992 cities around the world. *European Journal of Geography*, 13(2), 127–141. Advance online publication. DOI: 10.48088/ejg.a.bar.13.2.127.138

Basak, S., Kar, S., Saha, S., & Khaidema, L. (2019). Predicting the direction of stock market prices using tree-based classifiers. *The North American Journal of Economics and Finance*, 47, 552–567. DOI: 10.1016/j.najef.2018.06.013

Basiouka, S., Potsiou, C., & Bakogiannis, E. (2015). OpenStreetMap for cadastral purposes: An application using VGI for official processes in urban areas. *Survey Review*, 47(344), 333–341. Advance online publication. DOI: 10.1179/1752270615Y.0000000011

Bathia, U. (2021, September 23). *Conversational AI Examples: How Siri, Alexa & Google Assistant have Human-like Conversations*. CFTE. https://blog.cfte.education/conversational-ai-examples-how-siri-alexa-google-assistant-have-human-like-conversations/

Bauweraerts, J. (2016). Predicting bankruptcy in private frms: Towards a stepwise regression procedure. *Int J Financ Res*, 7(2), 147–153. DOI: 10.5430/ijfr.v7n2p147

Beaver, W. H. (1966). Financial Ratios As Predictors of Failure. *Journal of Accounting Research*, 4, 71–111. DOI: 10.2307/2490171

Benedikt Frey, C., & Osborne, M. (2023). *Generative AI and the Future of Work: a reappraisal – The Brown Journal of World Affairs*. The Brown Journal of World Affairs. https://bjwa.brown.edu/30-1/generative-ai-and-the-future-of-work-a-reappraisal/

Bergeron, F., Giroux, S., Lachiver, G., & Bouchard, B. (2018). Tracking objects within a smart home. ResearchGate. https://www.researchgate.net/publication/326162154_Tracking_objects_within_a_smart_home

Bharath, S. T., & Shumway, T. (2008). Forecasting Default with the Merton Distance to Default Model. *Review of Financial Studies*, 21(3), 1339–1369. DOI: 10.1093/rfs/hhn044

Bhattacharya, R., Nagwani, N. K., & Tripathi, S. (2023). Detecting influential nodes with topological structure via Graph Neural Network approach in social networks. *International Journal of Information Technology : an Official Journal of Bharati Vidyapeeth's Institute of Computer Applications and Management*, 15(4), 2233–2246. DOI: 10.1007/s41870-023-01271-1 PMID: 37256031

Black, F., & Scholes, M. (1973). The Pricing of Options and Corporate Liabilities. *Journal of Political Economy*, 81(3), 637–654. DOI: 10.1086/260062

Blömker, J., & Albrecht, C. M. (2024). Psychographic Segmentation of multichannel customers: Investigating the influence of individual differences on channel choice and switching behavior. *Journal of Retailing and Consumer Services*, 79, 103806. DOI: 10.1016/j.jretconser.2024.103806

Bockting, C. L., Van Dis, E. M., Van Rooij, R., Zuidema, W., & Bollen, J. (2023). Living guidelines for generative AI — Why scientists must oversee its use. *Nature*, 622(7984), 693–696. DOI: 10.1038/d41586-023-03266-1 PMID: 37857895

Bommarito, M. J.II, Katz, D. M., & Detterman, E. M. (2021). LexNLP: Natural language processing and information extraction for legal and regulatory texts. In *Research handbook on big data law* (pp. 216–227). Edward Elgar Publishing. DOI: 10.4337/9781788972826.00017

Borgs, C., Brautbar, M., Chayes, J., Lucier, B., & Sridhar, V. (2014). Maximizing social influence in nearly optimal time. In *Proceedings of the twenty-fifth annual ACM-SIAM symposium on Discrete algorithms* (pp. 966-977). Society for Industrial and Applied Mathematics. https://doi.org/DOI: 10.1137/1.9781611973402.70

Botha, J.G., & Pieterse, H. (2020). Fake news and deepfakes: A dangerous threat for 21st century information security.

Boyd, D. (2014). *It's complicated: The social lives of networked teens*. Yale University Press.

Brady, D. (2024, February 8). *What developers need to know about generative AI - The GitHub Blog*. The GitHub Blog. https://github.blog/2023-04-07-what-developers-need-to-know-about-generative-ai

Brown, P. F., Cocke, J., Della Pietra, S. A., Della Pietra, V. J., Jelinek, F., Lafferty, J., Mercer, R. L., & Roossin, P. S. (1990). *A statistical approach to machine translation*. ACL Anthology. https://aclanthology.org/J90-2002

Brown, T. B., Mann, B., Ryder, N., Subbiah, M., Kaplan, J., Dhariwal, P., Neelakantan, A., Shyam, P., Sastry, G., Askell, A., Agarwal, S., Herbert-Voss, A., Krueger, G., Henighan, T., Child, R., Ramesh, A., Ziegler, D. M., Wu, J., Winter, C., . . . Amodei, D. (2020). Language Models are Few-Shot Learners. arXiv (Cornell University). https://arxiv.org/pdf/2005.14165.pdf

Brown, B. (2022). *The gifts of imperfection: Let go of who you think you're supposed to be and embrace who you are*. Simon and Schuster.

Caballero, B. (2023, August 17). 8 best AI Chatbots for Customer Service. *Fit Small Business*. https://fitsmallbusiness.com/best-ai-chatbots-for-customer-service/

Callister, P. D. (2020). Law, artificial intelligence, and natural language processing: A funny thing happened on the way to my search results. *Law Library Journal*, 112, 161. DOI: 10.31228/osf.io/dw29y

Camacho, D., Luzon, M., & Cambria, E. (2020). New research methods & algorithms in social network analysis. *Future Generation Computer Systems*, 114, 290–293. Advance online publication. DOI: 10.1016/j.future.2020.08.006

Cambria, E., & White, B. (2014). Jumping NLP curves: A review of natural language processing research. *IEEE Computational Intelligence Magazine*, 9(2), 48–57. DOI: 10.1109/MCI.2014.2307227

Candelon, F., Gupta, A., Krayer, L., & Zhukov, L. (2023, May 25). *The CEO's guide to the Generative AI Revolution*. BCG Global. https://www.bcg.com/publications/2023/ceo-guide-to-ai-revolution

Cao, Y., Li, S., Liu, Y., Yan, Z., Dai, Y., Yu, P. S., & Sun, L. (2023). A Comprehensive Survey of AI-Generated Content (AIGC): A History of Generative AI from GAN to ChatGPT. *arXiv (Cornell University)*. https://doi.org//arxiv.2303.04226DOI: 10.48550

Capacity, T. (2023, April 11). *Everything you need to know about Helpdesk Chatbots (2023) - capacity*. Capacity. https://capacity.com/learn/helpdesk-chatbots/

Castelnovo, W. (2016). *Co-production makes cities smarter: Citizens' participation in smart city initiatives*. SpringerBriefs in Applied Sciences and Technology. DOI: 10.1007/978-3-319-30558-5_7

Cero, I., Luo, J., & Falligant, J. M. (2024). Lexicon-Based Sentiment Analysis in Behavioral Research. *Perspectives on Behavior Science*, 1–28. PMID: 38660506

Chamlertwat, W., Bhattarakosol, P., Rungkasiri, T., & Haruechaiyasak, C. (2012). Discovering consumer insight from Twitter via sentiment analysis. *Journal of Universal Computer Science*, 18(8), 973–992.

Charalambakis, E. C., & Garrett, I. (2019). On corporate fnancial distress prediction: What can we learn from private frms in a developing economy? Evidence from Greece. *Review of Quantitative Finance and Accounting*, 52(2), 467–497. DOI: 10.1007/s11156-018-0716-7

Charitou, A., Dionysiou, D., Lambertides, N., & Trigeorgis, L. (2013). Alternative bankruptcy prediction models using option-pricing theory. *Journal of Banking & Finance*, 37(7), 2329–2341. DOI: 10.1016/j.jbankfin.2013.01.020

Chen, W.-S., & Du, Y.-K. (2009). Using neural networks and data mining techniques for the financial distress prediction model. In *Expert Systems with Applications* (Vol. 36, Issue 2, pp. 4075–4086). DOI: 10.1016/j.eswa.2008.03.020

Chen, D., Xu, G., Yang, P., & Meng, L. (2023). TSIFIM: A three-stage iterative framework for influence maximization in complex networks. *Expert Systems with Applications*, 212, 118702. DOI: 10.1016/j.eswa.2022.118702

Cheng, C.-B., Chen, C.-L., & Fu, C.-J. (2006, February). Financial Distress Prediction by a Radial Basis Function Network with Logit Analysis Learning. *Computers & Mathematics with Applications (Oxford, England)*, 51(3-4), 579–588. DOI: 10.1016/j.camwa.2005.07.016

Chen, J. (2023). Analysis of Bitcoin price prediction using machine learning. *Journal of Risk and Financial Management*, 16(1), 51. DOI: 10.3390/jrfm16010051

Chen, T., Yan, S., Guo, J., & Wu, W. (2023). ToupleGDD: A fine-designed solution of influence maximization by deep reinforcement learning. *IEEE Transactions on Computational Social Systems*.

Chen, W., Lu, W., & Zhang, N. (2021). Time-Critical Influence Maximization in Social Networks with Time-Delayed Diffusion Process. In *Proceedings of the AAAI Conference on Artificial Intelligence* (Vol. 26, No. 1, pp. 591-598). https://doi.org/ DOI: 10.1609/aaai.v26i1.8204

Chen, Y. (2024). Urban Green Space Maps Based on GeoAI. In *Handbook of Geospatial Approaches to Sustainable Cities* (pp. 75–93). CRC Press. DOI: 10.1201/9781003244561-7

Chin, C., Ramiah, S., & Razali, N. F. (2023). Inventory Management Systems (IMS). 2600-7304. https://www.researchgate.net/publication/372653862_Inventory_Management_Systems_IMS

Chiu, M. C., Tsai, C. Z., & Huang, Y. C. (2024). Integrating object detection and natural language processing models to build a personalized attraction recommendation agent in a smart product service system. *Advanced Engineering Informatics*, 61, 102484. DOI: 10.1016/j.aei.2024.102484

Christison, C. (2024, April 8). *The 15 best AI tools for Social Media in 2024*. Social Media Marketing & Management Dashboard. https://blog.hootsuite.com/best-ai-tools-for-social-media/

Chui, M., Roberts, R., & Yee, L. (2022). *Generative AI is here: How tools like ChatGPT could change your business*. https://www.mckinsey.com/~/media/mckinsey/business%20functions/quantumblack/our%20insights/generative%20ai%20is%20here%20how%20tools%20like%20chatgpt%20could%20change%20your%20business/generative-ai-is-here-how-tools-like-chatgpt-could-change-your-business.pdf

Chu-Ke, C., & Dong, Y. (2024). *Misinformation and Literacies in the Era of Generative Artificial Intelligence: A Brief Overview and a Call for Future Research*. Emerging Media. DOI: 10.1177/27523543241240285

Cielen, A., Peeters, L., & Vanhoof, K. (2004). Bankruptcy prediction using a data envelopment analysis. *European Journal of Operational Research*, 154(2), 526–532. DOI: 10.1016/S0377-2217(03)00186-3

Clifford, C. (2018, February 1). *Google CEO: A.I. is more important than fire or electricity*. CNBC. https://www.cnbc.com/2018/02/01/google-ceo-sundar-pichai-ai-is-more-important-than-fire-electricity.html#:~:text=%E2%80%9CAI%20is%20one%20of%20the,the%20potential%20of%20artificial%20intelligence

Colicev, A., Kübler, R., & Pauwels, K. (2017). Social Media's Impact on Consumer Mindset: When to Use Which Sentiment Extraction Tool? *MSI Working Paper Series*.

Collier, J. R., & Van Duyn, E. (2022). Fake news by any other name: Phrases for false content and effects on public perceptions of US news media. *Journal of Applied Communication Research*, 51(4), 424–443. DOI: 10.1080/00909882.2022.2148487

Communications, H. (2023, September 6). HubSpot unveils HubSpot AI and new sales hub at INBOUND 2023. *HubSpot*. https://www.hubspot.com/company-news/inbound-2023-hubspot-ai-sales-hub

Critien, J. V., Gatt, A., & Ellul, J. (2022). Bitcoin price change and trend prediction through Twitter sentiment and data volume. *Financial Innovation*, 8(1), 1–20. DOI: 10.1186/s40854-022-00352-7

Cút, S. (2014). Prediction of Company Financial Distress Using Neural Network Based on the Radial Basis Function. *Lecture Notes in Management Science - 2014 2nd International Conference in Humanities, Social Sciences and Global Business Management (ISSGBM 2014).*

Cybenko, A. K., & Cybenko, G. V. (2018). AI and Fake News. *IEEE Intelligent Systems*, 33(5), 1–5. DOI: 10.1109/MIS.2018.2877280

D'Angelo, G., Poddar, D., & Vinci, C. (2023). Better bounds on the adaptivity gap of influence maximization under full-adoption feedback. *. *Artificial Intelligence*, 318, 103895. DOI: 10.1016/j.artint.2023.103895

Dang, N. C., Moreno-García, M. N., & De la Prieta, F. (2020). Sentiment analysis based on deep learning: A comparative study. *Electronics (Basel)*, 9(3), 483. DOI: 10.3390/electronics9030483

Daudt, H. M., van Mossel, C., & Scott, S. J. (2013). Enhancing the scoping study methodology: A large, interprofessional team's experience with Arksey and O'Malley's Framework. *BMC Medical Research Methodology*, 13(1), 48. DOI: 10.1186/1471-2288-13-48 PMID: 23522333

Davenport, T. H. (2023, August 15). *How generative AI is changing creative work.* Harvard Business Review. https://hbr.org/2022/11/how-generative-ai-is-changing-creative-work?utm_medium=paidsearch&utm_source=google&utm_campaign=intlcontent_bussoc&utm_term=Non-Brand&tpcc=intlcontent_bussoc&gad_source=1&gclid=Cj0KCQjw6auyBhDzARIsALIo6v80evxQNjT6M-JEvCAm_-InTpL3Y-1i8TE5yXus9ri1dFgphEmXHQAaAiuREALw_wcB

Davis, K., Drey, N., & Gould, D. (2009). What are scoping studies? A review of the nursing literature. *International Journal of Nursing Studies*, 46(10), 1386–1400. DOI: 10.1016/j.ijnurstu.2009.02.010 PMID: 19328488

Deakin, E. B. (1972). A Discriminant Analysis of Predictors of Business Failure. *Journal of Accounting Research*, 10(1), 167–179. DOI: 10.2307/2490225

Deep, A. I. (2020, June 25). *Autoregessive model.* DeepAI. https://deepai.org/machine-learning-glossary-and-terms/autoregressive-model

Devi, S. S., & Radhika, Y. (2018). A survey on machine learning and statistical techniques in bankruptcy prediction. *International Journal of Machine Learning and Computing*, 8(2), 133–139. DOI: 10.18178/ijmlc.2018.8.2.676

Dey, L., Chakraborty, S., Biswas, A., Bose, B., & Tiwari, S. (2016). Sentiment analysis of review datasets using naive bayes and k-nn classifier. *arXiv preprint arXiv:1610.09982*.

Dey, S. (2024, February 20). Which ethical implications of generative AI should companies focus on? *Forbes*. https://www.forbes.com/sites/forbestechcouncil/2023/10/17/which-ethical-implications-of-generative-ai-should-companies-focus-on/

Din, D. (2010). Sensor technology for smart homes. ResearchGate. https://www.researchgate.net/publication/51087784_Sensor_technology_for_smart_homes

Ding, Y., Song, X., & Zen, Y. (2008). Forecasting financial condition of Chinese listed companies based on support vector machine. *Expert Systems with Applications*, 34(4), 3081–3089. DOI: 10.1016/j.eswa.2007.06.037

Dizaji, S., Patil, K., & Avrachenkov, K. (2024). Influence maximization in dynamic networks using reinforcement learning. *SN Computer Science*, 5(1), 169. Advance online publication. DOI: 10.1007/s42979-023-02453-1

Doumpos, M., & Zopounidis, C. (1999). A Multicriteria Discrimination Method for the Prediction of Financial Distress: The Case of Greece. *Multinational Finance Journal*, 3(2), 71–101. DOI: 10.17578/3-2-1

Drus, Z., & Khalid, H. (2019). Sentiment analysis in social media and its application: Systematic literature review. *Procedia Computer Science*, 161, 707–714. DOI: 10.1016/j.procs.2019.11.174

Ejegwa, P. A., Onyeke, I. C., Kausar, N., & Kattel, P. (2023). A new partial correlation coefficient technique based on intuitionistic fuzzy information and its pattern recognition application. *International Journal of Intelligent Systems*, 2023, 1–14. Advance online publication. DOI: 10.1155/2023/5540085

Ejegwa, P. A., Onyeke, I. C., Terhemen, B. T., Onoja, M. P., Ogiji, A., & Opeh, C. U. (2022). Modified Szmidt and Kacprzyk's intuitionistic fuzzy distances and their applications in decision-making. *J. Niger. Soc. Phys. Sci.*, 4, 175–182. DOI: 10.46481/jnsps.2022.530

Ekechi, C. C., Chukwurah, E. G., Oyeniyi, L. D., & Okeke, C. D. (2024). AI-infused chatbots for customer support: A cross-country evaluation of user satisfaction in the USA and the UK. *International Journal of Management & Entrepreneurship Research*, 6(4), 1259–1272. DOI: 10.51594/ijmer.v6i4.1057

Ellingrud, K., Sanghvi, S., Dandona, G. S., Madgavkar, A., Chui, M., White, O., & Hasebe, P. (2023). Generative AI and the future of work in America. In *McKinsey & Company*. https://www.mckinsey.com/mgi/our-research/generative-ai-and-the-future-of-work-in-america

Explained: generative AI. (2023, November 9). MIT News | Massachusetts Institute of Technology. https://news.mit.edu/2023/explained-generative-ai-1109

Fan, C., Wang, Z., Zhang, J., Zhao, J., & Rui, X. (2023). Influence maximization in social networks based on discrete harris hawks optimization algorithm. *Computing*, 106(2), 1–25. DOI: 10.1007/s00607-023-01207-4

Farooq, U., & Qamar, M. A. J. (2019). Predicting multistage financial distress: Reflections on sampling, feature and model selection criteria. In *Journal of Forecasting* (Vol. 38, Issue 7, pp. 632–648). DOI: 10.1002/for.2588

Farooq, U., Jibran Qamar, M. A., & Haque, A. (2018). A three-stage dynamic model of financial distress. *Managerial Finance*, 44(9), 1101–1116. DOI: 10.1108/MF-07-2017-0244

Fawzi, N. S., Kamaluddin, A., & Sanusi, Z. M. (2015). *Monitoring Distressed Companies through Cash Flow Analysis. 7th International Conference on Financial Criminology 2015. 28*. ScienceDirect., DOI: 10.1016/S2212-5671(15)01092-8

Fisher, R. (1936). The use of multiple measurements in taxonomic problems. *Annals of Eugenics*, 7(2), 179–188. DOI: 10.1111/j.1469-1809.1936.tb02137.x

Florath, J., Chanussot, J., & Keller, S. (2024a). Rapid natural hazard extent estimation from twitter data: Investigation for hurricane impact areas. *Natural Hazards*, 120(7), 6775–6796. Advance online publication. DOI: 10.1007/s11069-024-06488-2

Florath, J., Chanussot, J., & Keller, S. (2024b). Utilizing Volunteered Geographic Information for Real-Time Analysis of Fire Hazards: Investigating the Potential of Twitter Data in Assessing the Impacted Areas. *Fire (Basel, Switzerland)*, 7(1), 6. Advance online publication. DOI: 10.3390/fire7010006

Floridi, L., Cowls, J., Beltrametti, M., Chatila, R., Chazerand, P., Dignum, V., Luetge, C., Madelin, R., Pagallo, U., Rossi, F., Schäfer, B., Valcke, P., & Vayena, E. (2018). AI4People—An Ethical Framework for a Good AI Society: Opportunities, risks, principles, and recommendations. *Minds and Machines*, 28(4), 689–707. DOI: 10.1007/s11023-018-9482-5 PMID: 30930541

Fung, A. (2015). Putting the Public Back into Governance: The Challenges of Citizen Participation and Its Future. *Public Administration Review*, 75(4), 513–522. Advance online publication. DOI: 10.1111/puar.12361

Gandhi, V. (2015). Interfacing brain and Machine. In *Elsevier eBooks* (pp. 7–63). DOI: 10.1016/B978-0-12-801543-8.00002-8

Gao, P., & Verrecchia, R. E. (2012). Economic consequences of idiosyncratic Information in diversified markets. *Social Science Research Network*. DOI: 10.2139/ssrn.2008364

Gao, Y., Janssen, M., & Zhang, C. (2023). Understanding the evolution of open government data research: Towards open data sustainability and smartness. *International Review of Administrative Sciences*, 89(1), 59–75. Advance online publication. DOI: 10.1177/00208523211009955

Garg, H., & Kumar, K. (2018). A novel correlation coefficient of intuitionistic fuzzy sets based on the connection number of set pair analysis and its application. *Scientia Iranica*, 25(4), 2373–2388.

Generative AI and cybersecurity: Strengthening both defenses and threats. (2023, September 18). Bain. https://www.bain.com/insights/generative-ai-and-cybersecurity-strengthening-both-defenses-and-threats-tech-report-2023/

Generative AI and the labor market: A case for techno-optimism. (2024, March 25). Deloitte Insights. https://www2.deloitte.com/us/en/insights/economy/generative-ai-impact-on-jobs.html

Generative AI for Software Engineering - Capgemini USA. (2023, August 11). Capgemini USA. https://www.capgemini.com/us-en/solutions/generative-ai-for-software-engineering/

Generative AI in marketing. (n.d.). https://www.marketsandmarkets.com/industry-practice/GenerativeAI/genai-marketing

Generative AI in sourcing and procurement operations. (n.d.). Deloitte United States. https://www2.deloitte.com/us/en/blog/business-operations-room-blog/2023/generative-ai-in-procurement.html

Giacosa, E., Alam, G. M., Culasso, F., & Crocco, E. (2023). Stress-inducing or performance-enhancing? Safety measure or cause of mistrust? The paradox of digital surveillance in the workplace. *Journal of Innovation & Knowledge*, 8(2), 100357. DOI: 10.1016/j.jik.2023.100357

Giuffrida, N., Le Pira, M., Inturri, G., & Ignaccolo, M. (2019). Mapping with stakeholders: An overview of public participatory GIS and VGI in transport decision-making. *ISPRS International Journal of Geo-Information*, 8(4), 198. Advance online publication. DOI: 10.3390/ijgi8040198

Global, C. E. E. 2023. (2024, January 18). *KPMG*. https://kpmg.com/xx/en/home/insights/2024/01/global-cee-2023.html

Goh, K. Y., Heng, C. S., & Lin, Z. (2013). Social media brand community and consumer behavior: Quantifying the relative impact of user-and marketer-generated content. *Information Systems Research*, 24(1), 88–107. DOI: 10.1287/isre.1120.0469

Goksu, M., & Cavus, N. (2019). Fake News Detection on Social Networks with Artificial Intelligence Tools: Systematic Literature Review.

Goldberg, Y. (2016). A primer on neural network models for natural language processing. *Journal of Artificial Intelligence Research*, 57, 345–420. DOI: 10.1613/jair.4992

Gomez-Uribe, C. A., & Hunt, N. T. (2015). The Netflix Recommender system. *ACM Transactions on Management Information Systems*, 6(4), 1–19. DOI: 10.1145/2843948

Gong, M., Yan, J., Shen, B., Ma, L., & Cai, Q. (2016). Influence Maximization in Social Networks Based on Discrete Particle Swarm Optimization. *Information Sciences*, 367, 600–614. Advance online publication. DOI: 10.1016/j.ins.2016.07.012

Gong, X., Wang, H., Wang, X., Chen, C., Zhang, W., & Zhang, Y. (2024). Influence maximization on hypergraphs via multi-hop influence estimation. *Information Processing & Management*, 61(3), 103683. https://doi.org/10. DOI: 10.1016/j.ipm.2024.103683

González-Ramiro, A., Gonçalves, G., Sánchez-Ríos, A., & Jeong, J. S. (2016). *Using a VGI and GIS-based multicriteria approach for assessing the potential of rural tourism in Extremadura (Spain)*. Sustainability., DOI: 10.3390/su8111144

Goodchild, M. F. (2007). Citizens as sensors: The world of volunteered geography. In *GeoJournal*. DOI: 10.1007/s10708-007-9111-y

Goodfellow, I., Pouget-Abadie, J., Mirza, M., Xu, B., Warde-Farley, D., Ozair, S., Courville, A., & Bengio, Y. (2020). Generative adversarial networks. *Communications of the ACM*, 63(11), 139–144. DOI: 10.1145/3422622

Gorbach, J. (2018). Not Your Grandpa's Hoax: A Comparative History of Fake News. *American Journalism*, 35(2), 236–249. DOI: 10.1080/08821127.2018.1457915

Goyal, A., Lu, W., & Lakshmanan, L. V. S. (2011). CELF++: optimizing the greedy algorithm for influence maximization in social networks. In *Proceedings of the 20th international conference companion on World wide web (WWW '11)* (pp. 47-48). Association for Computing Machinery. https://doi.org/DOI: 10.1145/1963192.1963217

Gregova, E., Valaskova, K., Adamko, P., Tumpach, M., & Jaros, J. (2020). Predicting financial distress of slovak enterprises: Comparison of selected traditional and learning algorithms methods. *Sustainability (Basel)*, 12(10), 3954. DOI: 10.3390/su12103954

Guille, A., Hacid, H., Favre, C., & Zighed, D. (2013). Information Diffusion in Online Social Networks: A Survey. *SIGMOD Record*, 42(2), 17–28. DOI: 10.1145/2503792.2503797

Guo, X., Shen, Z., Zhang, Y., & Wu, T. (2019). Review on the application of artificial intelligence in smart homes. Smart Cities, 2(3), 402-420. https://www.mdpi.com/2624-6511/2/3/25

Guo, C., Li, W., Liu, F., Zhong, K., Wu, X., Zhao, Y., & Jin, Q. (2023). Influence maximization algorithm based on group trust and local topology structure. *Neurocomputing*, 564, 126936. DOI: 10.1016/j.neucom.2023.126936

Guo, C., Li, W., Wang, J., Yu, X., Liu, X., Luvembe, A., Wang, C., & Jin, Q. (2024). Heterogeneous network influence maximization algorithm based on multi-scale propagation strength and repulsive force of propagation field. *. *Knowledge-Based Systems*, 291, 111580. DOI: 10.1016/j.knosys.2024.111580

Guo, J., Zhang, P., Zhou, C., Cao, Y.-N., & Guo, L. (2013). Personalized influence maximization on social networks. In *International Conference on Information and Knowledge Management, Proceedings* (pp. 199-208). DOI: 10.1145/2505515.2505571

Guo, Z., Schlichtkrull, M., & Vlachos, A. (2022). A survey on automated fact-checking. *Transactions of the Association for Computational Linguistics*, 10, 178–206. DOI: 10.1162/tacl_a_00454

Gursoy, F., & Gunnec, D. (2018). Influence maximization in social networks under Deterministic Linear Threshold Model. *Knowledge-Based Systems*, 161, 111–123. DOI: 10.1016/j.knosys.2018.07.040

Gu, X., Wang, Z., Chen, X., Lu, P., Du, Y., & Tang, M. (2023). Influence maximization in social networks using role-based embedding. *Networks and Heterogeneous Media*, 18(4), 1539–1574. DOI: 10.3934/nhm.2023068

Hafiene, N., Karoui, W., & Romdhane, L. (2019). Influential Nodes Detection in Dynamic Social Networks. In Lecture Notes in Computer Science (pp. 62-73). DOI: 10.1007/978-3-030-20482-2_6

Hafiz, A., Lukumon, O., Muhammad, B., Olugbenga, A., Hakeem, O., & Sa-heed, A. (2015) Bankruptcy prediction of construction businesses: towards a big data analytics approach. In: Big data computing service and applications (Big-DataService), 2015 IEEE first international conference, Redwood City, CA, USA, 347–352. DOI: 10.1109/BigDataService.2015.30

Haldorai, A., Murugan, S., & Balakrishnan, M. (2024). Harnessing Intelligent AI to Elevate Business Modeling: A Perspective. In *Artificial Intelligence for Sustainable Development* (pp. 429–440). Springer Nature Switzerland. DOI: 10.1007/978-3-031-53972-5_22

Handler, I., & Kawaminami, J. (2023). Why do Japanese people visit hot springs during a pandemic? A psychographic segmentation analysis. *Journal of Outdoor Recreation and Tourism*, 41, 100530. DOI: 10.1016/j.jort.2022.100530

Hao, K. (2022, January 10). The messy, secretive reality behind OpenAI's bid to save the world. *MIT Technology Review*. https://www.technologyreview.com/2020/02/17/844721/ai-openai-moonshot-elon-musk-sam-altman-greg-brockman-messy-secretive-reality/

Haritha, G., & Sahana, N. (2023, February). Cryptocurrency price prediction using Twitter sentiment analysis. In *CS & IT conference proceedings* (Vol. 13, No. 3). CS & IT Conference Proceedings. https://csitcp.com/paper/13/133csit02.pdf

Harkness, L., Robinson, K., Stein, E., & Wu, W. (2023, December 5). How generative AI can boost consumer marketing. McKinsey & Company. https://www.mckinsey.com/capabilities/growth-marketing-and-sales/our-insights/how-generative-ai-can-boost-consumer-marketing#/

Hasselwander, M., Kiko, M., & Johnson, T. (2022). Digital civic engagement, open data, and the informal sector: A think piece. *Transportation Research Interdisciplinary Perspectives*, 16, 100700. Advance online publication. DOI: 10.1016/j.trip.2022.100700

Heath, D., & Heath, C. (2007). *Made to Stick: Why Some Ideas Survive and Others Die. Random House*. Random House.

Henebry, K. L. (1996). Do cash flow variables improve the predictive accuracy of a Cox proportional hazards model for bank failure? *The Quarterly Review of Economics and Finance*, 36(3), 395–409. DOI: 10.1016/S1062-9769(96)90023-X

Heneghan, L., & Henninger, P. (2023, April 17). Generative AI models – the risks and potential rewards in business. *KPMG*. https://kpmg.com/xx/en/home/insights/2023/04/generative-ai-models-the-risks-and-potential-rewards-in-business.html

He, X., & Kempe, D. (2018). Stability and Robustness in Influence Maximization. *ACM Transactions on Knowledge Discovery from Data*, 12(6), 66. Advance online publication. DOI: 10.1145/3233227

Hickman, L., Thapa, S., Tay, L., Cao, M., & Srinivasan, P. (2022). Text preprocessing for text mining in organizational research: Review and recommendations. *Organizational Research Methods*, 25(1), 114–146. DOI: 10.1177/1094428120971683

Ho, J., Jain, A., & Abbeel, P. (2020). Denoising diffusion probabilistic models. *Neural Information Processing Systems, 33*, 6840–6851. https://proceedings.neurips.cc/paper/2020/file/4c5bcfec8584af0d967f1ab10179ca4b-Paper.pdf

Hoffmann, M. T., Ostapowicz, K., Bartoń, K., Ibisch, P. L., & Selva, N. (2024). Mapping roadless areas in regions with contrasting human footprint. *Scientific Reports*, 14(1), 4722. Advance online publication. DOI: 10.1038/s41598-024-55283-3 PMID: 38413813

Hossam, A., Ramadan, A., Magdy, M., Abdelwahab, R., Ashraf, S., & Mohamed, Z. (2024). Revolutionizing Retail Analytics: Advancing Inventory and Customer Insight with AI. arXiv preprint arXiv:2405.00023. DOI: 10.1109/ICMISI61517.2024.10580424

Hosseini-Pozveh, M., Zamanifar, K., & Naghsh-Nilchi, A. (2018). Assessing information diffusion models for influence maximization in signed social networks. *Expert Systems with Applications*, 119, 476–490. Advance online publication. DOI: 10.1016/j.eswa.2018.07.064

HOTOSM. (2023a). *Hot's Response To The Turkey & Syria Earthquake*. https://www.hotosm.org/projects/join-the-turkey-and-syria-earthquake-response/

HOTOSM. (2023b). *What We Do | Humanitarian OpenStreetMap Team*. https://www.hotosm.org/what-we-do

Huang, K., Wang, S., Bevilacqua, G., Xiao, X., & Lakshmanan, L. (2017). Revisiting the stop-and-stare algorithms for influence maximization. *Proceedings of the VLDB Endowment International Conference on Very Large Data Bases*, 10(9), 913–924. DOI: 10.14778/3099622.3099623

Hua, Y. (2020). Bitcoin price prediction using ARIMA and LSTM. In *E3S web of conferences* (Vol. 218, p. 01050). EDP Sciences. DOI: 10.1051/e3sconf/202021801050

Hugging Face – The AI community building the future. (n.d.). https://huggingface.co/

Jalayer, M., Azheian, M., & Agha Mohammad Ali Kermani, M. (2018). A hybrid algorithm based on community detection and multi attribute decision making for influence maximization. *Computers & Industrial Engineering*, 120, 234–250. DOI: 10.1016/j.cie.2018.04.049

James, G., Witten, D., Hastie, T., & Tibshirani, R. (2013). *An introduction to statistical learning* (Vol. 112, p. 18). New York: springer.

Jaouadi, M., & Romdhane, L. (2019). Influence Maximization Problem in Social Networks: An Overview. In 2019 IEEE/ACS 16th International Conference on Computer Systems and Applications (AICCSA) (pp. 1-8). DOI: 10.1109/AICCSA47632.2019.9035366

Jarke, J. (2021). Co-creating digital public services. In *Public Administration and Information Technology* (Vol. 6). DOI: 10.1007/978-3-030-52873-7_3

Jia, H., Ding, S., Xu, X., & Nie, R. (2014). The latest research progress on spectral Clustering. *Neural Computing & Applications*, 24(7-8), 1477–1486. DOI: 10.1007/s00521-013-1439-2

Job Openings and Labor Turnover Survey News Release. (n.d.). U.S. Bureau of Labor Statistics. https://www.bls.gov/news.release/jolts.htm

Jones, S. (2016). A Cash Flow Based Model of Corporate Bankruptcy in Australia. *JAMAR, 14*(1).

Jones, S., & Hensher, D. A. (2004). Predicting Firm Financial Distress: A Mixed Logit Model. *The Accounting Review*, 79(4), 1011–1038. DOI: 10.2308/accr.2004.79.4.1011

Juluru, K., Shih, H. H., Keshava Murthy, K. N., & Elnajjar, P. (2021). Bag-of-words technique in natural language processing: A primer for radiologists. *Radiographics*, 41(5), 1420–1426. DOI: 10.1148/rg.2021210025 PMID: 34388050

Juul, J. L., & Ugander, J. (2021). Comparing information diffusion mechanisms by matching on cascade size. *Proceedings of the National Academy of Sciences of the United States of America*, 118(46), e2100786118. DOI: 10.1073/pnas.2100786118 PMID: 34750252

Ju, W., Chen, Y., Chen, L., & Li, B. (2024). Locating Influence Sources in Social Network by Senders and Receivers Spaces Mapping. *Expert Systems with Applications*, 248, 123327. DOI: 10.1016/j.eswa.2024.123327

Kameswari, U. S., & Babu, I. R. (2015). Sensor data analysis and anomaly detection using predictive analytics for process industries. 2015 IEEE Workshop on Computational Intelligence: Theories, Applications and Future Directions (WCI), Kanpur, India. https://ieeexplore.ieee.org/document/7495528

Kang, C.-M., Wang, M.-C., & Lin, L. (2022). Financial Distress Prediction of Cooperative Financial Institutions—Evidence for Taiwan Credit Unions. In *International Journal of Financial Studies* (Vol. 10, Issue 2, p. 30). DOI: 10.3390/ijfs10020030

Karami, M., Mosallanezhad, A., Sheth, P., & Liu, H. (2022). Estimating topic exposure for under-represented users on social media. *arXiv preprint arXiv:2208.03796.*

Karas, M., & Režňáková, M. (2015). Are the bankruptcy predictors specific for a given time or branch? the evidence from the czech manufacturing and construction companies. *Finance and Performance of Firms in Science, Education and Practice.* researchgate. Retrieved from https://www.researchgate.net/publication/275353210

Karoui, W., Hafiene, N., & Romdhane, L. (2022). Machine learning-based method to predict influential nodes in dynamic social networks. *Social Network Analysis and Mining*, 12(1), 108. Advance online publication. DOI: 10.1007/s13278-022-00942-4

Kashyap, S., & Bansal, R. (2019). Modeling Financial Distress Prediction of Indian Companies. *International Journal of Recent Technology and Engineering (IJRTE), 8*(1C2).

Kaur, H. (2024, February 2). Types of Generative AI Models and Its Applications. *Signity Solutions.* https://www.signitysolutions.com/blog/types-of-generative-ai -models

Kazmi, H., Mehmood, F., & Amayri, M. (2017). Smart home futures: Algorithmic challenges and opportunities. 2017 14th International Symposium on Pervasive Systems, Algorithms and Networks & 2017 11th International Conference on Frontier of Computer Science and Technology & 2017 Third International Symposium of Creative Computing. https://ieeexplore.ieee.org/abstract/document/8121811

Keikha, M. M., Rahgozar, M., Asadpour, M., & Faghih Abdollahi, M. (2019). Influence Maximization across Heterogeneous Interconnected Networks based on Deep Learning. *Expert Systems with Applications*, 140, 112905. DOI: 10.1016/j.eswa.2019.112905

Kempe, D., Kleinberg, J., & Tardos, É. (2005). Influential nodes in a diffusion model for social networks. In Lecture Notes in Computer Science (pp. 1127-1138). DOI: 10.1007/11523468_91

Khademolqorani, S., Hamadani, A. Z., & Rafiei, F. M. (2015). A Hybrid Analysis Approach to Improve Financial Distress Forecasting: Empirical Evidence from Iran. In *Mathematical Problems in Engineering* (Vol. 2015, pp. 1–9). DOI: 10.1155/2015/178197

Khoja, L., Chipulu, M., & Jayasekera, R. (2019). Analysis of financial distress cross countries: Using macroeconomic, industrial indicators and accounting data. *International Review of Financial Analysis*, 66(101379), 101379. DOI: 10.1016/j.irfa.2019.101379

Khoo, C. S., & Johnkhan, S. B. (2018). Lexicon-based sentiment analysis: Comparative evaluation of six sentiment lexicons. *Journal of Information Science*, 44(4), 491–511. DOI: 10.1177/0165551517703514

Khosravi Kazazi, A., Rahmani, Y., Amiri, F., & Hosseinali, F. (2024). A School of Thought on VGI Challenges: A Literature Review. *Papers in Applied Geography*, 10(1), 53–68. Advance online publication. DOI: 10.1080/23754931.2023.2256344

Kietzmann, J. H., Hermkens, K., McCarthy, I. P., & Silvestre, B. S. (2011). Social media? Get serious! Understanding the functional building blocks of social media. *Business Horizons*, 54(3), 241–251. DOI: 10.1016/j.bushor.2011.01.005

Kim, C., Lee, S., Park, S., & Lee, S.-g. (2013). Influence Maximization Algorithm Using Markov Clustering. In Lecture Notes in Computer Science (pp. 112-126). DOI: 10.1007/978-3-642-40270-8_10

Kim, D., Hyeon, D., Oh, J., Han, W., & Yu, H. (2017). Influence maximization based on reachability sketches in dynamic graphs. *Information Sciences*, 394, 217–231. Advance online publication. DOI: 10.1016/j.ins.2017.02.023

Kim, S., Shin, W., & Kim, H. W. (2024). Predicting online customer purchase: The integration of customer characteristics and browsing patterns. *Decision Support Systems*, 177, 114105. DOI: 10.1016/j.dss.2023.114105

Kimura, M., & Saito, K. (2006). Tractable models for information diffusion in social networks. *Lecture Notes in Computer Science*, 4213, 259–271. DOI: 10.1007/11871637_27

Kiritchenko, S., & Mohammad, S. M. (2017). The effect of negators, modals, and degree adverbs on sentiment composition. *arXiv preprint arXiv:1712.01794.*

Kitsios, F., & Kamariotou, M. (2023). Digital innovation and entrepreneurship transformation through open data hackathons: Design strategies for successful start-up settings. *International Journal of Information Management*, 69, 102472. Advance online publication. DOI: 10.1016/j.ijinfomgt.2022.102472

Klein, E., & Patrick, S. (2024, March 21). *Envisioning a global regime complex to govern artificial intelligence*. Carnegie Endowment for International Peace. https://carnegieendowment.org/research/2024/03/envisioning-a-global-regime-complex-to-govern-artificial-intelligence?lang=en

Kodali, R. K., Jain, V., Bose, S., & Boppana, L. (2016). IoT based smart security and home automation system. International Conference on Computing, Communication and Automation, IEEE, 1286-1289, April 29-30. https://ieeexplore.ieee.org/abstract/document/7813916

Kolak, M., Steptoe, M., Manprisio, H., Azu-Popow, L., Hinchy, M., Malana, G., & Maciejewski, R. (2020). *Extending Volunteered Geographic Information (VGI) with Geospatial Software as a Service: Participatory Asset Mapping Infrastructures for Urban Health*. DOI: 10.1007/978-3-030-19573-1_11

Koto, F., & Adriani, M. (2015). A comparative study on twitter sentiment analysis: Which features are good? *Natural Language Processing and Information Systems: 20th International Conference on Applications of Natural Language to Information Systems, NLDB 2015, Passau, Germany, June 17-19, 2015Proceedings*, 20, 453–457.

Ko, Y., Cho, K.-J., & Kim, S.-W. (2018). Efficient and Effective Influence Maximization in Social Networks: A Hybrid-Approach. *Information Sciences*, 465, 144–161. Advance online publication. DOI: 10.1016/j.ins.2018.07.003

Kumar Singh, P., Setta, S., Singh, A. K., & Singh, A. P. (2024). Machine Learning and Sentiments Analysis: Analyzing Customer Reviews. In *Human-Centered Approaches in Industry 5.0: Human-Machine Interaction, Virtual Reality Training, and Customer Sentiment Analysis* (pp. 248-264). IGI Global.

Kumar, S., Gupta, A., & Khatri, I. (2022). CSR: A community based spreaders ranking algorithm for influence maximization in social networks. *World Wide Web (Bussum)*, 25(6), 2303–2322. Advance online publication. DOI: 10.1007/s11280-021-00996-y

Kumar, S., Tiwari, P., & Zymbler, M. (2019). Internet of Things is a revolutionary approach for future technology enhancement: A review. *Journal of Big Data*, 6(1), 111. DOI: 10.1186/s40537-019-0268-2

Kumar, V., Ashraf, A. R., & Nadeem, W. (2024). AI-powered marketing: What, where, and how? *International Journal of Information Management*, 77, 102783. DOI: 10.1016/j.ijinfomgt.2024.102783

Kurniasari, L., & Setyanto, A. (2020, February). Sentiment analysis using recurrent neural network. In *Journal of Physics: Conference Series* (Vol. 1471, No. 1, p. 012018). IOP Publishing. https://iopscience.iop.org/article/10.1088/1742-6596/1471/1/012018/pdf

Lahiri, K., & Zhao, Y. (2016). Determinants of consumer sentiment over business cycles: Evidence from the US surveys of consumers. *Journal of Business Cycle Research*, 12(2), 187–215. DOI: 10.1007/s41549-016-0010-5

Lamon, C., Nielsen, E., & Redondo, E. (2017). Cryptocurrency price prediction using news and social media sentiment. *SMU Data Sci. Rev, 1*(3), 1-22. https://cs229.stanford.edu/proj2017/final-reports/5237280.pdf

Lehr, D., & Ohm, P. (2017). Playing with the data: What legal scholars should learn about machine learning. *UCD Law Review*, 51, 653.

Li, J., & Du, W. (2011). An Empirical Study on the Corporate Financial Distress Prediction Based on Logistic Model: Evidence from China's Manufacturing Industry. *International Journal of Digital Content Technology and its Applications, 5*(6). DOI: 10.4156/jdcta.vol5.issue6.44

Li, S., Shi, W., Wang, J., & Zhou, H. (2021). A Deep Learning-Based Approach to Constructing a Domain Sentiment Lexicon: a Case Study in Financial Distress Prediction. In *Information Processing & Management* (Vol. 58, Issue 5, p. 102673). DOI: 10.1016/j.ipm.2021.102673

Liang, C., Wu, Z., Huang, W., & Giles, C. (2015). Measuring Prerequisite Relations Among Concepts. *InProceedings of the 2015 conference on empirical methods in natural language processing*, (pp. 1668-1674). DOI: 10.18653/v1/D15-1193

Liang, J., Gong, Y., Wu, X., & Li, Y. (2023). Customized influence maximization in attributed social networks: Heuristic and meta-heuristic algorithms. *Complex & Intelligent Systems*, 10(1), 1409–1424. Advance online publication. DOI: 10.1007/s40747-023-01220-2

Li, H., Xu, M., Bhowmick, S. S., Rayhan, J. S., Sun, C., & Cui, J. (2023). PIANO: Influence Maximization Meets Deep Reinforcement Learning. *IEEE Transactions on Computational Social Systems*, 10(3), 1288–1300. DOI: 10.1109/TCSS.2022.3164667

Li, H., Yang, S., Xu, M., Bhowmick, S. S., & Cui, J. (2023). Influence Maximization in Social Networks. *Survey (London, England)*.

Li, K., Zhang, L., & Huang, H. (2018). Social Influence Analysis: Models, Methods, and Evaluation. *Engineering (Beijing)*, 4(1), 40–46. Advance online publication. DOI: 10.1016/j.eng.2018.02.004

Li, M., Wang, X., Gao, K., & Zhang, S. (2017). A Survey on Information Diffusion in Online Social Networks: Models and Methods. *Information (Basel)*, 8(4), 118. DOI: 10.3390/info8040118

Lin, Y., & Lui, J. (2014). Algorithmic design for competitive influence maximization problems. arXiv preprint arXiv:1410.8664.

Lin, Y., Zhang, X., Xia, L., Ren, Y., & Li, W. (2019). A hybrid algorithm for influence maximization of social networks. In 2019 IEEE Intl Conf on Dependable, Autonomic and Secure Computing, Intl Conf on Pervasive Intelligence and Computing, Intl Conf on Cloud and Big Data Computing, Intl Conf on Cyber Science and Technology Congress (DASC/PiCom/CBDCom/CyberSciTech) (pp. 427-431). Fukuoka, Japan. DOI: 10.1109/DASC/PiCom/CBDCom/CyberSciTech.2019.00087

Lingam, G., Rout, R., & Somayajulu, D. (2019). Adaptive deep Q-learning model for detecting social bots and influential users in online social networks. *Applied Intelligence*, 49(11), 3947–3964. Advance online publication. DOI: 10.1007/s10489-019-01488-3

Liu, X., Meng, X., Chen, X., Su, W., Zhang, X., Li, L., & Su, X. (2020). The Research on Assistance Information Acquisition for Earthquake Emergency Response by Aggregating VGI Data and GF-2 Images. *Remote Sensing Technology and Application*, 34(6), 1286–1295.

Liu, Y., Hu, Y., Yu, S., Zhou, X., & Li, K. (2024). Conformity-aware adoption maximization in competitive social networks. *Neurocomputing*, 573, 127224. DOI: 10.1016/j.neucom.2023.127224

Li, W., Bai, Q., Liang, L., Yang, Y., Hu, Y., & Zhang, M. (2021). Social influence minimization based on context-aware multiple influences diffusion model. *Knowledge-Based Systems*, 227, 107233. DOI: 10.1016/j.knosys.2021.107233

Li, W., Li, Y., Liu, W., & Wang, C. (2022). An influence maximization method based on crowd emotion under an emotion-based attribute social network. *Information Processing & Management*, 59(2), 102818. DOI: 10.1016/j.ipm.2021.102818

Li, X., Cheng, X., Su, S., & Sun, C. (2017). Community-Based Seeds Selection Algorithm for Location Aware Influence Maximization. *Neurocomputing*, 275, 1601–1613. Advance online publication. DOI: 10.1016/j.neucom.2017.10.007

Li, Y., Fan, J., Wang, Y., & Tan, K.-L. (2018). Influence Maximization on Social Graphs: A Survey. *IEEE Transactions on Knowledge and Data Engineering*, 30(10), 1852–1872. DOI: 10.1109/TKDE.2018.2807843

Li, Y., Lu, T., Li, W., & Zhang, P. (2024). HCCKshell: A heterogeneous cross-comparison improved Kshell algorithm for Influence Maximization. *Information Processing & Management*, 61(3), 103681. DOI: 10.1016/j.ipm.2024.103681

Lo, A. W., & Ross, J. (2024). Generative AI from Theory to Practice: A Case Study of Financial Advice. *An MIT Exploration of Generative AI*. DOI: 10.21428/e4baedd9.a1f6a281

Lotf, J. J., Azgomi, M. A., & Ebrahimi Dishabi, M. R. (2022). An improved influence maximization method for social networks based on genetic algorithm. *. *Physica A*, 586, 126480. DOI: 10.1016/j.physa.2021.126480

Lv, J., Guo, J., Yang, Z., Zhang, W., & Jocshi, A. (2014). Improved algorithms OF CELF and CELF++ for influence maximization. *Journal of Engineering Science and Technology Review*, 7(3), 32–38. DOI: 10.25103/jestr.073.05

Maheshwari, Y., & Agrawal, K. (2016). Predicting financial distress: Revisiting the option based model. *South Asian Journal of Global Business Research.*

Mahmood, R., Wang, G., Kalra, M., & Yan, P. (2023). Fact-Checking of AI-Generated Reports. *ArXiv, abs/2307.14634*.

Maji, P. K., Roy, A. R., & Biswas, R. (2002). An application of soft sets in a decision making problem. *Computers & Mathematics with Applications (Oxford, England)*, 44(8-9), 1077–1083. DOI: 10.1016/S0898-1221(02)00216-X

Ma, L., Shao, Z., Li, X., Lin, Q., Li, J., Leung, V. C. M., & Nandi, A. K. (2023). Influence Maximization in Complex Networks by Using Evolutionary Deep Reinforcement Learning. *IEEE Transactions on Emerging Topics in Computational Intelligence*, 7(4), 995–1009. DOI: 10.1109/TETCI.2021.3136643

Malakauskas, A., & Lakštutienė, A. (2021). Financial distress prediction for small and medium enterprises using machine learning techniques. *The Engineering Economist*, 32(1), 4–14. DOI: 10.5755/j01.ee.32.1.27382

Malaviya, C., Wu, S., & Cotterell, R. (2019). A simple joint model for improved contextual neural lemmatization. *arXiv preprint arXiv:1904.02306*. DOI: 10.18653/v1/N19-1155

Mamatzakis, E., & Koutsomanoli-Filippaki, A. (2009, November). Performance and Merton-type default risk of listed banks in the EU: A panel VAR approach. *Journal of Banking & Finance*, 33(11), 2050–2061. DOI: 10.1016/j.jbankfin.2009.05.009

Mao, Y., Liu, Q., & Zhang, Y. (2024). Sentiment analysis methods, applications, and challenges: A systematic literature review. *Journal of King Saud University. Computer and Information Sciences*, 36(4), 102048. DOI: 10.1016/j.jksuci.2024.102048

Markham, A., Buchanan, E., & AOIR Ethics Working Committee. (2012). Ethical decision-making and internet research: Recommendations from the AoIR Ethics Working Committee (Version 2.0).

Marti, C. L., Liu, H., Kour, G., Bilgihan, A., & Xu, Y. (2024). Leveraging artificial intelligence in firm-generated online customer communities: A framework and future research agenda. *Journal of Service Management*, 35(3), 438–458. DOI: 10.1108/JOSM-10-2023-0443

Martin, D. (1977). Early warning of bank failure: A logit regression approach. *Journal of Banking & Finance*, 1(3), 249–276. DOI: 10.1016/0378-4266(77)90022-X

Martineau, K. (2024, May 1). *What is generative AI?* IBM Research. https://research.ibm.com/blog/what-is-generative-AI

Mast, G. (2023, June 20). *Toyota Research Institute unveils new generative AI technique for vehicle design - Toyota USA Newsroom.* Toyota USA Newsroom. https://pressroom.toyota.com/toyota-research-institute-unveils-new-generative-ai-technique-for-vehicle-design/

Mathew, L., & Bindu, V. R. (2020, March). A review of natural language processing techniques for sentiment analysis using pre-trained models. In *2020 Fourth international conference on computing methodologies and communication (ICCMC)* (pp. 340-345). IEEE. DOI: 10.1109/ICCMC48092.2020.ICCMC-00064

Mays, N., Roberts, E., & Popay, J. (2001). Synthesising research evidence. In Fulop, N., Allen, P., Clarke, A., & Black, N. (Eds.), *Studying the organisation and delivery of health services: research methods*. Routledge.

McCarty, L. T. (2018). Finding the right balance in artificial intelligence and law. In *Research Handbook on the Law of Artificial Intelligence* (pp. 55–87). Edward Elgar Publishing. DOI: 10.4337/9781786439055.00013

Meftah, M., Ounacer, S., & Azzouazi, M. (2024). Enhancing Customer Engagement in Loyalty Programs Through AI-Powered Market Basket Prediction Using Machine Learning Algorithms. In *Engineering Applications of Artificial Intelligence* (pp. 319–338). Springer Nature Switzerland. DOI: 10.1007/978-3-031-50300-9_18

Merritt, R. (2024, January 24). *What are foundation models? | NVIDIA Blogs.* NVIDIA Blog. https://blogs.nvidia.com/blog/what-are-foundation-models/

Merton, R. C. (1974, December 28-30). On the pricing of corporate debt: The risk structure of interest rates. *The Journal of Finance*, 449-470.

Mikel. (2011). Haiti OSM Mapping for Japan and Libya. In *Humanitarian OpenStreetMap Team*. https://hot.openstreetmap.org/weblog/2011/03/haiti-osm-mapping-for-japan-and-libya/\nhttp://hot.openstreetmap.org/weblog

Minghini, M., Sarretta, A., & Napolitano, M. (2022). OpenStreetMap Contribution to Local Data Ecosystems in COVID-19 Times: Experiences and Reflections from the Italian Case. *Data*, 7(4), 39. Advance online publication. DOI: 10.3390/data7040039

Mittal, A. (2024, January 9). *The Plagiarism Problem: How generative AI models reproduce copyrighted content*. Unite.AI. https://www.unite.ai/the-plagiarism-problem-how-generative-ai-models-reproduce-copyrighted-content/

Mittelstadt, B. D., Allo, P., Taddeo, M., Wachter, S., & Floridi, L. (2016). The ethics of algorithms: Mapping the debate. *Big Data & Society*, 3(2), 2053951716679679. DOI: 10.1177/2053951716679679

Monti, N. E., & Garcia, R. M. (2010). A Statistical Analysis to Predict Financial Distress. In *Journal of Service Science and Management* (Vol. 03, Issue 03, pp. 309–335). DOI: 10.4236/jssm.2010.33038

Morgan Stanley. (n.d.). Launch of AI @ Morgan Stanley Debrief | Morgan Stanley. https://www.morganstanley.com/press-releases/ai-at-morgan-stanley-debrief-launch#:~:text=The%20new%20AI%20%40%20Morgan%20Stanley,meetings%20and%20surfaces%20action%20items

Mukhopadhyay, S., Vijayalakshmi, A., & Jain, S. P. (2023). Understanding consumers in-store behavior: The dual role of episode-specific motive adjustment and motive selection. *Journal of Retailing*, 99(3), 460–479. DOI: 10.1016/j.jretai.2023.08.004

Nabatchi, T., Sancino, A., & Sicilia, M. (2017). Varieties of Participation in Public Services: The Who, When, and What of Coproduction. *Public Administration Review*, 77(5), 766–776. Advance online publication. DOI: 10.1111/puar.12765

Nair, N. P., & Suvarna, S. (2023, April 13). Xiaomi Smart TV X Pro Series with Google TV, Dolby Vision IQ Launched in India: All details. *Gadgets 360*. https://www.gadgets360.com/tv/news/xiaomi-smart-tv-x-pro-43-50-55-inch-price-india-rs-32999-41999-47999-launch-sale-date-april-19-specifications-features-3945522

Nair, J., & Sachdeva, J. (2018, March). Financial distress indicators of indian manufacturing companies–A factor analysis approach. *Global Journal for Research Analysis*, 7(3), 66–69.

Navigate the minefields of generative AI. (2023, August 15). *KPMG*. https://kpmg.com/my/en/home/insights/2023/08/navigate-the-minefields-of-generative-ai.html

Neri, F., Aliprandi, C., Capeci, F., & Cuadros, M. (2012, August). *Sentiment analysis on social media. In 2012 IEEE/ACM international conference on advances in social networks analysis and mining*. IEEE.

Nguyen, A. T., Kharosekar, A., Krishnan, S., Krishnan, S., Tate, E., Wallace, B. C., & Lease, M. (2018). Believe it or not: Designing a Human-AI Partnership for Mixed-Initiative Fact-Checking. *Proceedings of the 31st Annual ACM Symposium on User Interface Software and Technology*. DOI: 10.1145/3242587.3242666

Nguyen, D.-L., Nguyen, T.-H., Do, T.-H., & Yoo, M. (2017). Probability-Based Multi-hop Diffusion Method for Influence Maximization in Social Networks. *Wireless Personal Communications*, 93(4), 903–916. DOI: 10.1007/s11277-016-3939-8

Niemann, M., Schmidt, J. H., & Neukirchen, M. (2008). Improving performance of corporate rating prediction models by reducing financial ratio heterogeneity. *Journal of Banking & Finance*, 32(3), 434–446. DOI: 10.1016/j.jbankfin.2007.05.015

Ninan, N., Roy, J. C., & Thomas, M. R. (2019). Benefits of Cross - Training: Scale Development and Validity. In *Prabandhan: Indian Journal of Management* (Vol. 12, Issue 6, p. 41). DOI: 10.17010/pijom/2019/v12i6/144935

Noronha, M. R., & Mehta, P. U. (2012). Corporate governance and investment decisions. *Prabandhan: Indian Journal of Management*, 5(5), 41. DOI: 10.17010/pijom/2012/v5i5/60202

Novet, J. (2023, April 10). *Microsoft's $13 billion bet on OpenAI carries huge potential along with plenty of uncertainty*. CNBC. https://www.cnbc.com/2023/04/08/microsofts-complex-bet-on-openai-brings-potential-and-uncertainty.html

Nyitrai, T. (2019). Dynamization of bankruptcy models via indicator variables. *Benchmarking*, 26(1), 317–332. Advance online publication. DOI: 10.1108/BIJ-03-2017-0052

Ogundipe, D. O., Babatunde, S. O., & Abaku, E. A. (2024). AI and product management: A theoretical overview from idea to market. *International Journal of Management & Entrepreneurship Research*, 6(3), 950–969. DOI: 10.51594/ijmer.v6i3.965

Ohlson, J. A. (1980). Financial Ratios and the Probabilistic Prediction of Bankruptcy. *Journal of Accounting Research*, 18(1), 109–131. https://www.jstor.org/stable/2490395. DOI: 10.2307/2490395

Olteanu-Raimond, A. M., See, L., Schultz, M., Foody, G., Riffler, M., Gasber, T., Jolivet, L., le Bris, A., Meneroux, Y., Liu, L., Poupée, M., & Gombert, M. (2020). Use of automated change detection and VGI sources for identifying and validating urban land use change. *Remote Sensing (Basel)*, 12(7), 1186. Advance online publication. DOI: 10.3390/rs12071186

Open, A. I., & Pilipiszyn, A. (2021, March 25). GPT-3 powers the next generation of apps. *OpenAI.* https://openai.com/index/gpt-3-apps/

openstreetmap. (2023). *Why use OpenStreetMap?*https://welcome.openstreetmap.org/why-openstreetmap/

Orimaye, S. O. (2013). Learning to classify subjective sentences from multiple domains using extended subjectivity lexicon and subjective predicates. *Information Retrieval Technology: 9th Asia Information Retrieval Societies Conference, AIRS 2013, Singapore, December 9-11, 2013 Proceedings*, 9, 191–202.

Ortakci, Y., & Seker, H. (2024). Optimising customer retention: An AI-driven personalised pricing approach. *Computers & Industrial Engineering*, 188, 109920. DOI: 10.1016/j.cie.2024.109920

Ozen, C., & Tas, Z. M. (2024). The Role of Artificial Intelligence (AI) in Improving Product Development Efficiency. In Digital Technologies in Modeling and Management: Insights in Education and Industry (pp. 313-335). IGI Global.

Oz, I. O., & Yelkenci, T. (2017). A theoretical approach to financial distress prediction modeling. *Managerial Finance*, 43(2), 212–230. DOI: 10.1108/MF-03-2016-0084

Palacio-Niño, J. O., & Berzal, F. (2019). Evaluation metrics for unsupervised learning algorithms. *arXiv preprint arXiv:1905.05667.*

Pant, D. R., Neupane, P., Poudel, A., Pokhrel, A. K., & Lama, B. K. (2018, October). Recurrent neural network based bitcoin price prediction by twitter sentiment analysis. In *2018 IEEE 3rd international conference on computing, communication and security (ICCCS)* (pp. 128-132). IEEE. DOI: 10.1109/CCCS.2018.8586824

Paredes, D. G. (2023). *Strategies based on artificial intelligence for the detection of fake news.* AWARI. DOI: 10.47909/awari.58

Pasupulati, R., Mahendran, J., & Majumdar, A. (2024). Predictive Analysis of Digital Consumer Behaviour: Key to Understanding Success and Failure of a Brand. In Enhancing and Predicting Digital Consumer Behavior with AI (pp. 238-267). IGI Global.

Pathak, P. (2024, January 25). *GenAI vs. LLMs vs. NLP: A Complete Guide*. Scribble Data. https://www.scribbledata.io/blog/genai-vs-llms-vs-nlp-a-complete-guide/

Pavaloaia, V. D., Teodor, E. M., Fotache, D., & Danileţ, M. (2019). Opinion mining on social media data: Sentiment analysis of user preferences. *Sustainability (Basel)*, 11(16), 4459. DOI: 10.3390/su11164459

Peng, S., Zhou, Y., Cao, L., Yu, S., Niu, J., & Jia, W. (2018). Influence analysis in social networks: A survey. *Journal of Network and Computer Applications*, 106, 17–32. DOI: 10.1016/j.jnca.2018.01.005

Perry, L. G., Henderson, G. V.Jr, & Cronan, T. P. (1984). Multivariate analysis of corporate bond ratings and industry classifications. *Journal of Financial Research*, 7(1), 27–36. DOI: 10.1111/j.1475-6803.1984.tb00351.x

Pham, M. T., Rajic, A., Greig, J. D., Sargeant, J. M., Papadopoulos, A., & McEwen, S. A. (2014). A scoping review of scoping reviews: Advancing the approach and enhancing the consistency. *Research Synthesis Methods*, 5(4), 371–385. DOI: 10.1002/jrsm.1123 PMID: 26052958

Piriyakul, I., Kunathikornkit, S., & Piriyakul, R. (2024). Evaluating brand equity in the hospitality industry: Insights from customer journeys and text mining. *International Journal of Information Management Data Insights*, 4(2), 100245. DOI: 10.1016/j.jjimei.2024.100245

Ponti, M., & Craglia, M. (2020). *Citizen-generated data for public policy*. European Commission.

Provost, F., & Fawcett, T. (2013). *Data Science for Business: What you need to know about data mining and data-analytic thinking*. O'Reilly Media, Inc.

Psillaki, M., Tsolas, I., & Margaritis, D. (2010). Evaluation of credit risk based on firm performance. *European Journal of Operational Research*, 201(3), 873–881. DOI: 10.1016/j.ejor.2009.03.032

Purba, K. R., Asirvatham, D., & Murugesan, R. K. (2021). Influence maximization algorithm: Review on current approaches and limitations. *Engineering and Applied Science Research*, 48(2), 221–229.

Qin, Y., Ma, J., & Gao, S. (2017). Efficient influence maximization under TSCM: A suitable diffusion model in online social networks. *. *Soft Computing*, 21(4), 827–838. Advance online publication. DOI: 10.1007/s00500-016-2068-3

Rahwan, I. (2018). Society-in-the-loop: Programming the algorithmic social contract. *Ethics and Information Technology*, 20(1), 5–14. DOI: 10.1007/s10676-017-9430-8

Raji, M. A., Olodo, H. B., Oke, T. T., Addy, W. A., Ofodile, O. C., & Oyewole, A. T. (2024). E-commerce and consumer behavior: A review of AI-powered personalization and market trends. GSC Advanced Research and Reviews, 18(3), 66-77.

Rajpurohit, P. D., & Rijwani, P. R. (2020). Financial reporting practices and internal corporate governance mechanisms in emerging markets. *Prabandhan: Indian Journal of Management*, 13(8-9), 8. DOI: 10.17010/pijom/2020/v13i8-9/155230

Ranjan, N., Mundada, K., Phaltane, K., & Ahmad, S. (2016). A Survey on Techniques in NLP. *International Journal of Computer Applications*, 134(8), 6–9. DOI: 10.5120/ijca2016907355

Rao, A., Shetty, A., Uphade, A., Thawani, P., & Rl, P. (2020). A Proposal for a novel approach to analyse and detect the fake news using AI techniques. *2020 3rd International Conference on Intelligent Sustainable Systems (ICISS)*, 582-589.

Rao, G., Wang, Y., Chen, W., Li, D., & Wu, W. (2020). Matching Influence Maximization in Social Networks. *Theoretical Computer Science*, 857, 71–86. Advance online publication. DOI: 10.1016/j.tcs.2020.12.040

Rao, N. V., Atmanathan, G., Shankar, M., & Ramesh, S. (2013). Analysis of bankruptcy prediction models and their effectiveness: An indian perspective. *Great Lakes Herald*, 7(2), 3–17.

Rash, W. (2024, May 9). Generative AI exposes users to new security risks. *Forbes*. https://www.forbes.com/sites/waynerash/2024/02/07/generative-ai-exposes-users-to-new-security-risks/?sh=452246c52dfe

Rathan, K., Sai, S. V., & Manikanta, T. S. (2019, April). Crypto-currency price prediction using decision tree and regression techniques. In 2019 3rd International Conference on Trends in Electronics and Informatics (ICOEI) (pp. 190-194). IEEE. DOI: 10.1109/ICOEI.2019.8862585

Ray, S. (2011). Assessing Corporate Financial Distress in Automobile Industry of India: An Application of Altman's Model. *Research Journal of Finance and Accounting, 2*(3).

Rezvanian, A., Vahidipour, S. M., & Meybodi, M. R. (2023). A new stochastic diffusion model for influence maximization in social networks. *Scientific Reports*, 13(1), 6122. DOI: 10.1038/s41598-023-33010-8 PMID: 37059847

Risteska Stojkoska, B. L., & Trivodaliev, K. V. (2017). A review of Internet of Things for smart homes: Challenges and solutions. ScienceDirect. https://www.sciencedirect.com/science/article/abs/pii/S095965261631589X

Ritchie, J., & Spencer, L. (1994). Qualitative data analysis for applied policy research. In Bryman, A., & Burgess, R. G. (Eds.), *Analysing qualitative data*. Routledge. DOI: 10.4324/9780203413081_chapter_9

Robaldo, L., Villata, S., Wyner, A., & Grabmair, M. (2019). Introduction for artificial intelligence and law: Special issue "natural language processing for legal texts". *Artificial Intelligence and Law*, 27(2), 113–115. DOI: 10.1007/s10506-019-09251-2

Rogers, R., & Niederer, S. (2020). The Politics of Social Media Manipulation. *The Politics of Social Media Manipulation*.

Rosário, A. T. (2024). Artificial Intelligence in the Consumer Behavior Process in Business. In Enhancing and Predicting Digital Consumer Behavior with AI (pp. 39-73). IGI Global. DOI: 10.4018/979-8-3693-4453-8.ch004

Sadiku, M.N., Eze, T.P., & Musa, S.M. (2018). Fake news and misinformation.

Samad, T. (2020). A review of IoT devices and applications. ScienceDirect. https://www.sciencedirect.com/science/article/pii/S1389128620312068

Samanta, H., & Golui, K. (2023). Smart inventory control: Proposed framework on basis of IoT, RFID, and supply chain management. ResearchGate. https://www.researchgate.net/publication/370272042_Smart_Inventory_Control_Proposed_Framework_on_Basis_of_IoT_RFID_and_Supply_Chain_Management

Sam, C. Y. (2011). Corporate governance and financial crises: Lessons from Singapore's public sector. *Prabandhan: Indian Journal of Management*, 4(4), 3. DOI: 10.17010/pijom/2011/v4i4/62403

Sathianarayanan, M., Hsu, P. H., & Chang, C. C. (2024). Extracting disaster location identification from social media images using deep learning. *International Journal of Disaster Risk Reduction*, 104, 104352. Advance online publication. DOI: 10.1016/j.ijdrr.2024.104352

Sawang, S., Lee, C. C., Chou, C. Y., Vighnesh, N. V., & Chandrashekar, D. (2023). Understanding post-pandemic market segmentation through perceived risk, behavioural intention, and emotional wellbeing of consumers. *Journal of Retailing and Consumer Services*, 75, 103482. DOI: 10.1016/j.jretconser.2023.103482

Saxena, B., & Kumar, P. (2019). A node activity and connectivity-based model for influence maximization in social networks. *Social Network Analysis and Mining*, 9(1), 40. Advance online publication. DOI: 10.1007/s13278-019-0586-6

Schott, M., Zell, A., Lautenbach, S., Sumbul, G., Schultz, M., Zipf, A., & Demir, B. (2024). Analyzing and Improving the Quality and Fitness for Purpose of OpenStreetMap as Labels in Remote Sensing Applications. In *Volunteered Geographic Information*. DOI: 10.1007/978-3-031-35374-1_2

Schwartz, E. H. (2022, July 1). *Alexa beats Google Assistant and Siri in voice assistant popularity as voice AI market expands - Voicebot.ai*. Voicebot.ai. https://voicebot.ai/2022/06/24/alexa-beats-google-assistant-and-siri-in-voice-assistant-popularity-as-voice-ai-market-expands/

Senbet, L. W., & Wang, T. Y. (2012). Corporate fnancial distress and bankruptcy: A survey. *Found Trends Finance*, 5(4), 243–335. DOI: 10.1561/0500000009

Sepasgozar, S., Karimi, R., Farahzadi, L., Moezzi, F., Shirowzhan, S., Ebrahimzadeh, M. S., Hui, F., & Aye, L. (2020). A systematic content review of artificial intelligence and the Internet of Things applications in smart home. *Applied Sciences (Basel, Switzerland)*, 10(9), 3074. DOI: 10.3390/app10093074

Settles, B. (2022, July 6). *How we learn how you learn*. Duolingo Blog. https://blog.duolingo.com/how-we-learn-how-you-learn/

Shae, Z., & Tsai, J. J. (2019). AI Blockchain Platform for Trusting News. *2019 IEEE 39th International Conference on Distributed Computing Systems (ICDCS)*, 1610-1619.

Shah, P. (2020). Sentiment analysis using textblob. *Diambil kembali dari Medium:*https://towardsdatascience.com/my-absolute-go-to-for-sentimentanalysis-textblob-3ac3a11d524

Shaik Vadla, M. K., Suresh, M. A., & Viswanathan, V. K. (2024). Enhancing Product Design through AI-Driven Sentiment Analysis of Amazon Reviews Using BERT. *Algorithms*, 17(2), 59. DOI: 10.3390/a17020059

Shakeel, N., & Dwivedi, R. K. (2022a). A comparative analysis of the techniques for influence maximization in social networks. In 2022 9th International Conference on Computing for Sustainable Global Development (INDIACom) (pp. 674-678). New Delhi, India. DOI: 10.23919/INDIACom54597.2022.9763226

Shakeel, N., & Dwivedi, R. K. (2022b). A learning based influence maximization across multiple social networks. In 2022 12th International Conference on Cloud Computing, Data Science & Engineering (Confluence) (pp. 53-58). Noida, India. DOI: 10.1109/Confluence52989.2022.9734145

Shang, J., Zhou, S., Li, X., Liu, L., & Wu, H. (2016). CoFIM: A community-based framework for influence maximization on large-scale networks. *Knowledge-Based Systems*, 10. Advance online publication. DOI: 10.1016/j.knosys.2016.09.029

Shankar, M. S. S. N., Kishore, G. D. K., Anudeep, N. L. N. C., Bhavya, R., & Sankar, R. S. E-commerce Chatbot For Price Negotiation Using NLP.

Shao, Y., Chen, L., Chen, Y., Liu, W., & Dai, C. (2023). Identifying multiple influence sources in social networks based on latent space mapping. *Information Sciences*, 635, 375–397. DOI: 10.1016/j.ins.2023.01.127

Sharma, N., Arahna, V. P., Saxena, S., Ateef, M., & Samuel, A. J. (2022). Scoping out the scope of scoping reviews in neonatal and pediatric pain management: A scoping review methodological framework. In *Journal of Pediatric Surgery* (Vol. 57, Issue 8, pp. 1599–1608). DOI: 10.1016/j.jpedsurg.2021.11.012

Sharma, A., Sharma, S., & Gupta, D. (2021). A review of sensors and their application in Internet of Things (IOT). *International Journal of Computer Applications*, 174(24), 27–34. DOI: 10.5120/ijca2021921148

Sharma, H., & Bhardwaj, S. (2024). *Ethical Explorations: AI's Role in Shaping the Journalism Ecosystem Across Asia*. New Global Studies., DOI: 10.1515/ngs-2024-0002

Shaw, F. X. (2023, November 17). Microsoft Ignite 2023: AI transformation and the technology driving change - The Official Microsoft Blog. The Official Microsoft Blog. https://blogs.microsoft.com/blog/2023/11/15/microsoft-ignite-2023-ai-transformation-and-the-technology-driving-change/

Shayaa, S., Ainin, S., Jaafar, N. I., Zakaria, S. B., Phoong, S. W., Yeong, W. C., Al-Garadi, M. A., Muhammad, A., & Zahid Piprani, A. (2018). Linking consumer confidence index and social media sentiment analysis. *Cogent Business & Management*, 5(1), 1509424. DOI: 10.1080/23311975.2018.1509424

Sheshar, S., Srivastva, S., Verma, M., & Kumar, R. (2021). Influence Maximization Frameworks, Performance, Challenges and Directions on Social Network: A Theoretical Study. *Journal of King Saud University. Computer and Information Sciences*, 34(9), 7570–7603. Advance online publication. DOI: 10.1016/j.jksuci.2021.08.009

Shetty, M., & Shetty, A. (2023). Automated Method for Fake News Detection Using Machine Learning. *2023 International Conference on Network, Multimedia and Information Technology (NMITCON)*, 1-6. DOI: 10.1109/NMITCON58196.2023.10275880

Shilpa, N., & Amulya, M. (2017). Corporate Financial Distress: Analysis of Indian Automobile Industry. *SDMIMD Journal of Management*, 8(1), 85. Advance online publication. DOI: 10.18311/sdmimd/2017/15726

Shivdas, A., & Chandrasekhar, J. (2016). Sustainability through frugal innovations: An application of Indian spiritual wisdom. *Prabandhan: Indian Journal of Management*, 9(5), 7–23. DOI: 10.17010/pijom/2016/v9i5/92567

Shukla, N. (2024, April 11). *Evolution of Language models: From Rules-Based Models to LLMs*. Appy Pie. https://www.appypie.com/blog/evolution-of-language-models

Shumway, T. (2001, January). Forecasting Bankruptcy More Accurately: A Simple Hazard Model. *The Journal of Business*, 74(1), 101–124. DOI: 10.1086/209665

Singh, A. K., & Kailasam, L. (2021). Link prediction-based influence maximization in online social networks. *Neurocomputing*, 453, 151–163. DOI: 10.1016/j.neucom.2021.04.084

Singh, S. S., Kumar, A., Singh, K., & Biswas, B. (2019). C2IM: Community based context-aware influence maximization in social networks. *Physica A*, 514, 796–818. DOI: 10.1016/j.physa.2018.09.142

Skalic, M., Jiménez, J., Sabbadin, D., & De Fabritiis, G. (2019). Shape-Based Generative Modeling for de Novo Drug Design. *Journal of Chemical Information and Modeling*, 59(3), 1205–1214. DOI: 10.1021/acs.jcim.8b00706 PMID: 30762364

Školkay, A., & Filin, J. (2019). *A Comparison of Fake News Detecting and Fact-Checking AI Based Solutions*. Studia Medioznawcze. DOI: 10.33077/uw.24511617.ms.2019.4.187

Souitaris, V., & Balabanis, G. (2007). Tailoring online retail strategies to increase customer satisfaction and loyalty. *Long Range Planning*, 40(2), 244–261. DOI: 10.1016/j.lrp.2006.11.006

Srivastava, S., & Bagga, T. (2014). A Comparative Study on the Usage of HRIS in the IT/ITES, Services, and Manufacturing Sectors in the Indian Scenario. In *Prabandhan: Indian Journal of Management* (Vol. 7, Issue 6, p. 21). DOI: 10.17010/pijom/2014/v7i6/59325

Stacy. (2021, October 28). Adding that extra 'You' to your discovery: Oskar Stål, Spotify vice president of personalization, explains how it works — Spotify. Spotify. https://newsroom.spotify.com/2021-10-13/adding-that-extra-you-to-your-discovery-oskar-stal-spotify-vice-president-of-personalization-explains-how-it-works/

Stanton, C., & Higgins, M. (2023, December). *Harvard Business School.* https://www.hbs.edu/faculty/Pages/item.aspx?num=65160#:~:text=Generative%20AI%20seemed%20poised%20to,Creation%2C%20and%20the%20labor%20market. https://hbsp.harvard.edu/product/824130-PDF-ENG

Stenqvist, E., & Lönnö, J. (2017). Predicting Bitcoin price fluctuation with Twitter sentiment analysis. https://www.diva-portal.org/smash/get/diva2:1110776/FULLTEXT01.pdf

Stone, D., Patton, B., & Heen, S. (2023). *Difficult conversations: How to discuss what matters most*. Penguin.

Sucharew, H., & Macaluso, M. (2019). Methods for research evidence synthesis: The scoping review approach. *Journal of Hospital Medicine*, 14(7), 416–418. DOI: 10.12788/jhm.3248 PMID: 31251164

Sun, J., Fujita, H., Zheng, Y., & Ai, W. (2021). Multi-class financial distress prediction based on support vector machines integrated with the decomposition and fusion methods. In *Information Sciences* (Vol. 559, pp. 153–170). DOI: 10.1016/j.ins.2021.01.059

Sun, B., & Li, J. (2008). The Recognition of new and old banknotes Based on SVM. *Second International Symposium on Intelligent Information Technology Application*, 2, 95-98. DOI: 10.1109/IITA.2008.157

Sun, J., Li, H., Huang, Q.-H., & He, K.-Y. (2013). Predicting financial distress and corporate failure: A review from the state-of-the-art definitions, modeling, sampling, and featuring approaches. *Knowledge-Based Systems*.

Sun, L., Huang, W., Yu, P., & Chen, W. (2018). Multi-Round Influence Maximization. In *Proceedings of the 27th ACM International Conference on Information and Knowledge Management* (pp. 2249-2258). https://doi.org/DOI: 10.1145/3219819.3220101

Surden, H. (2021). Machine learning and law: An overview. *Research Handbook on Big Data Law*, 171-184.

Surden, H. (2019). Artificial intelligence and law: An overview. *Georgia State University Law Review*, 35, 19–22.

Susskind, D. (2020). A world without work: technology, automation and how we should respond. https://openlibrary.org/books/OL28086183M/World_Without_Work

Sykora, M., Elayan, S., Hodgkinson, I. R., Jackson, T. W., & West, A. (2022). The power of emotions: Leveraging user generated content for customer experience management. *Journal of Business Research*, 144, 997–1006. DOI: 10.1016/j.jbusres.2022.02.048

Szudarski, P. (2023). *Collocations, corpora and language learning*. DOI: 10.1017/9781108992602

Tabassum, A., & Patil, R. R. (2020). A survey on text preprocessing & feature extraction techniques in natural language processing. *International Research Journal of Engineering and Technology*, 7(06), 4864–4867.

Talukder, A. (2017). An approach of cost optimized influence maximization in social networks. In 2017 19th Asia-Pacific Network Operations and Management Symposium (APNOMS) (pp. 354-357). Seoul, Korea (South). DOI: 10.1109/APNOMS.2017.8094146

Tang, S., & Yuan, J. (2020). Influence maximization with partial feedback. *Operations Research Letters*, 48(1), 24–28. DOI: 10.1016/j.orl.2019.10.013

Tang, W., Luo, G., Wu, Y., Tian, L., Zheng, X., & Cai, Z. (2019). A second-order diffusion model for influence maximization in social networks. *. *IEEE Transactions on Computational Social Systems*, 6(4), 702–714. DOI: 10.1109/TCSS.2019.2921422

Tan, K. L., Lee, C. P., Anbananthen, K. S. M., & Lim, K. M. (2022). RoBERTa-LSTM: A hybrid model for sentiment analysis with transformer and recurrent neural network. *IEEE Access : Practical Innovations, Open Solutions*, 10, 21517–21525. DOI: 10.1109/ACCESS.2022.3152828

Tavra, M., Lisec, A., Galešić Divić, M., & Cetl, V. (2024). Unpacking the role of volunteered geographic information in disaster management: focus on data quality. In *Geomatics, Natural Hazards and Risk* (Vol. 15, Issue 1). DOI: 10.1080/19475705.2023.2300825

Team, S. (2023, September 21). *Maximizing SEO Impact with ChatGPT: A Comprehensive Guide*. Semrush Blog. https://www.semrush.com/blog/chatgpt-seo/

Tejesh, B. S. S., & Kumar, T. R. (2018). Warehouse inventory management system using IoT and open-source framework. ResearchGate._https://www.researchgate.net/publication/329230209_Warehouse_inventory_management_system_using_IoT_and_open_source_framework

Teng, L., & Wang, Y. (2024). Topic-aware Most Influential Community Search in Social Networks. ArXiv, abs/2402.07601.

Teng, Y.-W., Tai, C.-H., Yu, P., & Chen, M.-S. (2018). Revenue Maximization on the Multi-grade Product. .DOI: 10.1137/1.9781611975321.65

Terenggana, C. A. (2024). The Influence of Artificial Intelligence on Customer Experience (Study of Maxim Users in Surabaya, East Java). *Economics Studies and Banking Journal*, 1(1), 37–45. DOI: 10.62207/jhctec97

Text generation models. (n.d.). OpenAI; OPenAI. https://platform.openai.com/docs/guides/text-generation/json-mode

The CEO's Guide to Generative AI. (2024, April). IBM Institute for Business Value. Retrieved May 18, 2024, from https://www.ibm.com/thought-leadership/institute-business-value/en-us/report/ceo-generative-ai

The startling power generative AI is bringing to software development. (n.d.). KPMG. https://kpmg.com/us/en/articles/2023/generative-artificial-intelligence.html

The state of AI in 2023: Generative AI's breakout year. (2023, August 1). McKinsey & Company. https://www.mckinsey.com/capabilities/quantumblack/our-insights/the-state-of-ai-in-2023-generative-AIs-breakout-year

The Transformer model family. (n.d.). https://huggingface.co/docs/transformers/en/model_summary

Thielen, A. (2018). Fake News – Två ord, två betydelser: En statsvetenskaplig begreppsstudie på traditionella medieartiklar.

Thiruvenkatasuresh, M. P., Venkatachalam, V., & Revathy, G. (2020). Brain Tumor Detection in MRI Images, using Fuzzy C-Means and Cuckoo Search Algorithm. *Solid State Technology*, 63(5), 6759–6769.

Thorne, J., & Vlachos, A. (2018). Automated Fact Checking: Task Formulations, Methods and Future Directions. *ArXiv, abs/1806.07687*.

Throop, A. W. (1992). Consumer sentiment: Its causes and effects. *Federal Reserve Bank of San Francisco Economic Review*, 1, 35–59.

Tian, S., Mo, S., Wang, L., & Peng, Z. (2020). Deep Reinforcement Learning-Based Approach to Tackle Topic-Aware Influence Maximization. *Data Science and Engineering*, 5(1), 5. DOI: 10.1007/s41019-020-00117-1

Tinoco, H. M., & Wilson, N. (2013, February 15). Financial distress and bankruptcy prediction among listed companies using accounting, market and macroeconomic variables. *International Review of Financial Analysis*, 30, 394–419. Advance online publication. DOI: 10.1016/j.irfa.2013.02.013

Toews, R. (2024, February 20). The next generation of large language models. *Forbes*. https://www.forbes.com/sites/robtoews/2023/02/07/the-next-generation-of-large-language-models/?sh=1e80997e18db

Tripathi, A. K., Agrawal, S., & Gupta, R. D. (2020). Cloud enabled SDI architecture: a review. In *Earth Science Informatics* (Vol. 13, Issue 2). DOI: 10.1007/s12145-020-00446-9

Trujillo-Ponce, A., Samaniego-Medina, R., & Cardone-Riportella, C. (n.d.). Examining What Best Explains Corporate Credit Risk: Accounting-Based versus Market-Based Models. In SSRN *Electronic Journal*. DOI: 10.2139/ssrn.2072176

Tsai, C.-W., & Liu, S.-J. (2019). SEIM: Search economics for influence maximization in online social networks. *Future Generation Computer Systems*, 93, 1055–1064. DOI: 10.1016/j.future.2018.08.033

Tsarapatsanis, D., & Aletras, N. (2021). On the ethical limits of natural language processing on legal text. *arXiv preprint arXiv:2105.02751*. DOI: 10.18653/v1/2021.findings-acl.314

Tseng, F.-M., & Hu, Y.-C. (2010). Comparing four bankruptcy prediction models: Logit, quadratic interval logit, neural and fuzzy neural networks. *Expert Systems with Applications*, 37(3), 1846–1853. DOI: 10.1016/j.eswa.2009.07.081

Tzavella, K., Skopeliti, A., & Fekete, A. (2022). Volunteered geographic information use in crisis, emergency and disaster management: A scoping review and a web atlas. *Geo-Spatial Information Science*. Advance online publication. DOI: 10.1080/10095020.2022.2139642

Urrea, G., & Yoo, E. (2023). The role of volunteer experience on performance on online volunteering platforms. *Production and Operations Management*, 32(2), 416–433. Advance online publication. DOI: 10.1111/poms.13879

Uthayasuriyan, A., Hema Chandran, G., Kavvin, U., Mahitha, S. H., & Jeyakumar, G. (2023). A comparative analysis on influence maximization models in social networks. In *2023 International Conference on Advancement in Computation & Computer Technologies (InCACCT)* (pp. 1-6). Gharuan, India. DOI: 10.1109/InCACCT57535.2023.10141772

Valchanov, I. (2024, April 23). *35 Best ChatGPT Prompts for SEO in 2024 [Expert Sourced]*. Team-GPT. https://team-gpt.com/blog/chatgpt-prompts-for-seo/

Valencia, F., Gómez-Espinosa, A., & Valdés-Aguirre, B. (2019). Price movement prediction of cryptocurrencies using sentiment analysis and machine learning. *Entropy (Basel, Switzerland)*, 21(6), 589. DOI: 10.3390/e21060589 PMID: 33267303

Varghese, S., & Agrawal, M. (2021). Impact of social media on Consumer Buying Behavior. *Saudi Journal of Business and Management Studies*, 6(3), 51–55. DOI: 10.36348/sjbms.2021.v06i03.001

Vaswani, A., Shazeer, N., Parmar, N., Uszkoreit, J., Jones, L., Gomez, A. N., Kaiser, L., & Polosukhin, I. (2017). Attention is All you Need. *arXiv (Cornell University), 30*, 5998–6008. https://arxiv.org/pdf/1706.03762v5

Velentzas, J., Broni, G., & Professor, A. (2020). *Fake News and the Role of Semantics and AI (Artificial Intelligence)*. How to Recognise the Falsity of the Information.

Verplanke, J., McCall, M. K., Uberhuaga, C., Rambaldi, G., & Haklay, M. (2016). A Shared Perspective for PGIS and VGI. *The Cartographic Journal*, 53(4), 308–317. Advance online publication. DOI: 10.1080/00087041.2016.1227552

Von Luxburg, U., Belkin, M., & Bousquet, O. (2008). Consistency of spectral Clustering. *Annals of Statistics*, 555–586.

Waddington, H., Sonnenfeld, A., Finetti, J., Gaarder, M., John, D., & Stevenson, J. (2019). Citizen engagement in public services in low- and middle-income countries: A mixed-methods systematic review of participation, inclusion, transparency and accountability (PITA) initiatives. In *Campbell Systematic Reviews* (Vol. 15, Issues 1–2). DOI: 10.1002/cl2.1025

Walters, R., & Novak, M. (2021). Artificial Intelligence and Law. In *Cyber Security, Artificial Intelligence, Data Protection & the Law* (pp. 39–69). Springer Singapore. DOI: 10.1007/978-981-16-1665-5_3

Wang, G. (2016). Predicting financial distress for privately held firms in the European Union. *University of Sydney Business School*, 240.

Wang, H., Zhao, Y., Gao, X., & Gao, B. (2021). Collaborative decision-making for urban regeneration: A literature review and bibliometric analysis. *Land Use Policy*, 107, 105479. Advance online publication. DOI: 10.1016/j.landusepol.2021.105479

Wang, P., Li, K., Du, Q., & Wang, J. (2024). Customer experience in AI-enabled products: Scale development and validation. *Journal of Retailing and Consumer Services*, 76, 103578. DOI: 10.1016/j.jretconser.2023.103578

Wang, T., Wu, X., Garg, H., Liu, Q., & Chen, G. (2023). A prospect theory-based MABAC algorithm with novel similarity measures and interactional operations for picture fuzzy sets and its applications. *Engineering Applications of Artificial Intelligence*, 126, 106787. DOI: 10.1016/j.engappai.2023.106787

Wang, W., & Street, W. N. (2018). Modeling and maximizing influence diffusion in social networks for viral marketing. *Applied Network Science*, 3(6), 6. DOI: 10.1007/s41109-018-0062-7 PMID: 30839789

Wang, Y., Dong, W., & Dong, X. (2018). A novel ITÖ Algorithm for influence maximization in the large-scale social networks. *Future Generation Computer Systems*, 88, 755–763. DOI: 10.1016/j.future.2018.04.026

Wankhade, M., Rao, A. C. S., & Kulkarni, C. (2022). A survey on sentiment analysis methods, applications, and challenges. *Artificial Intelligence Review*, 55(7), 5731–5780. DOI: 10.1007/s10462-022-10144-1

Waqas, H., & Md-Rus, R. (2018). Predicting fnancial distress: Importance of accounting and frm-specifc market variables for Pakistan's listed frms. *Cogent Economics & Finance*, 6(1), 1545739. DOI: 10.1080/23322039.2018.1545739

Weizenbaum, J. (1966). ELIZA—A computer program for the study of natural language communication between man and Machine. *Communications of the ACM*, 9(1), 36–45. DOI: 10.1145/365153.365168

West, S. M., Whittaker, M., & Crawford, K. (2019). *Discriminating Systems: Gender, Race and Power in AI.* AI Now Institute.

What are Autoregressive Models? - AR Models Explained - AWS. (n.d.). Amazon Web Services, Inc. https://aws.amazon.com/what-is/autoregressive-models/

What are Foundation Models? - Foundation Models in Generative AI Explained - AWS. (n.d.). Amazon Web Services, Inc. https://aws.amazon.com/what-is/foundation-models/

What is generative AI? (2024, April 2). McKinsey & Company. https://www.mckinsey.com/featured-insights/mckinsey-explainers/what-is-generative-ai

What is generative AI? (2024a). In *McKinsey & Company*. McKinsey Global Publishing.

What is Generative AI? (n.d.). Darktrace. https://darktrace.com/cyber-ai-glossary/generative-ai

What is GPT AI? - Generative Pre-Trained Transformers Explained - AWS. (n.d.). Amazon Web Services, Inc. https://aws.amazon.com/what-is/gpt/

Wittenberg, C., Epstein, Z., Berinsky, A. J., & Rand, D. G. (2024). Labeling AI-Generated Content: Promises, Perils, and Future Directions. *An MIT Exploration of Generative AI*. DOI: 10.21428/e4baedd9.0319e3a6

Wu, B., Tian, F., Zhang, M., Zeng, H., & Zeng, Y. (2020). Cloud services with big data provide a solution for monitoring and tracking sustainable development goals. *Geography and Sustainability*, 1(1), 25–32. Advance online publication. DOI: 10.1016/j.geosus.2020.03.006

Wu, P., & Pan, L. (2017). Scalable influence blocking maximization in social networks under competitive independent cascade models. *Computer Networks*, 123, 38–50. Advance online publication. DOI: 10.1016/j.comnet.2017.05.004

Xiaomi Smart TV X Series 2023 Edition. (n.d.). Xiaomi. https://www.mi.com/in/product/xiaomi-smart-tv-x-series-2023/

Xiao, Z., Yang, X., Pang, Y., & Dang, X. (2012). The prediction for listed companies' financial distress by using multiple prediction methods with rough set and Dempster–Shafer evidence theory. *Knowledge-Based Systems*, 26, 196–206. DOI: 10.1016/j.knosys.2011.08.001

Xie, M., Yang, Q., Wang, Q., Cong, G., & Melo, G. (2015). DynaDiffuse: A Dynamic Diffusion Model for Continuous Time Constrained Influence Maximization. *Proceedings of the AAAI Conference on Artificial Intelligence*, 29(1). Advance online publication. DOI: 10.1609/aaai.v29i1.9203

Xu, M., & Zhang, C. (2009). Bankruptcy prediction: The case of Japanese listed companies. *Review of Accounting Studies*, 14(4), 534–558. DOI: 10.1007/s11142-008-9080-5

Yang, A., Fan, H., Jia, Q., Ma, M., Zhong, Z., Li, J., & Jing, N. (2024). How do contributions of organizations impact data inequality in OpenStreetMap? *Computers, Environment and Urban Systems*, 109, 102077. Advance online publication. DOI: 10.1016/j.compenvurbsys.2024.102077

Yang, J., & Liu, J. (2018). Influence Maximization-Cost Minimization in Social Networks Based on a Multiobjective Discrete Particle Swarm Optimization Algorithm. *IEEE Access : Practical Innovations, Open Solutions*, 6, 2320–2329. DOI: 10.1109/ACCESS.2017.2782814

Yang, L., Zhang, Z., Song, Y., Hong, S., Xu, R., Zhao, Y., Zhang, W., Cui, B., & Yang, M. H. (2023). Diffusion models: A comprehensive survey of methods and applications. *ACM Computing Surveys*, 56(4), 1–39. DOI: 10.1145/3626235

Yang, S., Du, Q., Zhu, G., Cao, J., Chen, L., Qin, W., & Wang, Y. (2024). Balanced influence maximization in social networks based on deep reinforcement learning. *Neural Networks*, 169, 334–351. DOI: 10.1016/j.neunet.2023.10.030 PMID: 37922716

Ye, Y., Chen, Y., & Han, W. (2022). Influence maximization in social networks: Theories, methods and challenges. *Array (New York, N.Y.)*, 16, 100264. DOI: 10.1016/j.array.2022.100264

Yuan, B. (2024). Influence max: An efficient algorithm for maximizing influence in networks. *International Journal of IoT*, 4(1), 1–4.

Zadeh, L. A. (1996). On fuzzy algorithms. In *fuzzy sets, fuzzy logic, and fuzzy systems: selected papers By Lotfi A Zadeh* (pp. 127-147).

Zareie, A., & Sakellariou, R. (2024). Fuzzy Influence Maximization in Social Networks. *ACM Transactions on the Web*, 18(3), 32. Advance online publication. DOI: 10.1145/3650179

Zaremba, W., Brockman, G., & Open, A. I. (2021, August 10). OpenAI Codex. *OpenAI*. https://openai.com/index/openai-codex/

Zhang, J., Dong, B., & Yu, P. S. (2018). FakeDetector: Effective Fake News Detection with Deep Diffusive Neural Network. *2020 IEEE 36th International Conference on Data Engineering (ICDE)*, 1826-1829.

Zhang, R., Zhang, Z., Wang, D., & Du, M. (2022). Financial Distress Prediction with a Novel Diversity-Considered GA-MLP Ensemble Algorithm. In *Neural Processing Letters* (Vol. 54, Issue 2, pp. 1175–1194). DOI: 10.1007/s11063-021-10674-9

Zhang, W., Yang, Z., Wang, Y., Shen, Y., Li, Y., Wang, L., & Cui, B. (2021). Grain: Improving data efficiency of graph neural networks via diversified influence maximization. arXiv preprint arXiv:2108.00219.

Zhang, Z., Song, Y., & Qi, H. (2017). Age Progression/Regression by Conditional Adversarial Autoencoder. *arXiv (Cornell University)*. /arxiv.1702.08423DOI: 10.1109/CVPR.2017.463

Zhou, Y., Ejegwa, P. A., & Johnny, S. E. (2023). Generalized similarity operator for intuitionistic fuzzy sets and its applications based on recognition principle and multiple criteria decision making technique. *International Journal of Computational Intelligence Systems*, 16(1), 85. DOI: 10.1007/s44196-023-00245-2

Zimmer, M. (2020). “But the data is already public”: on the ethics of research in Facebook. In *The ethics of information technologies* (pp. 229–241). Routledge. DOI: 10.4324/9781003075011-17

Zlatanova-Pazheva, E. (2024). Model of consumer behavior when applying ai at every point of consumer contact. *International Journal of Engineering Technologies and Management Research*, 11(3), 35–46. DOI: 10.29121/ijetmr.v11.i3.2024.1433

Zmijewski, M. E. (1984). Methodological Issues Related to the Estimation of Financial Distress Prediction Models. *Journal of Accounting Research*, 22, 59–82. DOI: 10.2307/2490859

About the Contributors

Arul Kumar Natarajan currently serves as an Assistant Professor in the Department of Computer Science at the Samarkand International University of Technology in Uzbekistan. He earned his Doctor of Philosophy degree in Computer Science from Bharathidasan University, India, in 2017. Concurrently, he is engaged in postdoctoral research in Generative AI for Cybersecurity at the Singapore Institute of Technology, Singapore. Throughout his 14-year teaching career, Dr. Arul has held esteemed positions at various institutions, including Christ University, Bishop Heber College in India, and Debre Berhan University in Ethiopia. Dr. Arul has made significant contributions to academia, specializing in cybersecurity and artificial intelligence, as evidenced by his portfolio of scholarly works. He has authored 50 peer-reviewed and internationally indexed publications and delivered 35 conference presentations. Additionally, he has edited and published 04 books with IGI Global, USA, which are indexed in Scopus and focus on Artificial Intelligence and Cybersecurity. He also has 04 more books in the processing stage with IGI Global, Wiley, and Springer. In addition to his academic pursuits, Dr. Arul is a prolific innovator. He has 17 patents granted in India and 1 granted in the United Kingdom, spanning diverse fields such as communication and computer science. His latest work involves 1 copyrighted research (Govt. of India) in Artificial Intelligence and Machine Learning, specifically focused on segmenting, classifying, and tracking issue nuclei in images. Dr. Arul also exhibits notable proficiency in networking and cybersecurity, having completed the CCNA Routing and Switching Exam from CISCO and the Networking Fundamentals exam from Microsoft. He continues to demonstrate a strong interest in Generative AI for Cybersecurity.

Mohammad Gouse Galety is a seasoned professional in computer science, is currently a Professor at the computer science department of Samarkand International University of Technology, Samarkand, Uzbekistan. His research interests encompass a wide range of computer and information science, focusing on Web Mining, Computer Vision, IoT, Machine Learning, and AI. His impactful research has led to the (co) authorship of over 50 journal papers and international conference proceedings indexed by Springer, Web of Science, and Scopus, the holding of four patents, and the writing of six books. His standing in the academic community is further solidified by his role as a Fellow of the IEEE and ACM. With a career spanning over two decades, Mohammad Gouse Galety has served in many national and international organizations, solidifying his expertise in the field. His teaching experience includes roles at Sree Vidyanikethan Degree College, India; Emeralds Degree College, Tirupati, India; Brindavan College of Engineering, India; Kuwait Educational Center, Kuwait; Ambo University, Ethiopia; Debre Berhan University, Ethiopia; Lebanese French University, Iraq; and Catholic University in Erbil, Iraq. He imparts his knowledge to undergraduate and postgraduate students, teaching various courses in computer science and information technology/science engineering.

Celestine Iwendi is an IEEE Brand Ambassador. He has a PhD in Electronics Engineering, is a Past ACM Distinguished Speaker, a Senior Member of IEEE, a Seasoned Lecturer, and a Chartered Engineer. A highly motivated researcher and teacher with an emphasis on communication, hands-on experience, willingness to learn, and 23 years of technical expertise. He has developed operational, maintenance, and testing procedures for electronic products, components, equipment, and systems; provided technical support and instruction to staff and customers regarding equipment standards, assisting with specific, difficult in-service engineering; Inspected electronic and communication equipment, instruments, products, and systems to ensure conformance to specifications, safety standards, and regulations. He is a wireless sensor network Chief Evangelist, AI, ML, and IoT expert and designer. Celestine is a Reader (Professor) at the University of Bolton, United Kingdom. He is also the IEEE University of Bolton, Student Branch Counselor and former Board Member of the IEEE Sweden Section, a Fellow of The Higher Education Academy, United Kingdom, and a fellow of the Institute of Management Consultants to add to his teaching, managerial, and professional experiences. Celestine is an Ambassador in the prestigious Manchester Conference Ambassador Programme, a Visiting Professor to five Universities, and an IEEE Humanitarian Philanthropist. Celestine has received the prestigious recognition of the Royal Academy of Engineering through the Exceptional Talent Scheme, acknowledging his substantial contributions to Artificial Intelligence and its medical applications. Additionally, he takes pride in his three-year inclusion in Elsevier's publication, featuring the World's Top 2% Influential Scientists. Celestine is the Chair of the Election Committee of IEEE Computer Society Worldwide.

Deepthi Das, an Associate Professor and Associate Dean at the School of Sciences, CHRIST (Deemed to be University) in Bangalore, India, is a dedicated educator with over two decades of teaching experience. With a profound commitment to fostering academic excellence and ethical values among students, Dr. Deepthi creates inclusive learning environments that promote critical thinking and social responsibility. Holding a Ph.D. in Computer Science and Engineering from CMR University, Bangalore, she actively integrates innovative teaching methods to deliver dynamic educational experiences. Dr. Deepthi's research interests primarily focus on Artificial Intelligence and Machine Learning, with notable contributions to customer churn prediction in the motor insurance sector. Alongside her academic responsibilities, she has served in various leadership roles, including Head of the Department of Statistics and Data Science and Programme Coordinator for Undergraduate Programmes. Dr. Deepthi's dedication to education, research, and leadership acumen make her a valuable asset to the academic community.

Achyut Shankar is currently working as a Postdoc Research Fellow at the University of Warwick, United Kingdom, and was recently appointed as a visiting Associate Professor at the University of Johannesburg, South Africa. He obtained his PhD in Computer Science and Engineering, majoring in wireless sensor networks, from VIT University, Vellore, India.

* * *

Priyanka Aggarwal is a research scholar in Department of Management Studies, Jamia Millia Islamia, New Delhi. She received her Bachelor of Commerce from Delhi University and MBA from Amity University, Noida. She holds a working experience of more than ten years and her current research interest includes the areas of artificial intelligence, e commerce, and service marketing.

Munir Ahmad is a seasoned professional in the realm of Spatial Data Infrastructure (SDI), Geo-Information Productions, Information Systems, and Information Governance, boasting over 25 years of dedicated experience in the field. With a PhD in Computer Science, Dr. Ahmad's expertise spans Spatial Data Production, Management, Processing, Analysis, Visualization, and Quality Control. Throughout his career, Dr. Ahmad has been deeply involved in the development and deployment of SDI systems specially in the context of Pakistan, leveraging his proficiency in Spatial Database Design, Web, Mobile & Desktop GIS, and Geo Web Services Architecture. His contributions to Volunteered Geographic Information (VGI) and Open Source Geoportal & Metadata Portal have significantly enriched the geospatial community. As a trainer and researcher, Dr. Ahmad has authored over 50 publications, advancing the industry's knowledge base and fostering innovation in Geo-Tech, Data Governance, and Information Infrastructure, and Emerging Technologies. His commitment to Research and Development (R&D) is evident in his role as a dedicated educator and mentor in the field.

M. Bangaru Lakshmi is currently working as an Assistant Professor in the Department of CSE, GITAM Deemed to be University, Visakhapatnam, Andhra Pradesh, India. She is Pursuing her Ph.D. in CSE from GITAM Deemed to be Un University, Visakhapatnam, India. She has more than 14 years of teaching and research experience as an assistant professor at Vignan Engineering College and GITAM deemed to be a university, and she received a B.Tech. Degree: Computer science and engineering from JNTUK affiliated College and MTech from JNTU Kakinada. I am pursuing a Ph.D. in computer science and engineering from GITAM deemed to be a university. Her research interests include Image Processing, AI and Machine Learning Software engineering, and soft computing Techniques.

Disha Batra is a seasoned academician with over 8 years of experience in teaching media students. With a Doctorate in Business Journalism and a repertoire of reputed publications, their expertise in media and communications is unparalleled. Having worked previously with Punjabi University, Patiala, and Chandigarh University, Kharar, Dr. Batra has successfully merged academic knowledge with industry experience as the Head of Branding and PR at Humans of Digital Marketing firm. She is currently aligned with Microsoft, Gurugram as a Mentor for their CWB initiative as she possesses expertise in the realm of Generative AI, chatbots, and other AI tools. Her in-depth knowledge in this field has led her to explore the potential of ChatGPT and unleash its power in the context of media. Dr. Batra has shared her insights and findings in a book dedicated to harnessing the capabilities of ChatGPT in the media industry, solidifying her position as a thought leader in the intersection of AI and journalism.

Saket Kumar Bhardwaj is currently working as an Assistant Professor at the Centre for Journalism and Mass Communication, Hemwati Nandan Bahuguna Garhwal University, Uttarakhand.

Ipsa Bharti is a Research Scholar at Jamia Millia Islamia, New Delhi. She holds an MBA from the Indian Institute of Management, Udaipur (IIMU), and an M.Com and B.Com from Jamia Millia Islamia, New Delhi. With about three years of corporate experience, her research focuses on the impact of artificial intelligence on various industries and its transformative effects on people's lives. Ipsa has published an article in a UGC CARE (list 1) journal, contributing to her field's academic discourse.

Kavita Chauhan is a Professor at the Department of Management Studies, Jamia Millia Islamia, New Delhi. She graduated from Lady Shri Ram College, Delhi University, and obtained her MBA in E-commerce from GGSIP University. She holds a Doctorate in E-commerce (Indian Tourism Marketing). She has a rich experience of nineteen years in academics and teaching postgraduate students. She has published research papers in Scopus/Web of Science indexed research papers in Springer, Taylor, and Francis and Inderscience publication. Her research interests are in the areas of customer relationship marketing, and digital marketing.

Sudheep Elayidom is the Hod and Professor, in the Department Of Computer Science And Engineering, School Of Engineering, Cochin University Of Science And Technology, Kerala. India.

Reynal Lavita Fernandes, a Ph.D. scholar in Commerce at CHRIST University. Her research areas are in Accounting and Finance with respect to the prediction of financial distress and the Restructuring of Companies. She has presented papers at several national and international conferences. Her area of interest is to find a solution for bankrupt companies through Machine learning techniques. Based on her research interest, she is currently working on a restructuring model for financially distressed companies in India through the application of Machine Learning Algorithms in her Ph.D.

Minnu Elizabeth Ittan is a research scholar in the area of Social Media Analytics in School Of Engineering, Cochin University Of Science And Technology, Kerala, India.

Jayapriya J. completed Ph.D. in National Institute of Technology, Tiruchirappalli. Published seven papers in a reputed journal that includes SCIE and Scopus. Presented in seven International Conferences and won Best Paper Award which was held in Singapore. Acted as Resource Person for various workshops and seminars. Completed various national FDP programmes. Qualified GATE and NET exam.

Ambeshwar Kumar is working as an Assistant Professor in GITAM University, Visakhapatnam, India, He has obtained his Ph.D. in Machine learning and deep learning from SASTRA Deemed University, Thanjavur, India. He has received his Bachelor of Engineering and Master of Technology Degree from Visvesvaraya Technological University Belagavi and Dayananda Sagar University Bangalore respectively. He possesses two year of academic and four years of research experience in the field of Computer Science and Engineering. He has more than 30 research contributions to his credit, which are published in referred and indexed journals, book chapters and conferences. He has knowledge in the field of Machine Learning and Big Data Analytics. His contemporary research interests include Machine Learning, Deep Learning, React.js, IoT and Health Care Applications. He is a reviewer in Hindawi Journal.

Bala Lakhendra is a senior professor in the Department of Journalism and Mass Communication at Banaras Hindu University, India.

A. Manimaran is an Assistant Professor and dedicated researcher in the Department of Mathematics at VIT-AP University, Andhra Pradesh. He has 10 years of teaching experience and 4 years of research experience. He has published 25 research papers in reputed national and international journals and conferences. He pursued a Bachelor of Science from Bharathidasan University, Tamil Nadu, in 2006, a Master of Computer Applications from Anna University in 2009, and a Ph.D. from Bharathidasan University in 2018. His main research work focuses on machine learning, deep learning, NLP, and data science and he is committed to mentoring the next generation of scholars. In addition to his professional endeavors, he serves as a reviewer for prestigious journals in the fields of data science and artificial intelligence. He is also an active member of the Association for Computing Machinery (ACM) and the Computer Society of India (CSI), where he engages in networking opportunities and stays abreast of the latest developments in the field.

Midhun P. Mathew is a research scholar working in the area of artificial intelligence in School Of Engineering, Cochin University Of Science And Technology, Kerala, India.

Sachin Mishra is a dedicated and passionate scholar currently pursuing his Master's degree in Data Science from VIT-AP University. He holds a bachelor's degree in computer application from Guru Gobind Singh Indraprastha University. With a strong foundation in machine learning, statistics, regression analysis, and natural language processing (NLP), Sachin has developed a keen interest in and expertise in these areas. During his academic journey, Sachin gained practical experience through an internship at Doubtnut, where he worked as a content developer. This experience allowed him to hone his skills and contribute to the field of educational content development. This chapter marks Sachin's first publication, showcasing his commitment to advancing knowledge and contributing to academic discourse. Apart from his academic pursuits, Sachin enjoys staying active and spends his free time at the gym, emphasizing the importance of maintaining a healthy lifestyle.

Manikandan Ramachandran obtained his Ph.D. in VLSI Physical design from SASTRA Deemed University, India in 2014. He received his Bachelor of Engineering in Computer Science from Bharathidasan University in 1996 and Master of Technology in VLSI Design from SASTRA Deemed University in 2002. He possesses three decades of academic and 15 years of research experience in the field of Computer Science and Engineering. He has more than 200 research contributions to his credit, which are published in referred and indexed journals, book chapters and conferences. He is presently working as Senior Assistant Professor at SASTRA Deemed University for the last 17 years. He has delivered many lectures and has attended and presented in International Conferences in India and Abroad. He has edited more than 200 research articles to his credit, which includes his editorial experience across refereed and indexed journals, conferences and book chapters at national and international levels. His contemporary research interests include Big Data, Data Analytics, VLSI Design, IoT and Health Care Applications.

Harshwardhani Sharma is currently working as an Assistant Professor in the School of Media Studies and Humanities at Manav Rachna International Institute of Research and Studies, Faridabad, India. She worked as an assistant professor in the Department of Journalism and Mass Communication at Manipal University Jaipur, Jaipur, India from 2022 to 2023. She defended her doctoral thesis at the Department of Mass Communication and Journalism, Tezpur University, Assam, India in 2023. She completed a project titled "An exploratory study of contemporary Estonian animation films through the semiotics of Yuri Lotman" at Tallinn University from 11-09-2021 to 11-02-2022, which is funded by the Education and Youth Board of the Republic of Estonia under the programme of the Academic studies of Estonian Language and Culture Abroad (the E.K.K.A.V. programme). She was a visiting doctoral student at Tallinn University, Estonia, from 1st-11-2019 to 31st-01-2021, under the Dora Plus mobility grant, funded by the European Regional Development Fund.

Veerta Tantia is working as Associate Professor at the Department of Commerce, CHRIST (Deemed to be University), Hosur Road, Bangalore, India. Her research areas are in Accounting and Finance. She has researched on organisational culture, Occupational stress on self-efficacy on Teachers, Corporate Performance of India Companies . She has presented and participated in several International Conferences.

Akshay Upmanyu is a qualified and experienced Assistant Professor with over 3 years of teaching experience. He is UGC NET qualified and has completed his master's degree from Aligarh Muslim University, Aligarh. He is pursuing his PhD from Sharda University and working as an Assistant Professor at Chandigarh University. He is a passionate and dedicated teacher committed to providing his students with the best possible education.

Index

N

O

P

R

S

T

U

V